621.43
ENG

Engines

JOHN DEERE

FUNDAMENTALS OF SERVICE
Engines
ISBN-0-86691-354-8

FOS3010NC (2009)   (ENGLISH)

A service, testing, and maintenance guide for engine systems
in off-road vehicles, trucks, and buses

D0817156

Deere & Company
LITHO IN U.S.A.

CIVIC CENTER

Introduction

Free Catalog — Call 1-800-522-7448

Check out all of our titles in the FUNDAMENTALS OF SERVICE series!

Here are a few of the titles in this series:

Air Conditioning
Service, testing, and maintenance guide for air conditioning systems in off-road vehicles, trucks, and buses.

Electronic and Electrical Systems
Service, testing, and maintenance guide for electronic and electrical systems in off-road vehicles, trucks, and buses.

Hydraulics
Service, testing, and maintenance guide for hydraulic systems in off-road vehicles, trucks, and buses.

Hydraulic Diagnostics Systems
Service, testing, and troubleshooting guide for hydraulic systems in off-road vehicles, trucks, and buses.

Identification of Parts Failures
A highly illustrated failure analysis guide for automotive and off-road vehicle parts.

Power Trains
Service, testing, and maintenance guide for power trains in off-road vehicles, trucks, and buses.

Shop Tools
A basic guide showing the right tool for each type of job and its proper use.

Welding
The fundamentals of welding, cutting, brazing, soldering, and surfacing of metals.

Our FUNDAMENTALS OF SERVICE series brings together all the technical information you need and combines it with clearly written and amply illustrated instructional aids for many types of mechanical systems, their components, the tools needed, and repair procedures.

There are many ways to order, to inquire into prices, or to receive our free catalog:

- Call 1-800-522-7448 to order using a credit card
- Search online from http://www.JohnDeere.com/publications

FUNDAMENTALS OF SERVICE (FOS)

Fundamentals of Service (FOS) is a series of manuals created by Deere & Company. Each book in the series is conceived, researched, outlined, edited, and published by Deere & Company, John Deere Publishing. Authors are selected to provide a basic technical manuscript that could be edited and rewritten by staff editors.

HOW TO USE THE MANUAL: This FOS manual can be used by anyone — experienced mechanics, shop trainees, vocational students, and lay readers. The instructions are written in simple language so that they can be easily understood.

Persons not familiar with the topics discussed in this book should begin with Chapter 1 and then study the chapters in sequence. The experienced person can find what is needed on the "Contents" page.

Each guide was written by Deere & Company, John Deere Publishing staff in cooperation with the technical writers, illustrators, and editors at Almon, Inc. — a full-service technical publications company headquartered in Waukesha, Wisconsin (www.almoninc.com).

This material is the property of Deere & Company, John Deere Publishing. All use and/or reproduction not specifically authorized by Deere & Company, John Deere Publishing is prohibited.

Engines

Engines is the definitive "how-to" book of both on- and off-road engines — from showing you how to diagnose problems and test components, to explaining how to repair each system. And when we say "show you," we mean just that! Our book is filled with illustrations to clearly demonstrate what must be done — photographs, drawings, pictorial diagrams, troubleshooting charts, and diagnostic charts.

Instructions are written in simple language so that they can be easily understood. This book can be used by anyone, from a novice to an experienced mechanic. And it can be used to work on many types of engines, from gasoline to LP-gas to diesel. By starting with the basics, the book builds your knowledge step-by-step, from how engines work to diagnosing and testing an engine. It even shows you how to perform an engine tune-up.

ACKNOWLEDGMENTS
John Deere gratefully acknowledges the following people for their contributions to this manual:

AUTHOR: *Mike Hall* is the original author. He has a B.S. degree in Industrial Technology from Pittsburg State University at Pittsburg, Kansas, and has over 25 years of experience writing and editing technical publication for agricultural equipment.

CONSULTING EDITOR: *Lon R. Shell Ed.D.*, Professor of Agricultural Systems Management, Department of Agriculture at Southwest Texas State University, is the original consulting editor. He has over 30 years of experience in research and teaching. His research specialty is tractor power delivery dealing with tractive efficiencies.

COPY EDITOR: *Tom Rader* is the original editor of the Fundamentals of Service (FOS) series of teaching materials during his career with Deere and Company. He was also a technical writer and later the manager of ⸱⸱⸱ining for Deere in Waterloo,

3 1232 00893 5761

Introduction

John Deere also acknowledges the following groups:

AW Dynamometer, Inc., Dana Corporation, Allen Electric and Equipment Co., Bacharach Industrial Instrument Co., Bendix Corp., Central Tool Co., Federal-Mogul Corp., F.W. Dwyer Co., Garrett Corp., General Motors Corp., J.H. Williams and Co., Kent-Moore Corp., K.O. Lee Co., Koppers Co., L.S. Starrett Co., Marquette Manufacturing Co., Marvel-Schebler, Division of Borg-Warner Corp., MVE, Inc., National LP-Gas Assn., Nuday Co., Owatonna Tool Co., Roosa Master, Hartford Division of Standard Screw Co., Taylor Dynamometer and Machine Co., Texaco, Inc., TRW Replacement Div., Union Carbide Corp., United Tool Process Corp., Westberg Manufacturing Co.

OUO1010,0000EAC -19-10JUL09-2/2

Contents

Continued on next page

*Original Instructions. All information, illustrations and specifications in this
manual are based on the latest information available at the time of publication.
The right is reserved to make changes at any time without notice.*

Continued on next page

100709
PN=2

100709
PN=4

How Engines Work

How Engines Work — Introduction

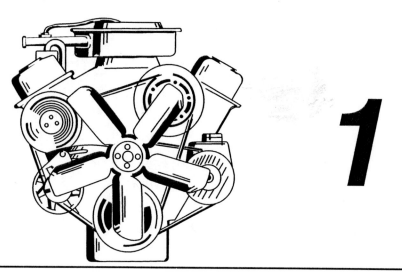

WHAT IS AN INTERNAL COMBUSTION ENGINE?

First let's see what the term internal combustion means:

- Internal means "inside" or "enclosed".
- Combustion is the "act of burning."

An internal combustion engine **burns** fuel **internally**.

To burn the fuel, the fuel must be first mixed with the air in the correct proportion, depending on the type of fuel being used.

So basically, the internal combustion engine is a container where a correct air-fuel mixture is ignited and burned.

The air-fuel mixture expands rapidly while burning and pushes outward. This push can be used to move a part of the engine, and transmitted to drive the machine.

In summary, an internal combustion engine is a device that converts heat energy into mechanical energy to do work.

TYPES OF INTERNAL COMBUSTION ENGINES

There are many types of internal combustion engines: reciprocating, rotary (pistonless), oscillating, and toroidal, to name a few. This book focuses primarily on the traditional reciprocating type engine and the more common fuels used to power it.

Continued on next page

OUO1010,0000E8B -19-23JUL09-1/2

ELEMENTS NEEDED FOR AN INTERNAL COMBUSTION ENGINE

The elements needed to construct a simple engine are as follows:

- **Air, Fuel, and Combustion**
- **Reciprocating and Rotary Motion**
- **Compression of Air-Fuel Mixture**
- **Engine Cycles**

Let's discuss these items one by one.

A—Rotary Motion
B—Reciprocating Motion
C—Air-Fuel Mixture
D—Compression
E—Combustion
F—Cycles: Two-Stroke or Four-Stroke

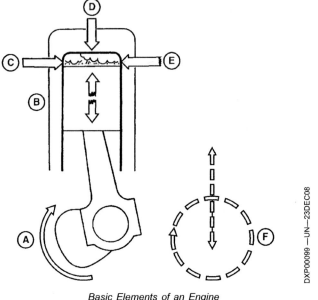

Basic Elements of an Engine

OUO1010,0000E8B -19-23JUL09-2/2

Engine Elements — Air, Fuel, and Combustion

Three basic elements are needed to produce heat energy in the engine:

- **Air**
- **Fuel**
- **Combustion**

AIR

Air is needed to combine with fuel and give it oxygen for fast burning. Air also has two other properties which affect the engine:

1. Air will compress; 1 cubic foot (28 L) of air can be packed into 1 cubic inch (16 cm^3) or less.

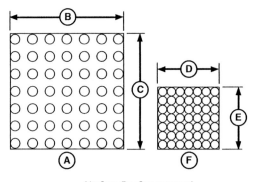

Air Can Be Compressed

A—Sea Level Air
B—12 in. (300 mm)
C—12 in. (300 mm)
D—1 in. (25 mm)
E—1 in. (25 mm)
F—Compressed Air

Continued on next page

OUO1010,0000E8C -19-10JUL09-1/4

100709
PN=10

2. Air heats when it is compressed. The molecules of air rub against each other and produce heat.

G—Atmospheric Air   **H—Compressed Air**

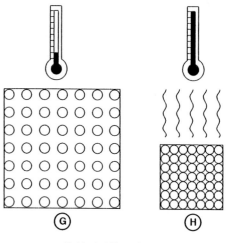

Air Heats When Compressed

OUO1010,0000E8C -19-10JUL09-2/4

FUEL

Fuel must mix readily with air and ignite easily. The four types of fuels we will cover are:

- Gasoline
- LP-Gas
- Natural Gas
- Diesel Fuel

These fuels ignite easily and are readily broken down or vaporized. Why do we want to vaporize the fuel? To help each particle of fuel contact enough air to burn fully.

DXP00102 —UN—23DEC08

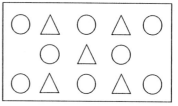

Fuel Must Mix Readily with Air and Ignite Easily

I— Air   **J—Fuel**

Continued on next page   OUO1010,0000E8C -19-10JUL09-3/4

COMBUSTION

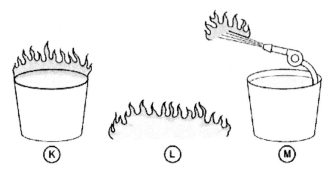

Vaporized Fuel Burns Faster

K—Solid Fuel in Container   **L—Solid Fuel Spread Out**   **M—Vaporized Fuel**

Combustion is the actual igniting and burning of the air-fuel mixture. It is the oxygen in the air that combines with the vaporized fuel that provides combustion.

What is important here is how fast the fuel burns, for this force must be "explosive" to get full power from the engine.

If a container of gasoline is ignited in calm outside air, it burns rather lazily. This is because the air contacts only the surface of the fuel. To make the fuel burn faster, two things can be done:

• Heat the fuel

• Vaporize the fuel

However, too powerful an explosion would destroy an engine, since combustion takes place in a closed container.

We can control the rate of burning by:

• How far we compress the air (and so heat it up)
• How much fuel is used
• How volatile the fuel is

OUO1010,0000E8C -19-10JUL09-4/4

Engine Elements — Reciprocating and Rotary Motion

The reciprocating engine uses two forms of motion to transmit energy:

• **Reciprocating Motion (up-and-down or back-and-forth motion)**
• **Rotary Motion (circular motion around a point)**

The engine converts reciprocating motion into rotary motion.

So the engine can perform this function, four basic parts are needed:

• Cylinder
• Piston
• Connecting Rod
• Crankshaft

The **piston and cylinder** are mated parts, fitted closely so that the piston glides easily in the cylinder but with little clearance at the sides. The top of the cylinder is

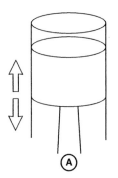 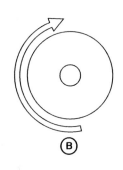

Reciprocating and Rotary Motion

A—Reciprocating Motion   **B—Rotary Motion**

closed off, usually by a component called the cylinder head. The cylinder head contains an extra space called a combustion chamber, where the combustion of the air-fuel mixture occurs.

Continued on next page   OUO1010,0000E8D -19-23JUL09-1/3

The **connecting rod** is the link that transmits the reciprocating motion of the piston to the crankshaft.

A simple **crankshaft** is a shaft that contains a section offset from the center line of the shaft. This offset is called the crank or "throw," and it is where the connecting rod attaches to provide the rotary motion (cranking) of the shaft.

The motion is basically the same as when you pedal a bicycle. Your leg is like the connecting rod while the pedal crank and sprocket are like the crankshaft.

C—Cylinder   E—Connecting Rod
D—Piston   F—Crankshaft

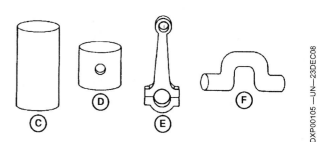

Basic Parts of the Engine

OUO1010,0000E8D -19-23JUL09-2/3

As a result, we now have a way of converting the reciprocating motion of the piston into useful rotary motion.

The stroke of the piston (how far it travels in the cylinder) is set by the "throw" of the crankshaft (how far it is offset).

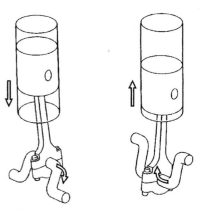

How Reciprocating Motion Is Transmitted to the Crank-shaft as Rotary Motion

OUO1010,0000E8D -19-23JUL09-3/3

Engine Elements — Compression of Air-Fuel Mixture

An average gasoline engine works best when about 14.7 parts of air are mixed with 1 part of fuel when measured by weight. That equals 14.7 lb (6.7 kg) of air to each 1.0 lb (0.45 kg) of fuel. This converts to approximately 1,361 gallons (5,152 L) of air to 0.16 gallon (0.6 L) of fuel. The illustrations to the right show how much greater the volume of air needed is than the volume of fuel.

The 14.7-to-1 air-fuel mixture is known as the stoichiometric ratio. It is a calculated ratio of air and fuel that, under near-perfect conditions, allows all the fuel to be burned and all oxygen to be used up during the combustion process. The stoichiometric ratio can vary and is dependent on the amount of octane and type of additives in the fuel, such as detergents and oxygenators.

A—Air   B—Fuel

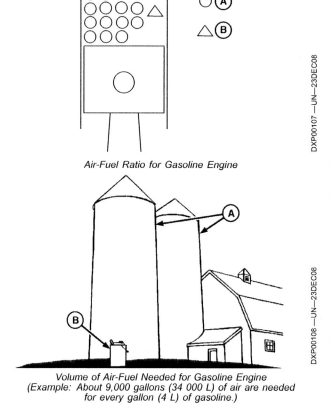

Air-Fuel Ratio for Gasoline Engine

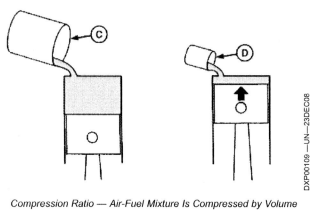

*Volume of Air-Fuel Needed for Gasoline Engine
(Example: About 9,000 gallons (34 000 L) of air are needed
for every gallon (4 L) of gasoline.)*

OUO1010,0000E8E -19-23JUL09-1/2

For proper combustion, compression of the air-fuel mixture is needed.

Compression ratios tell us how much the air-fuel mixture is compressed by volume. The illustration shows an imaginary case.

When the piston is at the bottom of its stroke, let's measure the amount of liquid the cylinder will hold. Let's say it takes 8 pints (3.75 L).

Now remove all the liquid and move the piston to the top of its stroke. Fill the cylinder full of liquid again. Let's say it now only holds 1 pint (0.5 L).

The ratio is then 8 to 1, which is the compression ratio.

In other words, air in this engine is compressed to one-eighth of its former volume by the moving piston. Later we'll see how compression affects the engine.

Compression Ratio — Air-Fuel Mixture Is Compressed by Volume

C—8 Parts   D—1 Part

OUO1010,0000E8E -19-23JUL09-2/2

Engine Elements — Cycles

For an engine to operate, a definite series of events must occur in sequence. They are:

1. Fill the cylinder with a combustible mixture.
2. Compress this mixture into a smaller space.
3. Ignite the mixture, causing it to expand and producing power.
4. Remove the burned gases from the cylinder.

This sequence is generally called:

- **Intake**
- **Compression**
- **Power**
- **Exhaust**

To produce sustained power, the engine must repeat this sequence over and over again. One complete series of these events in an engine is called a *cycle*.

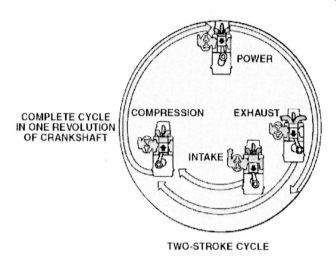

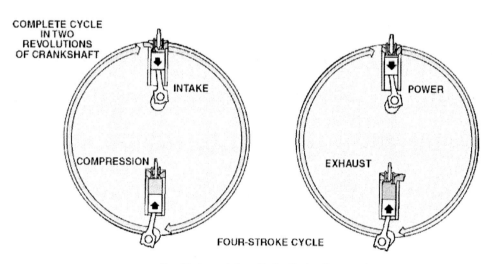

Two-Stroke and Four-Stroke Cycles Compared

Most engines have one of two types of cycles:

- **Two-Stroke Cycle**
- **Four-Stroke Cycle**

In the **two-stroke cycle** engine, there are two strokes of the piston, one up and one down, during each cycle. Then it starts over again on another cycle of the same two strokes. This whole cycle occurs during one revolution of the crankshaft.

In the **four-stroke cycle** engine, there are four strokes of the piston, two up and two down, during each cycle. Then it starts over again on another cycle of the same four strokes. This cycle occurs during two revolutions of the crankshaft. Most engines today operate on the four-stroke cycle.

Let's see how each type of cycle works in detail.

Continued on next page   OUO1010,0000E8F -19-23JUL09-1/4

DXP01076 —UN—11JUN09

TWO-STROKE CYCLE ENGINE

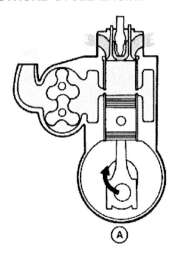 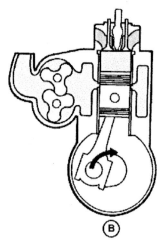 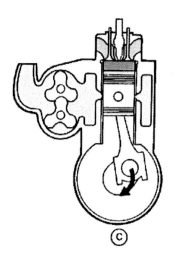

Two-Stroke Cycle Engine (Diesel Shown)

A—Intake and Exhaust   B—Compression   C—Power

In two-stroke cycle engines, the complete cycle of events — **intake**, **compression**, **power**, and **exhaust** — takes place during two piston strokes. Every other stroke is a power stroke, so each time the piston moves down it is a power stroke.

NOTE: *The illustration above shows a two-stroke diesel engine in operation, and the following text describes the events of a two-stroke diesel engine. The cycle events of a two-stroke gas engine are similar with the exception of fuel delivery and ignition.*

In the diesel engine shown, air alone is compressed in the cylinder. For a diesel two-stroke engine, a charge of fuel is then sprayed into the cylinder and ignites from the heat of compression.

In the two-stroke cycle engine, intake and exhaust take place during part of the compression and power strokes.

A blower is sometimes used to force air into the cylinder for expelling exhaust gases and to supply fresh air for combustion. The cylinder wall contains a row of ports that are above the piston when it is at the bottom of its stroke.

These ports admit air from the blower into the cylinder when they are uncovered (during intake).

The flow of air toward the exhaust valves pushes the exhaust gases out of the cylinders, leaving the cylinders full of clean air when the piston again rises to cover the ports (during compression).

At the same time, the exhaust valves close and the fresh air is compressed in the closed cylinders.

When the piston almost reaches the top of its compression stroke, fuel is sprayed into the combustion area as shown. The heat of compression ignites the fuel and the resulting pressure forces the piston down on its power stroke.

As the piston nears the bottom of its stroke, the exhaust valves are again opened and the burned gases escape.

The piston then uncovers the intake ports, and the cycle begins once more.

This entire cycle is completed in one revolution of the crankshaft or two strokes of the piston — one up and one down.

Continued on next page

OUO1010,0000E8F -19-23JUL09-2/4

DXP00111—UN—23DEC08

FOUR-STROKE CYCLE ENGINE

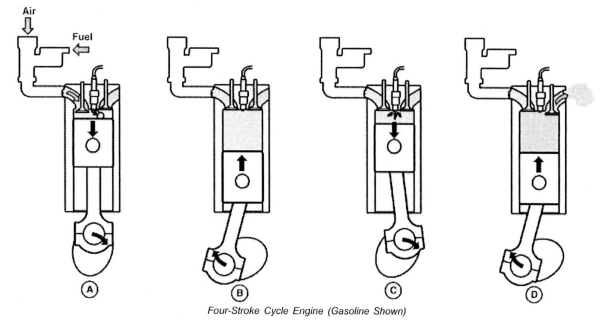

Four-Stroke Cycle Engine (Gasoline Shown)

A—Intake   B—Compression   C—Power   D—Exhaust

In four-stroke cycle engines, the same four operations occur — **intake, compression, power,** and **exhaust**. However, four strokes of the piston — two up and two down — are needed to complete the cycle. As a result, the crankshaft will rotate two complete turns before one cycle is completed.

- INTAKE STROKE: The intake stroke starts with the piston near the top and ends shortly after the bottom of its stroke. The intake valve is opened, allowing the cylinder to receive the air-fuel mixture as the piston moves down. The valve is then closed, sealing the cylinder.
- COMPRESSION STROKE: The compression stroke begins with the piston at bottom and rising up to compress the air-fuel mixture. Since the intake and exhaust valves are closed, there is no escape for the air-fuel mixture and it is compressed to a fraction of its original volume.
- POWER STROKE: The power stroke begins when the piston almost reaches the top of its stroke and the air-fuel mixture is ignited. As the mixture burns and expands, it forces the piston down on its power stroke. The valves remain closed so that all the force is exerted on the piston.

- EXHAUST STROKE: The exhaust stroke begins when the piston nears the end of its power stroke. The exhaust valve is opened and the piston rises, pushing out the burned gases. When the piston reaches the top, the exhaust valve is closed and the piston is ready for a new four-stroke cycle of intake, compression, power, and exhaust. As it completes the cycle, the crankshaft has gone all the way around twice.

TWO-STROKE CYCLE VERSUS FOUR-STROKE CYCLE ENGINES

It might seem that the two-stroke cycle engine can produce twice as much power as a four-stroke cycle engine.

However, this is not true. With the two-stroke cycle engine, some power may be used to drive the blower that forces the air-fuel charge into the cylinder under pressure. Also, the burned gases are not completely cleared from the cylinder, this results in less power per power stroke.

The actual gain in power with a two-stroke cycle engine is about 75% (over a four-stroke cycle engine of the same displacement).

Continued on next page   OUO1010,0000E8F -19-23JUL09-3/4

MULTIPLE-CYLINDER ENGINES

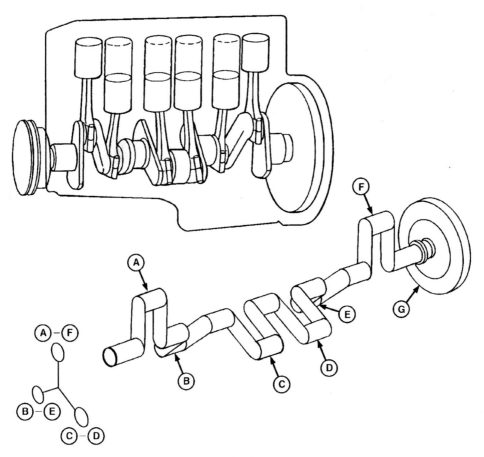

Crankshaft for a Six-Cylinder Engine

A—No. 1 Cylinder	C—No. 3 Cylinder	E—No. 5 Cylinder	G—Flywheel
B—No. 2 Cylinder	D—No. 4 Cylinder	F—No. 6 Cylinder	

So far we have covered only basic one-cylinder engines. A single cylinder gives only one power impulse every two revolutions of the crankshaft in a four-stroke cycle engine. Thus it is producing power only one-fourth of the time.

For a more continuous flow of power, modern engines use four, six, eight, or more cylinders. The same series of cycles takes place in each cylinder.

For example, in a typical four-stroke cycle engine having six cylinders, the cranks (offsets) on the crankshaft are set 120 degrees apart . The cranks for cylinders 1 and 6, 2 and 5, 3 and 4 are in line with each other as shown.

The cylinders normally fire and deliver their power strokes in the following order: 1-5-3-6-2-4. Thus the power strokes follow each other so closely that there is a fairly continuous and smooth delivery of power to the crankshaft.

A heavy flywheel, attached to the rear of the crankshaft, provides momentum to return the pistons to the top of the cylinders after each power stroke. Weights on the crankshaft are used to help balance the forces created by the reciprocating motion of the pistons and connecting rods.

For more information on construction of the basic engine, see Chapter 3.

Now let's look at the systems that help the engine to operate.

OUO1010,0000E8F -19-23JUL09-4/4

DXP00113 —UN—23DEC08

Engine Systems

Now that we've put together a basic engine, let's look at some other systems that are required for good operation:

- **Fuel System**
- **Intake and Exhaust System**

- **Lubrication System**
- **Cooling System**
- **Governing System**

Let's discuss these systems one by one.

OUO1010,0000E90 -19-12JUN09-1/1

Fuel Systems

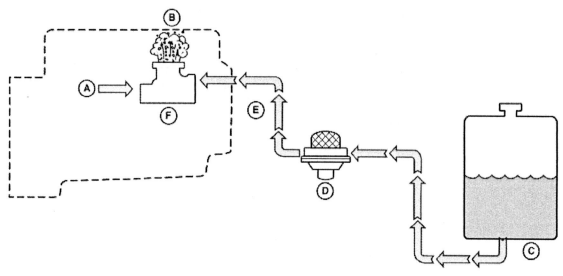

Gasoline Fuel System

A—Air Intake
B—Air-Fuel Mixture
C—Fuel Tank
D—Fuel Pump and Filter
E—Fuel Intake
F—Carburetor

A fuel system must deliver clean fuel, in the quantity required, to the fuel intake of an engine. It must provide for safe fuel storage and transfer.

The four fuel systems of concern to us are:

- **Gasoline**
- **LP-Gas**
- **Natural Gas**
- **Diesel**

GASOLINE FUEL SYSTEMS

The gasoline fuel system supplies a combustible mixture of air and fuel for the engine.

The basic gasoline fuel system has three parts.

- **Fuel Tank (stores fuel)**
- **Fuel Pump (moves fuel to carburetor)**
- **Carburetor (atomizes fuel and mixes with air)**

In operation, the fuel pump moves gasoline from the fuel tank to the carburetor bowl.

The carburetor is basically an air tube that atomizes fuel and mixes it with air by a difference in air pressure. It meters both the air and fuel for the engine.

On its intake stroke, the engine creates a partial vacuum. This allows outside air pressure to force the air-fuel vapor mixed in the carburetor into the engine cylinder.

FUEL SUPPLY SYSTEMS

Fuel can be supplied to the carburetor in two ways:

- **Gravity-Feed**
- **Force-Feed**

The gravity-feed system has the fuel tank placed above the level of the carburetor. This system does not use a fuel pump. Instead, the fuel flows by gravity to the carburetor.

The force-feed system allows the fuel tank to be located below the carburetor if necessary. A fuel pump moves the fuel from the tank to the carburetor as shown.

For more information on gasoline fuel systems, see Chapter 4.

Continued on next page

OUO1010,0000E91 -19-05AUG09-1/5

LP-GAS FUEL SYSTEMS

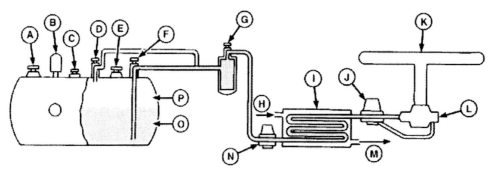

LP-Gas Fuel System

A—Filling Valve	E—Vapor Return Valve	I— Vaporizer	M—Return Water
B—Pressure Relief Valve	F—Liquid Line Valve	J— Low-Pressure Regulator	N—High-Pressure Regulator
C—80% Fill Valve	G—Strainer Valve	K—Engine Intake Manifold	O—Fuel
D—Vapor Line Valve	H—Hot Water From Engine	L—Carburetor	P—Vapor

The LP (Liquefied Petroleum) gas fuel system supplies a combustible mixture of vaporized air and fuel to the engine. LP-gas is made up mainly of propane and butane gases that have been liquefied by compressing many gallons of vapor into one gallon of liquid. However, LP-gas vaporizes at low temperatures. Thus the fuel tank must be a closed unit to prevent vapor from escaping.

To withdraw fuel from the tank, two methods are used:

• **Liquid Withdrawal**
• **Vapor Withdrawal**

The vapor withdrawal of fuel is used when starting the engine. The fuel system is later switched to liquid withdrawal after the engine has warmed up. This is because in a cold engine, the heat exchanger cannot change the liquid fuel to vapor, and the carburetor operates only on vapor.

The liquid and vapor line valves shown provide for safety and for selection of fuel — liquid or vapor. The filters

remove moisture and dirt. The pressure regulators keep a constant pressure of fuel at the carburetor for accurate fuel metering.

In the liquid withdrawal system, a heat exchanger converts the liquid fuel to vapor. The heat exchanger does this by circulating hot water from the engine cooling system around the fuel line. As the fuel heats up and pressure is reduced, it vaporizes. The liquid withdrawal system is most common today.

The LP-gas carburetor is simpler than the gasoline type, since the fuel is already vaporized. It meters the vapor and mixes it with the proper amount of air for the engine.

For more information on LP-gas fuel systems, see Chapter 5.

Continued on next page

OUO1010,0000E91 -19-05AUG09-2/5

NATURAL GAS FUEL SYSTEMS

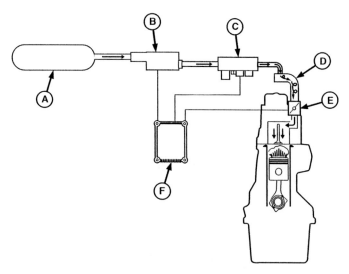

Natural Gas Fuel System

A—Fuel Tank
B—Pressure Regulator
C—Fuel Metering Valve
D—Air-Fuel Mixture
E—Throttle Body
F—Engine Control Unit (ECU)

A natural gas fuel system supplies a combustible mixture of vaporized fuel and air to the engine.

Natural gas is in a gaseous state at any temperature above −259°F (−126°C). Because of this, natural gas must be stored in a closed tank to prevent the gas from escaping.

Using natural gas to power an engine in a mobile application requires a fuel storage system that can be safely transported. The storage system must contain a sufficient supply of gas to provide the vehicle with an operating range comparable to that of a gasoline fuel system. To accomplish this, natural gas is stored in one of the following forms:

- **Compressed Natural Gas (CNG)**
- **Liquefied Natural Gas (LNG)**

To produce CNG, natural gas must be pressurized to 3000–3600 psi.

To produce LNG, natural gas must be cooled to at least −259°F (−126°C).

The major components of a natural gas fuel system as shown are:

- **Fuel Tank**
- **Vaporizer/Heat Exchanger (LNG systems only) (not shown)**
- **Pressure Regulator**
- **Fuel Metering Valve**
- **Air-Fuel Mixer**
- **Throttle Body**

FUEL TANKS — Compressed natural gas (CNG) is stored in high-strength fuel tanks at pressures up to 3600 psi.

The tanks are equipped with shutoff valves and pressure relief valves.

Liquefied natural gas (LNG) is stored in a high-strength insulated fuel tank. The insulated tank is necessary to maintain the low temperature of the liquefied gas. The tank normally includes a fuel pressure regulator, manual supply and vent valves, relief valves, and a fuel gauge. The tank may also include a vaporizer/heat exchanger.

VAPORIZER/HEAT EXCHANGER — The vaporizer/heat exchanger converts LNG into compressed natural gas for delivery to the engine. The heat exchanger does this by circulating warm engine coolant around the fuel line. The heat from the coolant vaporizes the liquefied gas.

PRESSURE REGULATOR — The pressure regulator reduces the high-pressure gas to the proper pressure required by the fuel system. Engine coolant heats the pressure regulator to prevent it from freezing as a result of the rapid drop in gas pressure.

FUEL METERING VALVE — The fuel metering valve delivers the correct amount of fuel to the engine. An electronic engine control unit (ECU) controls the output of the fuel metering valve to meet the operating conditions on most modern engines.

AIR-FUEL MIXER — The air-fuel mixer combines the vaporized fuel with the incoming air to form the air-fuel mixture for combustion. The air-fuel mixer is located upstream of the throttle body.

THROTTLE BODY — The throttle body controls the amount of air-fuel mixture that flows to the combustion chamber.

Continued on next page

OUO1010,0000E91 -19-05AUG09-3/5

For more information on natural gas fuel systems, see
Chapter 6.

OUO1010,0000E91 -19-05AUG09-4/5

DIESEL FUEL SYSTEMS

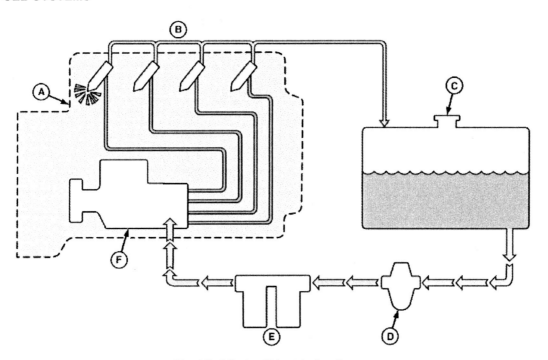

Diesel Fuel System Using Injection Pump

A—Combustion Chamber
B—Injection Nozzles
C—Fuel Tank
D—Fuel Transfer Pump
E—Fuel Filters
F—Injection Pump

In the diesel fuel system, fuel is sprayed directly into the engine combustion chamber where it mixes with hot compressed air and ignites. No electrical spark is used to ignite the mixture (as in gasoline, LP-gas, and natural gas engines).

Instead of a carburetor, a fuel injection pump and spray nozzle are used.

The major parts of a typical diesel fuel system are:

• **Fuel Tank**
• **Fuel Transfer Pump**
• **Fuel Filters**
• **Water Separator (optional)**
• **Injection Pump or High-Pressure Pump**
• **Injection Nozzles**

In operation, the fuel transfer pump moves fuel from the fuel tank, where the fuel is stored. The transfer pump pushes fuel through the fuel filters. The fuel filters remove dirt and particulates from the fuel. An optional water separator (not shown) removes water that may be present in the fuel.

Clean fuel free of water is very vital to the precision parts of the diesel injection system. Extra filters are often used to ensure clean fuel, but buying clean fuel and storing it properly are also very important.

Depending on the type of system, the fuel is then pushed on to one of the following:

• INJECTION PUMP: The injection pump meters the fuel, puts the fuel under high pressure, then delivers it to each injection nozzle in the proper sequence.
• HIGH-PRESSURE PUMP: The high-pressure pump (not shown) delivers fuel under high pressure to common rail type systems.

The nozzles each serve one cylinder; they atomize the fuel and spray it under controlled high pressure into the combustion chamber at the correct moment.

High-pressure fuel is needed at each nozzle to get a fine spray of fuel. This ensures good mixing of fuel with the hot compressed air for full combustion.

For more information on diesel fuel systems, see Chapter 7.

OUO1010,0000E91 -19-05AUG09-5/5

Intake and Exhaust Systems

The intake system carries the air-fuel mixture into the combustion chamber.

The exhaust system removes the exhaust gases after combustion.

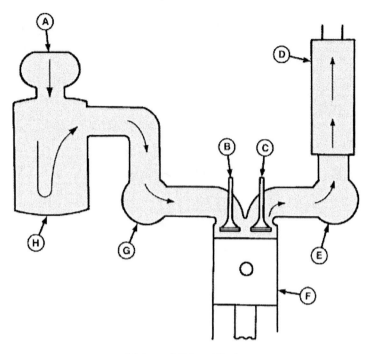

Intake and Exhaust Systems

A—Pre-Cleaner
B—Intake Valve

C—Exhaust Valve
D—Muffler

E—Exhaust Manifold
F—Cylinder

G—Intake Manifold
H—Air Cleaner

INTAKE SYSTEM

The intake system supplies the engine with clean air of the proper quantity, temperature, and fuel to provide the correct fuel mixture for good combustion.

NOTE: For diesel engines, fuel is not delivered by the intake system. Instead, it is injected directly into the combustion chamber.

The intake system has five parts:

- Air Cleaner
- Supercharger or Turbocharger (optional)
- Intake Manifold
- Carburetor/Air Inlet (gas engines)
- Intake Valves

Air cleaners remove dust and dirt from the air flowing to the intake manifold. A pre-cleaner (if equipped) prevents larger particles from reaching and plugging the air cleaner.

Superchargers pressurize the intake air. They can be used on two-stroke engines to force air into the cylinder and drive out exhaust gases. The supercharger can increase horsepower by packing more air (diesel engine) or air-fuel mixture (gas engine) into the engine cylinders than the engine could take in by natural aspiration. It is usually gear or belt driven and gets it power directly from the engine crankshaft.

Turbochargers also increase horsepower by packing more air (diesel engine) or air-fuel mixture (gas engine) into the engine cylinders than the engine could take in by natural aspiration. The turbocharger is driven by the engine's exhaust.

Intake manifolds route the air (diesel engine) or air-fuel mixture (gas engine) to the engine cylinders.

Carburetors (gas engines) mix incoming air with fuel in the proper proportion for combustion, and to control engine speed.

Intake valves allow air (diesel engine) or air-fuel mixture (gas engine) into the combustion chambers. They are usually opened and closed by mechanical linkage from the camshaft.

For more information on intake systems, see Chapter 8.

EXHAUST SYSTEM

The exhaust system collects the exhaust gases after combustion and carries them away. This is really three jobs:

- Removing heat
- Muffling engine sounds
- Carrying away burned and unburned gases

Continued on next page

OUO1010,0000E92 -19-05AUG09-1/2

PN=23

The exhaust system has these basic parts:

• Exhaust Valves
• Exhaust Manifold
• Muffler

The **exhaust valves** open to release the burned gases. This pertains to all four-cycle engines and some types of two-cycle engines. The valves are usually operated by a camshaft.

The **exhaust manifold** collects the exhaust gases and routes it away from the cylinder.

The **muffler** reduces the sound of the engine exhaust.

For more information on exhaust systems, see Chapter 8.

OUO1010,0000E92 -19-05AUG09-2/2

Lubrication Systems

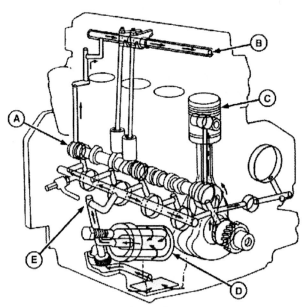

Lubrication System

A—Reduces Friction and Wear
B—Absorbs Heat
C—Seals Piston Rings
D—Cleans Parts
E—Deadens Noise

The lubrication system does these jobs for the engine:

• **Reduces friction between moving parts**
• **Absorbs and dissipates heat**
• **Seals the piston rings and cylinder walls**
• **Cleans and flushes moving parts**
• **Helps deaden the noise of the engine**

With lubricating oil, the system is able to do all these jobs at once. Without oil, the engine would soon wear out, burn up, or seize. Oil not only reduces friction by forming a film between parts; it also conducts heat away from these parts.

The lubrication system may work by splashing oil on the moving parts or it may feed oil under pressure to the parts via internal oil passages as shown. In some cases, both methods are used at the same time.

The engine crankcase forms an oil reservoir where oil is stored and cooled. The crankcase must be vented to prevent pressure buildup from the blow-by of gases past the pistons.

Most venting systems route crankcase vapors back to the intake system to reduce air pollution.

For more information on lubrication systems, see Chapter 9.

OUO1010,0000E93 -19-23JUL09-1/1

DXP00119 —UN—23DEC08

Cooling System

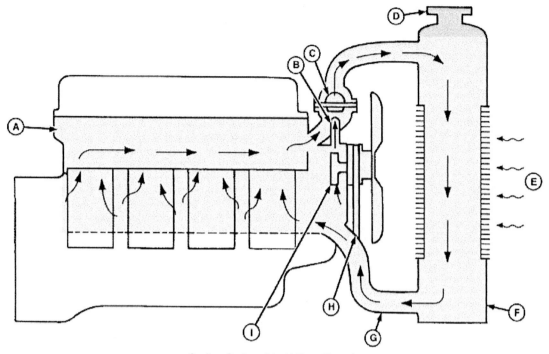

Cooling System (Liquid Type Shown)

A—Engine Water Jacket
B—Bypass
C—Thermostat
D—Pressure Cap
E—Cooling Air
F—Radiator
G—Hose
H—Fan Belt Drive
I— Water Pump

The cooling system prevents the engine from overheating. Although some heat is necessary for combustion, a working engine generates too much heat. Therefore, the cooling system carries off the excess heat.

Cooling systems are designed to use parts that are matched in capacity. A matched cooling system will provide adequate heat rejection. If one part is replaced that is under or over capacity, the effectiveness of the system will be decreased. Parts include the coolant pump, radiator, coolant, piping, thermostat, and fan.

TYPES OF COOLING SYSTEMS

Two types of cooling systems are used on modern engines:

• **Air Cooling**
• **Liquid Cooling**

AIR COOLING:

This method of cooling uses air passing around the engine to dissipate heat. It is used primarily on small engines

or aircraft. Metal baffles, ducts, and blowers are used to aid in distributing air.

LIQUID COOLING:

This method of cooling uses water or a water-based coolant solution to dissipate heat. The water or coolant solution circulates in a jacket around the cylinders and cylinder head. A large amount of engine heat is absorbed by the coolant, which then flows to the radiator. Air flowing through the radiator removes the heat from the coolant, then dissipates the heat into the air. The cooled coolant then recirculates into the engine to pick up more heat.

Water, by itself, acts as a very good medium for cooling, but cold weather can cause it to freeze. Because of this, a solution made of water and ethylene glycol is normally used. This solution prevents freezing during cold weather while still providing good cooling during hot weather.

For more information on cooling systems, see Chapter 10.

OUO1010,0000E94 -19-10JUL09-1/1

Governing Systems

The governing system keeps the engine speed at a constant level. It does this by varying the amount of fuel (diesel engine) or air-fuel mixture (gas engine) supplied to the engine, according to the demands of the load. The level of engine speed is controlled by the position of the speed control lever, connected by linkage to the governor.

The object is to get the engine's power to match the load at all times, to keep the speed at a steady level.

Governors can be either mechanical, hydraulic, or electrical. For more information, see Chapter 11.

OUO1010,0000E95 -19-23JUL09-1/1

Types of Engines

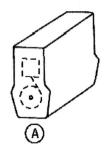
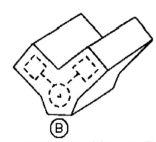
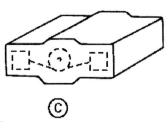

Arrangement of Cylinders (Three Types)

A—In-Line   B—V-Type   C—Opposed

Engines can be typed in three ways:

• **Cylinder Arrangement**
• **Valve Arrangement**
• **Type of Fuel Used**

CYLINDER ARRANGEMENT

Multi-cylinder engines are classified according to the arrangement of cylinders:

• **In-Line**
• **V-Type**
• **Opposed**

IN-LINE engines have all cylinders in a straight line above the crankshaft. This type engine is popular on farm and industrial machines.

V-TYPE engines have two banks of cylinders in a V-shape above the crankshaft. This type engine is common on automobiles, although its use in farm and industrial machines is increasing.

OPPOSED type engines have two rows of cylinders opposite the crankshaft. This type engine is limited primarily to small cars and aircraft.

The cylinders are normally numbered. With in-line models, the No. 1 cylinder is normally at the end opposite the flywheel. The others are numbered 2, 3, 4, etc., from front to rear. In V-type and opposed engines, the sequence varies with the manufacturer.

Continued on next page
OUO1010,0000E96 -19-23JUL09-1/4

VALVE ARRANGEMENT

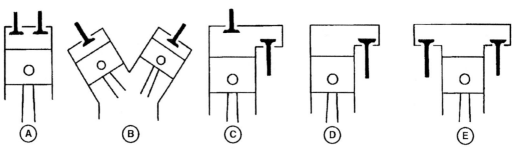

Valve Arrangement (Five Types)

A—I-Head
B—H-Head (2-Cycle)

C—F-Head
D—L-Head

E—T-Head

Engines can also be classified by the position and arrangement of the intake and exhaust valves. This normally depends on whether the valves are located in the cylinder block or the cylinder head. The most common types are the I-head, H-head (two-stroke cycle), F-head, L-head, and T-head. See Chapter 3 for more information.

FUEL TYPES OF ENGINES

The most common way to type engines is by the type of fuel used.

Four fuel types are most common:

- **Gasoline Engine**
- **Diesel Engine**
- **LP-Gas Engine**
- **Natural Gas Engine**

The basic operation of each engine is the same and we have already compared the methods of fueling. But now let's look at the overall performance of each one while comparing gasoline and diesel.

Gasoline and Diesel Engines — What Are the Main Differences?

- The method of supplying and igniting fuel
- The higher compression ratio in diesels
- The generally more rugged design of diesels
- The grade and type of fuel used

Let's look at each of these differences.

Continued on next page

OUO1010,0000E96 -19-23JUL09-2/4

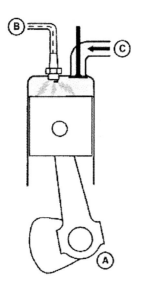 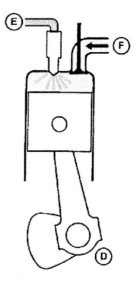

Methods of Supplying and Igniting Fuel (Power Stroke)

DXP00123 —UN—23DEC08

A—Gasoline   C—Air-Fuel Mixture   E—Fuel
B—Spark   D—Diesel   F—Air

METHODS OF SUPPLYING AND IGNITING FUEL

In gasoline engines, fuel and air are mixed outside the cylinders, in the carburetor and manifold. The mixture is forced into the cylinders due to the partial vacuum created as the piston moves downward on the intake stroke.

In diesel engines, there is no premixing of air and fuel outside the cylinder. Air only is taken into the cylinder through the intake manifold and compressed. Fuel is then sprayed into the cylinder and mixed with air as the piston nears the top of its compression stroke.

Gasoline engines use an electric spark to ignite the air-fuel mixture, while diesels use the heat of the compressed air for ignition.

Continued on next page

OUO1010,0000E96 -19-23JUL09-3/4

COMPRESSION RATIOS

Compression ratio compares the volume of air in the cylinder before compression with the volume after compression.

An 8-to-1 compression ratio is common for gasoline engines, while a 16-to-1 ratio is common for diesels.

The higher compression ratio of the diesel raises the temperature of the air high enough to ignite the fuel without a spark.

This also gives the diesel more efficiency because the higher compression results in greater expansion of gases in the cylinder following combustion. Result: a more powerful stroke.

The higher efficiency that results from diesel combustion must be offset by the need for sturdier, more expensive parts to withstand the greater forces of combustion.

DESIGN OF ENGINE PARTS

We have just touched on the next point: diesels must be built sturdier to withstand the greater forces of combustion. This is generally done by "beefing up" the pistons, pins, rods, and cranks, and by adding more bearings to support the crankshaft.

GRADES AND TYPES OF FUEL

Fuel energy is measured in standard heat units, "British Thermal Units" (Btu), or watts. This allows various types of fuels to be compared against each other when selecting an engine for a particular application. For example, a gallon of diesel fuel can produce more energy than a gallon of gasoline.

LP-Gas Engines

The LP-gas engine is similar to the gasoline engine, but requires special fuel handling and equipment.

LP-gas engines have higher compression ratios than gasoline engines, but not as high as diesels.

In areas where LP-gas fuel is available at low prices, these engines are very popular. However, in many areas, LP-gas fuel is not competitive with the other fuels.

Natural Gas Engines

The natural gas engine is similar to the LP-gas engine in that it requires special fuel handling and equipment.

Some advantages of natural gas fuel systems are:

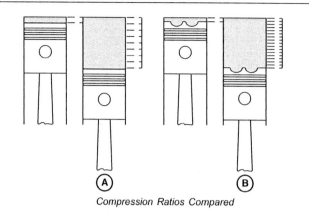

Compression Ratios Compared

A—Gasoline 8-to-1 Ratio   **B—Diesel 16- to-1 Ratio**

- Low fuel cost
- Secure fuel supply
- Reduced engine maintenance and extended engine life

Some disadvantages of natural gas fuel systems are:

- Reduction in operating range and power
- Increased weight to accommodate fuel storage
- Limited refueling infrastructure

Summary: Comparing Engines

The following chart compares gasoline, LP-gas, natural gas, and diesel engines. The comparisons assume that each fuel is available at reasonable prices. Performance is based on general applications that are suited to the engine and fuel type. It is also assumed that the engines are all in good condition.

COMPARING THE ENGINES				
	Gasoline	LP-Gas	Natural Gas	Diesel
Fuel Economy	Good	Good	Good	Best
Hours Before Maintenance	Fair	Good	Good	Good
Weight per Horsepower	Low	Low	Medium	High
Cold Weather Starting	Good	Fair	Good	Fair
Acceleration	Good	Good	Good	Fair
Continuous Duty	Fair	Fair	Good	Good
Lubricating Oil Contamination	Moderate	Lowest	Lowest	Low
Exhaust Emissions	Fair	Good	Good	Fair
Initial Cost of Engine	Moderate	Moderate	Moderate	Highest

OUO1010,0000E96 -19-23JUL09-4/4

Uses of Engines

There are two basic uses for engines:

• **Stationary**
• **Mobile**

STATIONARY ENGINES

Stationary engines supply power from a fixed location. Couplers, belts, chains and drive shafts transfer the engine power to other machines.

Because they are in a "fixed" location, they can be designed for one application. Stationary engines, or "power units," drive such machines as compressors, motor driven pumps, and generators.

MOBILE ENGINES

Mobile engines supply power on the move. They power a wide variety of vehicles from road graders to race cars. Mobile engines can be subdivided into two basic types:

• **Structural**
• **Nonstructural**

A structural engine is mounted to and becomes part of the vehicle frame. It helps to support and carry the load of the vehicle. The structural engine block must be strong enough to withstand the load and road stress.

Structural engines are used on some construction equipment and other off-road vehicles. An example of this type is a farm tractor engine that is bolted directly to the sides of the main machine frame. The engine helps support and hold the tractor together.

A non-structural engine is not a part of the vehicle frame. The vehicle frame in this example is just as strong with or without the engine. Nonstructural engine blocks do not have to withstand load and road stress like a structural engine. The nonstructural engine block can be made of lighter metals, such as aluminum.

Stationary Engine (Industrial Power Unit Shown)

These lighter engines are usually mounted on rubber pads on the vehicle frame. The main application for nonstructural engines is on-road vehicles, such as cars and trucks. The lighter engine block weight helps boost vehicle fuel economy and load carrying ability.

A nonstructural engine may supply the same performance and reliability of a comparable structural engine. However, when a new engine is used to replace original equipment, make sure you do not replace a structural engine with a nonstructural engine. Structural stress may severely damage the engine block.

Also, if you are replacing original equipment with a different type of engine, be sure the fuel and cooling systems are compatible. These systems must be matched with the engine performance and load carrying ability.

OUO1010,0000E97 -19-22JUL09-1/1

The Basics of Engine

We have seen how the engine works. Now let's look at some of the basics which go into the design and operation of engines.

First the laws of mechanics:

- **Matter**
- **Momentum**
- **Mass**
- **Torque**
- **Energy**
- **Work**
- **Mechanical Power**
- **Inertia**
- **Force**

Let's discuss the basics one by one.

MATTER

The substances we encounter in engines are these: **solids**, **liquids**, and **gases**; they are the three physical states of matter.

If you look at the illustration you will see the following truths:

- Solids have a definite volume and shape.
- Liquids have a definite volume but no definite shape.

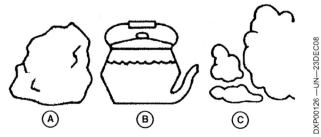

Three Physical States of Matter

A—Solid   C—Gas
B—Liquid

- Gases have no definite volume or shape.

All matter can be changed from one state to another by heating or cooling. Water is a liquid which can be changed to ice (solid) or steam (gas) by changing its temperature. However, if the temperature is returned to the original point, the water will become liquid again; the water has been subjected to physical change only, because its characteristics stayed the same.

In summary, matter can be changed, but it cannot be destroyed.

OUO1010,0000E98 -19-05AUG09-1/9

MASS

Mass is often confused with weight. **Mass is the measure of how much matter is in a body.** Weight is the measure of Earth's gravitational pull. A body has the same mass on Earth (A) as it has 2,000 miles (3200 km) (F) out in space, but its weight is much less out in space.

The illustration shows a man weighing 170 pounds (77 kg) (B) when standing on the Earth, but out in space he may weigh only 5 pounds (2.25 kg) (D). However, in either location he has the same amount of matter, or equal mass (C), in his body.

A—Earth   D—5 pounds (2.25 kg)
B—170 pounds (77 kg)   F—2,000 miles (3200 km)
C—Equal Mass

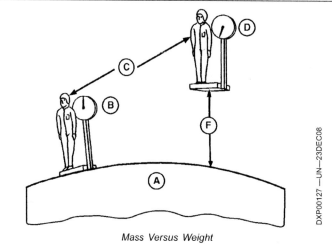

Mass Versus Weight

Continued on next page   OUO1010,0000E98 -19-05AUG09-2/9

ENERGY

Electricity, light, sound, and heat are forms of energy. They do not occupy space or have weight in the usual sense. It is energy that produces changes in matter.

DXP00128 —UN—23DEC08

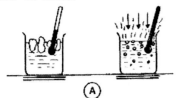 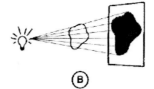 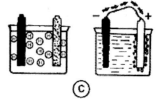

Energy at Work

A—Heat   B—Light   C—Chemical

The illustration shows heat energy converting water (A), light energy forming an image on film (B), and electrical energy working in a chemical cell (C).

Chemical energy heats your home and runs an engine, and mechanical energy does work.

OUO1010,0000E98 -19-05AUG09-3/9

INERTIA

Inertia is the tendency of a body at rest to remain at rest or a body in motion to stay in motion in a straight line unless acted on by an outside force.

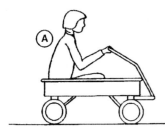 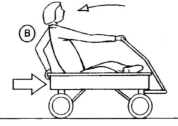 

Inertia

A—At Rest   B—Pitch Rearward [When   C—Pitch Forward [from Sudden
                     Pushed from Behind]   Stop]

If you're sitting in a wagon at rest (A) and someone gives the wagon a push from behind; your body will pitch rearward (B). Nothing actually pushed you rearward, your body just tried to stay at rest.

If someone stops the wagon while you are moving forward, you will pitch forward (C). This is because your body wants to keep moving at the same speed.

The larger the mass of your body, the more you will be affected by inertia.

Continued on next page

OUO1010,0000E98 -19-05AUG09-4/9

FORCE

A force is a push or pull which starts, stops, or changes the motion of a body.

From this we conclude that if all the forces acting on a body are equal from all directions, the body will be at rest. If any one of the forces is greater, the body will be set into motion in the direction of the force.

If the equal forces are applied where shown in the illustration, the box will retain its position. When one of the forces becomes greater than the others, the box will move in the direction of the greater force.

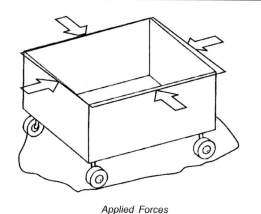

Applied Forces

OUO1010,0000E98 -19-05AUG09-5/9

MOMENTUM

When a body is in motion, it is said to have momentum, which is the product of its mass and velocity (speed). A body moving in a straight line will keep going in a straight line at the same speed forever if no other forces act upon it. The laws of momentum are equally effective when a body is rotating; it would continue to rotate. Momentum and inertia are sources of energy because of their mass.

Momentum Force

OUO1010,0000E98 -19-05AUG09-6/9

TORQUE

If the forces applied to a body do not all act at a single point, the body will tend to rotate. The turning effect of any force applied to a body is found by multiplying the amount of the force by the distance from the pivot point to the line of the force.

This turning effect of a force is called torque, and the distance mentioned is the torque arm.

Consider the torque wrench shown in the illustration. If we apply more force at the point shown or apply the force farther out, the torque will be increased. This increased torque will cause the wrench to either rotate faster or give us more turning force at the pivot point.

An engine crankshaft reacts to the pushing force of the piston and connecting rod in the same manner.

TORQUE (Tw)

FORCE (F)

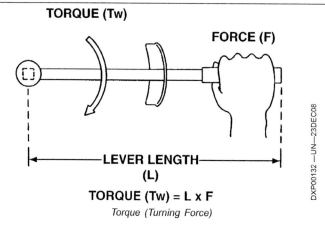

LEVER LENGTH (L)

TORQUE (Tw) = L x F

Torque (Turning Force)

Continued on next page   OUO1010,0000E98 -19-05AUG09-7/9

WORK

When you're pushing on a large rock and it fails to move, you feel like you're working hard. In Physics, work is accomplished only when the rock is moved by the pushing you're doing. Work is expressed as a force unit multiplied by a distance unit.

If you stand and hold a 20 lb (9 kg) weight at the 2-foot (600 mm) level as shown in the illustration, you aren't really doing any work, as the weight is stationary. However, if you move it to the 5-foot (1.5 m) level, you have moved the weight 3 feet (900 mm), and work was done. The path you take to get to the 5-foot (1.5 m) level is not important; the amount of work done is 20 lb (9 kg) times 3 feet (900 mm) = 60 pound feet (80 J), whether you take short path A or long path B as shown.

A—Short Path   B—Long Path

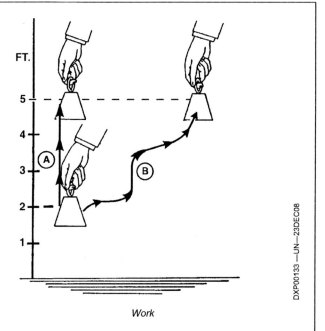

Work

OUO1010,0000E98 -19-05AUG09-8/9

MECHANICAL POWER

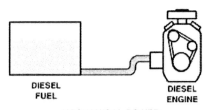

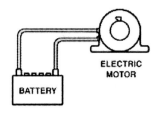

Mechanical and Electrical Power Compared

Power is a rate of doing work and the term mechanical is the energy method used. Other energy methods include chemical, electrical, heat, and sound.

Energy conversion allows us to compare different types of power. A diesel engine produces mechanical power by chemical change. Fuel (liquid) and air (gas) is burned (chemical change) in the combustion chamber and the engine crankshaft rotates.

When diesel fuel burns and produces heat at the rate of 2545 Btu per hour (746 watts), the fuel energy is being expended in the engine at the rate of 1 horsepower per hour.

If a battery delivers electricity to an electric motor for 1 hour at the rate of 746 watts, the electric motor will consume 746 watt-hours or 1 hp-hr of energy.

It is interesting to note that by simple conversion, 746 watts of electric power equals 2545 Btu per hour when converted to heat.

This completes our discussion of the basic laws of mechanics.

OUO1010,0000E98 -19-05AUG09-9/9

Engine Measurements and Performance

A number of engine measurement factors determine the ability of an engine to produce usable power. The engine manufacturer determines these measurements when designing the engine to provide the size of the engine suited for particular applications.

CYLINDER BORE

The cylinder bore of an engine is the diameter (distance across the cylinder at the center) of the engine's cylinders. The measurement is usually expressed in inches or millimeters.

The engine designer determines the size of the cylinder bore based on the number of cylinders, cylinder length, engine configuration, and other factors.

PISTON STROKE

The piston stroke is the distance traveled by the piston either from the bottom dead center (BDC) to its top dead center (TDC) position or TDC to BDC. The length of the stroke is usually expressed in inches or millimeters.

The design of the crankshaft determines the piston stroke. The connecting rod journals are offset from the crankshaft centerline, which causes the rod journals to orbit the crankshaft centerline as the crankshaft rotates. The distance from the centerline of the crankshaft to the center of the rod journal is exactly one-half the piston's stroke. This is sometimes called the crankshaft throw.

Long-stroke engines are usually slower speed engines, while short-stroke engines can be run faster because of the smaller orbit path of their rod journal around the crankshaft centerline Although both engines may have the same horsepower rating, the long-stroke engines will generally produce more torque and have better lugging ability. This is a result of its longer power stroke and the greater torque arm length of the crankshaft throw.

DISPLACEMENT

The displacement of an engine is determined by the cylinder bore diameter, the length of the stroke, and the number of cylinders. Engine displacement is expressed as cubic inches, cubic centimeters, or liters.

Each piston in the engine displaces a given amount or volume of air as it moves from the bottom to the top of its stroke. This is called piston displacement. The area of the piston head and the length of the stroke determines the piston displacement.

The area of the piston head can be determined by using the following formula: *Area = (Bore multiplied by Bore) multiplied by 0.7854.* Therefore, the piston head area of an engine that has a 5.0-inch bore would be: (5.0 multiplied by 5.0) = 25.0 multiplied by 0.7854 = 19.64 square inches.

The piston displacement can be determined by using the following formula: *Piston Displacement equals Piston*

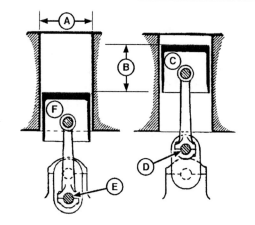

Two Basic Engine Measurements Are Cylinder Bore (Diameter) and Piston Stroke (Travel)

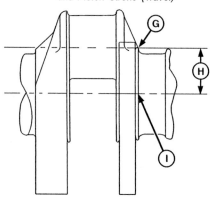

The Crankshaft Throw Determines the Piston Stroke

A—Bore	F—BDC
B—Stroke	G—Rod Journal Centerline
C—TDC	H—Crankshaft Throw
D—Crankshaft at TDC	I— Crankshaft Centerline
E—Crankshaft at BDC	

Area multiplied by Piston Stroke. The piston displacement of an engine that has a 5.0 inch bore and a 4.4 inch stroke would be: (5.0 multiplied by 5.0) = 25.0 multiplied by 0.7854 = 19.64 (piston area) multiplied by 4.4 (stroke) = 86.42 cubic inches.

The total displacement of an engine can be determined by multiplying the piston displacement by the number of cylinders. The total displacement of a six-cylinder engine that has a 5.0-inch bore and 4.4-inch stroke would be: 86.42 cubic inches (piston displacement) multiplied by 6 (number of cylinders) = 518.52 cubic inches.

In metric terms, the displacement of a six-cylinder engine with a 130.0 millimeter bore and a stroke of 110.0 millimeters would be calculated as follows: First, since metric displacement is expressed in cubic centimeters, it is necessary to convert the millimeter dimensions to centimeters. Divide the millimeter (mm) dimensions by 10 to convert the bore and stroke dimensions to centimeters (cm). Example: 130.0 mm divided by 10 equals 13.0 cm.

Continued on next page

OUO1010,0000E86 -19-14JUL09-1/6

The piston displacement and engine displacement in metric terms are determined using the same formulas as for the unified inch measurements. The piston area would be: (13.0 cm x 13.0 cm) multiplied by 0.7854 = 132.73 square centimeters. The piston displacement of one cylinder would be: (13.0 x 13.0) multiplied by 0.7854 = 132.73 multiplied by 11.0 cm = 1460 cubic centimeters. The total displacement of the engine would be: 1460 (piston displacement) multiplied by 6 (number of cylinders) = 8760 cubic centimeters or 8.760 liters.

The power that an engine produces depends in part on the total displacement of the engine. Engines with more displacement are able to take in a greater amount of air on each intake stroke and can therefore produce more power.

HORSEPOWER

Because we are mainly concerned with engines, our unit of power is the horsepower.

A Scotsman named James Watt is credited with defining and standardizing this unit of power in the late 18th century. Watt, an engineer and designer of steam engines, sought to accurately define a unit of power for use in rating his steam engine. Since the draft horse was the primary source of power at the time, it was logical to define the output of the steam engine in terms of how many horses it could replace.

As Watt tried to compare the power of a steam engine with the power of a horse, he found that an average-sized draft horse could pull 330 pounds a distance of 100 feet in one minute. By multiplying 330 pounds by 100 feet,

he determined that 33,000 pounds-feet of work (force multiplied by distance equals work) per minute was equal to one horsepower.

Therefore, it can be said that one horsepower is the ability to do 33,000 pounds-feet of work in 1 minute, or 550 pounds-feet of work in 1 second (33,000 divided by 60 seconds equals 550). This evaluation has been accepted since that time as the unit of measurement in rating engines and motors.

The formula for horsepower is:

$$\text{Horsepower} = \frac{\text{Force (pounds) x Distance (feet)}}{\text{Time (minutes) x 33,000}}$$

In metric terms, horsepower is measured in watts. One horsepower equals approximately 746 watts or 0.746 kilowatts (kW).

There are several categories of horsepower, all very necessary for the design of an efficient engine. To sum them up quickly, we talk about **theoretical horsepower** and **net horsepower (useful)** and of those in between.

The most common horsepower terms are:

- **Indicated (IHP)**
- **Friction (FHP)**
- **Flywheel or Brake (BHP)**
- **Drawbar**
- **Power Take-Off (PTO)**
- **Rated**

Let's see what each type of horsepower means.

Continued on next page   OUO1010,0000E86 -19-14JUL09-2/6

INDICATED HORSEPOWER (IHP):

Indicated horsepower (IHP) is the power that an engine is theoretically capable of producing. Indicated horsepower is determined by taking measurements in the combustion chamber of a cylinder using a special instrument. The instrument measures the actual gas pressure developed during the combustion process. Using this measurement, an engineer can calculate the amount of energy that is released in the cylinder.

However, we are more interested in the measurements that are more useful. Indicated horsepower neglects such things as the friction of the moving engine parts and the type of power that we actually need to do our work. Indicated horsepower is *theoretical* horsepower.

FRICTION HORSEPOWER (FHP):

Friction horsepower (FHP) is the power required to overcome the friction between engine parts such as the pistons and cylinder walls, the crankshaft and bearings, and the power needed for compression. Friction is a loss factor and a producer of heat.

Remember, energy cannot be destroyed, merely converted or divided. If the bearings are heating up while the mechanical energy is working, some of that mechanical energy is being converted to heat and is lost into the cooling system.

FHP then is the difference between indicated horsepower and usable horsepower. So it is a factor in engine *efficiency*.

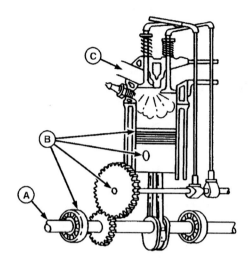

Indicated Horsepower Minus Friction Horsepower

A—Usable HP at Engine Output Shaft
B—Friction HP (Friction between Parts that Subtract from Indicated HP)
C—Indicated HP (Theoretical Power in the Cylinder)

Lubricating oils place a thin film between two surfaces to reduce friction. In most engines, bearings are lubricated with oil under pressure to make the shafts float on an oil film and so reduce friction.

Continued on next page

OUO1010,0000E86 -19-14JUL09-3/6

PN=37

FLYWHEEL HORSEPOWER OR BRAKE HORSEPOWER:

Now we come to the first really practical unit of measurement for an engine. This is the point where we can couple to the engine and actually draw power.

Flywheel or brake horsepower is the FHP (friction losses) subtracted from the IHP (theoretical horsepower). For simplicity, FHP is all the engine losses; friction, power needed for compression, etc. If the losses equal 10 horsepower (7.5 kW), and if IHP equals 100 horsepower (75 kW), the result is 100 minus 10 = 90 hp (75 minus 7.5 = 67.5 kW), which is the flywheel horsepower.

Flywheel horsepower is also called brake horsepower (BHP). It is the maximum usable horsepower the engine can produce without alteration.

Flywheel or brake horsepower is measured by a *Prony brake* or a *dynamometer*. Both test instruments apply a load to the engine that is measured in pounds (kilograms). Engine speed is measured with a tachometer in revolutions per minute (rpm).

The Prony brake uses a friction device with an arm attached to it to apply a load to an engine. The other end of the arm is attached to a scale. The arm deflects in proportion to the force applied to it and actuates the scale, which indicates the load in pounds (kilograms).

If the length of the arm is known in feet (meters), we can measure pound-feet (lb-ft) [Newton-meters (N·m)] of force as engine load. Example: If the length of the arm is 4 feet and the brake is applied to provide a reading on the scale of 60 pounds, the force or torque would be 60 pounds multiplied by 4 feet = 240 lb-ft (325 N·m) of torque.

If the engine speed under these conditions is 1800 rpm, the brake horsepower (BHP) can be calculated as follows: torque multiplied by rpm divided by a coefficient of 5252 equals BHP. Example: 240 lb-ft multiplied by 1800 rpm divided by 5252 equals 82.25 BHP. This can be converted to kilowatts (kW) by multiplying 82.25 by 0.746 = 61.36 kW.

Dynamometers may be classified as follows:

• Engine Dynamometer
• PTO Dynamometer
• Chassis Dynamometer (shown in illustration)

The engine dynamometer is coupled directly to the engine's flywheel, thus it measures flywheel or brake horsepower.

The PTO dynamometer is driven by the machine's power take-off (PTO) shaft. This type of dynamometer is typically used to measure the horsepower of agricultural tractors.

The chassis dynamometer applies a load to an engine via the vehicle's drive wheels. The chassis dynamometer

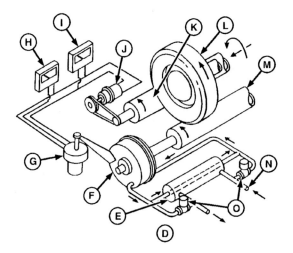

Measuring Engine Horsepower (Chassis Dynamometer Shown)

D—Chassis Dynamometer	J— Tachometer Generator
E—Heat Exchanger	K—Idle Roll
F—Power Absorption Unit	L—Power Source (Machine
G—Torque Bridge	Wheel)
H—Power Meter	M—Driven Roll
I— Speed Meter	N—Water Supply
	O—Solenoid Valves

gives a true indication of the engine's usable horsepower, because the engine's indicated horsepower will be reduced by the friction horsepower loss that occurs in the vehicle's power train.

Most dynamometers, whether of the engine type, PTO type, or chassis type, convert the torque and speed measurements automatically to a brake horsepower or road horsepower reading on a dial.

It is important to understand that no matter where you measure it or how you measure it, 1 horsepower is 33,000 lb-ft per minute or 746 watts.

For more information on dynamometer testing, see Chapter 14.

PTO (POWER TAKE-OFF) HORSEPOWER:

PTO horsepower is a function of torque and speed (rpm) and is measured at the machine's power take-off shaft.

A power take-off usually has some gear reduction between the engine and the PTO shaft. This reduction increases the torque value but reduces the speed. When measuring PTO horsepower, the speed is usually held constant at 1,000 rpm, so the horsepower can be read directly on a gauge measuring torque but having its scale calibrated in horsepower.

Continued on next page   OUO1010,0000E86 -19-14JUL09-4/6

DRAWBAR HORSEPOWER:

Drawbar horsepower is the measure of pulling power an engine can produce when mounted in a moving machine. The load is attached to the machine, and the horsepower required to move the machine is calculated by knowing the force required to move the load and the speed with which it is moved.

A—Load
B—Drawbar Horsepower

C—Machine

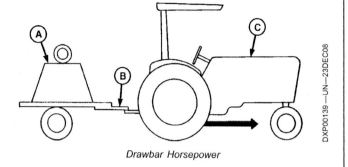

Drawbar Horsepower

OUO1010,0000E86 -19-14JUL09-5/6

RATED HORSEPOWER:

Rated horsepower is a value used by engine manufacturers to indicate the horsepower an engine should produce under normal operating conditions. This rating takes into account the maximum pressure forces in the engine as well as the speed and torsion forces. If these values are exceeded, the engine can be damaged.

Rated horsepower depends in part on the total cubic inches of piston displacement in the engine. From this, the engine manufacturer determines the maximum pressure stresses and rpm the engine can tolerate without internal damage. The manufacturer then tests and develops the engine for long life and reliability at a given rated horsepower.

Rated horsepower may not be the most efficient operating point for the best fuel consumption; this rating is expressed in terms of recommended operating horsepower and rpm.

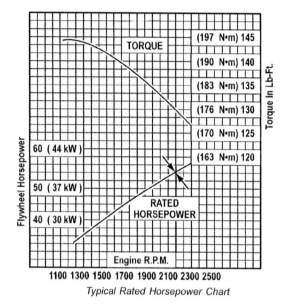

Typical Rated Horsepower Chart

OUO1010,0000E86 -19-14JUL09-6/6

Engine Efficiency

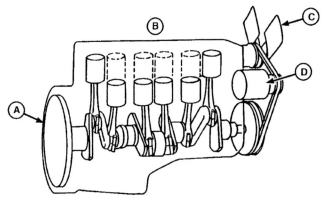

Mechanical Efficiency

A—Flywheel [Subtract 2 HP (1.5 kW)]

B—Engine Friction and Heat [Subtract 20 HP (15 kW)]

C—Fan and Water Pump [Subtract 5 HP (3.7 kW)]

D—Alternator or Generator [Subtract 1 HP (0.7 kW)]

The importance of efficiency can not be overstressed. Efficiency represents more than fuel economy; it also means the ability to do work at a constant rate with low maintenance.

Here are the prime efficiency factors in engines:

• **Mechanical Efficiency**
• **Volumetric Efficiency**
• **Thermal Efficiency**
• **Effective Pressure**
• **Fuel Consumption**
• **Compression Efficiency**
• **Load Effects**

Let's take a look at each type of engine efficiency.

MECHANICAL EFFICIENCY

We've mentioned some losses in horsepower between IHP and BHP — friction losses, etc. But in a functional engine other things also absorb horsepower:

• Fuel Pump

• Water Pump
• Cooling Fan
• Generator or Alternator
• Ignition System
• Valves
• Oil Pump
• Blowers and Superchargers
• Hydraulic Pumps (on standby)
• Air Conditioner Compressor

Driving all these extra devices draws power from the engine and reduces its efficiency.

Mechanical efficiency then takes into account all these losses as well as frictional losses. To get a true value for mechanical efficiency, the engine must be operating at its rated output, with all accessories performing their normal functions when we measure flywheel of brake horsepower (BHP). Then we divide the BHP by the IHP and multiply by 100 to get the mechanical efficiency of the engine.

Continued on next page

OUO1010,0000E87 -19-23JUL09-1/4

VOLUMETRIC EFFICIENCY

Most engines get intake air into the cylinders by creating a partial vacuum as the piston travels down on the intake stroke. However, the intake manifold, carburetor, air cleaner, and intake system restrict the amount of air that can actually get into the cylinder.

Volumetric efficiency is calculated by dividing the actual amount of engine air taken in the by the piston displacement and multiplying by 100.

Volumetric efficiency is one of the main factors governing the maximum torque output of an engine. The rpm at which an engine "breathes" the best will often determine the point of maximum torque.

THERMAL EFFICIENCY

The ratio of the work done by the gases in a cylinder (indicated work) to the heat energy (thermal energy) of the fuel is called thermal efficiency. Thermal efficiency is a laboratory value. We want something more practical — **brake thermal efficiency**.

Brake thermal efficiency is brake horsepower, converted to Btu, divided by the fuel heat input in Btu, and multiplied by 100.

Thus, the brake thermal efficiency tells us how effectively an engine converts heat energy into usable power.

Brake thermal energy takes into account all engine losses and is sometimes called overall efficiency.

The following table shows the formula applied to an engine that develops 100 flywheel or brake HP (75 kW) per hour while consuming 800,000 Btu (234 kW) of fuel per hour:

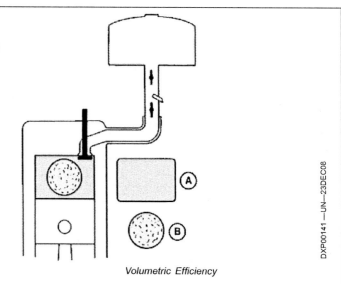

Volumetric Efficiency

A—Piston Placement   B—Actual Air Intake

Engine Brake HP (BHP) = 100 HP (75 kW)
Fuel Burned per Hour (Fuel Heat Input) = 800,000 Btu (234 kW)
NOTE: 1 Horsepower per Hour = 2544 Btu (0.75 kW)
2544 x 100 BHP = 254,400 Btu (75 kW)
BRAKE THERMAL EFFICIENCY = $\dfrac{254,400}{800,000}$ x 100 = 31.8%

Continued on next page   OUO1010,0000E87 -19-23JUL09-2/4

MEAN EFFECTIVE PRESSURE

Mean effective pressure consists of two types:

- **Indicated Mean Effective Pressure (IMEP)**
- **Brake Mean Effective Pressure (BMEP)**

The elements shown are:

- Beginning of compression stroke (A)
- Start of injection (B)
- Beginning of combustion (C)
- Peak firing pressure (D)
- Exhaust valve opens (E)
- Ignition delay (B to C) *
- Power stroke (D to E)

* Ignition delay is the period of time from the beginning of injection until the actual burning starts.

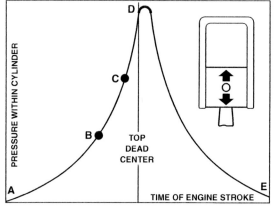

Typical Diesel Engine Pressure Indicator Tracing

These pressures are developed within the cylinder. Pressures increase from the beginning of the compression stroke, become maximum at peak firing pressure, and decrease as the power stroke progresses until the exhaust valve opens.

Indicated mean effective pressure (IMEP) is the average pressure in the cylinder. This is measured by an instrument which records a tracing on a calibrated chart. From this tracing the actual horsepower developed within the cylinder can be calculated.

Calculating the indicated mean effective pressure from the tracing is a complicated process requiring instruments not commonly available outside of engineering establishments. For those interested in more detail, refer to engineering publications on the subject.

IMEP is used to calculate the horsepower the engine develops without taking into consideration the losses due to friction and other losses (heat, volumetric efficiency, etc.). Obviously this calculated horsepower is much higher than the usable horsepower obtained from the engine.

Brake mean effective pressure (BMEP) is calculated from the actual horsepower developed by the engine as measured by a dynamometer.

COMPRESSION RATIO

Compression ratio is the ratio of the volume of the combustion chamber at the beginning of the upstroke of the piston to the volume at the end of that stroke.

Thus, if the volume of the combustion chamber at the end of the upstroke is one-tenth (1/10) of the volume at the bottom of the upstroke, the compression ratio of the engine would be ten to one — commonly expressed 10:1.

Continued on next page

OUO1010,0000E87 -19-23JUL09-3/4

FUEL CONSUMPTION

Fuel consumption is generally measured by pounds (kilograms) of fuel per horsepower-hour (kilowatt-hour).

In a gasoline engine, air-fuel ratio is considered best at 15 pounds (7 kg) of air to one pound (0.5 kg) of fuel. Pounds (kilograms) of fuel per horsepower-hour (kilowatt-hour) is the measure of engine efficiency.

Each engine manufacturer can furnish a chart that indicates fuel consumption at various speeds and horsepower outputs.

COMPRESSION EFFICIENCY

Increased compression ratio permits the air-fuel mixture to be compressed more, which in turn gives greater heat and expansion during combustion.

Also, the higher compression and the resulting extra heat and pressure give better burning and more energy.

There are limiting factors in the thermal efficiency produced by increasing the compression ratio. These

The Amount of Fuel Used Is Measured by Pounds

include mechanical stresses, temperatures, and combustion chamber pressures.

Increased compression ratio may reduce fuel consumption, but certain fuels do not permit high compression ratios.

OUO1010,0000E87 -19-23JUL09-4/4

Load Effects

Every phase of engine efficiency is affected by the nature of the load the engine drives.

A fluctuating load can set up serious vibration and, in extreme cases, "lugging" of the engine.

"Lugging" slows an engine while the throttle is still wide open. The result is excessive heat and firing pressures. This may cause the engine to be damaged.

Lean fuel mixtures result in high cylinder temperatures and possible detonation with certain fuels.

USES OF THERMODYNAMICS

Thermodynamics is the science that deals with heat and mechanical energy and their conversion one to the other.

The basics are as follows:

- **Laws of Thermodynamics**
- **Boyle's Law**
- **Charles' Law**
- **Boyle's and Charles' Law Combined**
- **Conservation of Energy**
- **Heat and Energy**
- **Adiabatic Compression**
- **Compression Ratio vs. Pressure**

Let's see what each of these means to the engine.

LAWS OF THERMODYNAMICS

Two basic laws of thermodynamics apply to engine fundamentals: *The first states that energy is conserved. The second deals with the irreversibility of this process.*

Mechanical energy can be converted completely to heat but heat energy can never be converted completely to mechanical energy.

These fundamental truths are based upon the fact that heat naturally travels to a lower-temperature object and seeks equilibrium.

An engine burns fuel and develops heat to produce mechanical energy for doing work. However, a major portion of the heat in the combustion chamber is absorbed by the cylinder walls or blown out the exhaust system; it is lost as far as mechanical energy resulting in BHP output is concerned.

Therefore, the efficiency of an engine is limited in part by the engine temperature range.

BOYLE'S LAW

Boyle's Law says that a gaseous mass can be compressed and that its volume is inversely proportional to the pressure on it, as long as the temperature stays the same.

This law is a governing factor in engine design, relative to piston displacement and compression ratio.

CHARLES' LAW

Charles' Law says that temperature changes on a gaseous mass result in direct changes of volume and pressure.

If the temperature of the gas is raised and the volume is held constant, the pressure will increase; raising the temperature and holding the pressure constant will increase the volume of the gas.

The heat energy in an engine combustion chamber causes the air-fuel mixture to expand. The combustion chamber is somewhat restricted in volume (compression ratio) and the change in volume is quite large over the entire power stroke. However, the change is not as great as the gas would need for full expansion at combustion, so the pressure is raised appreciably.

BOYLE'S AND CHARLES' LAWS COMBINED

The diesel engine takes advantage of the laws just discussed.

Air (a gas) is fed into the combustion chamber. Reducing the volume during the compression stroke raises the air temperature and the pressure.

Fuel injected into this mixture will then ignite due to the high temperatures of compression. Ignition further raises the temperature and the gases expand to force the piston down on the power stroke.

The pressure applied is dependent on the rate of burning and the heat retained in the gases and not lost through the cylinder walls and exhaust system.

This same principle is the cause of detonation or the preignition in gasoline engines. Gasoline will ignite at much lower temperatures than diesel fuel. If the temperature is raised too much by compression, spontaneous ignition will take place before the compression stroke is complete and the engine will tend to run backwards.

Engine design must, therefore, take into account all known conditions resulting from the three laws.

CONSERVATION OF ENERGY

Energy can neither be created nor destroyed. The different forms (work, heat, etc.) are mutually convertible. When, in some thermodynamic change, a quantity of one form disappears, an equivalent quantity must necessarily appear.

HEAT AND ENERGY

Heat is the greatest source of energy loss, as most engines operate in the varying temperature of the atmosphere.

Remember that heat energy lost to the engine cannot be recovered. This includes the normal losses of heat through the engine cooling and exhaust systems.

Continued on next page

OUO1010,0000E88 -19-14JUL09-1/2

100709
PN=44

COMPRESSION RATIO VERSUS PRESSURE

The compression ratio of an engine is fixed by the design engineer. The engineer then determines the combustion pressures that result under normal operation of the engine.

Tampering with the fuel rate or combustion rate can change the cylinder pressure and may damage the engine.

The higher the compression ratio of gasoline, natural gas and LP-gas engines, the more critical become fuel quality and operating temperatures.

Detonation is the stray, unwanted explosions that take place in the engine as a result of too-high compression or heat, or low-quality fuels. These explosions cause piston and cylinder damage by creating excess pressure.

"Lugging" a high-compression engine at low rpm can cause serious detonation as the charge is in the combustion chamber a longer interval.

The laws of thermodynamics cannot be ignored when efficient, trouble-free engine performance is required.

OUO1010,0000E88 -19-14JUL09-2/2

Test Yourself

Questions

1. (Fill in the blanks.) An internal combustion engine converts _____ energy into _____ energy.

2. What three basic elements are needed to produce heat energy in the engine cylinder?

3. Air is compressed in the engine cylinder. What else happens to this air?

4. Fuel must be in what form for good combustion in the engine?

5. Match the two items at left with the definitions at the right.

   a. Rotary Motion   1. Up-and-down movement
   b. Reciprocating Motion   2. Circular movement

6. (Fill in the blanks.) The engine converts _____ motion to _____ motion.

7. What three engine measurement factors determine the displacement of an engine?

8. Which of these statements is true?

   a. One horsepower is the amount of power required to raise 550 pounds (250 kg) a distance of 1 foot (30 cm) in 1 second.

   b. One horsepower is the amount of power required to raise 33,000 pounds (15,000 kg) a distance of 1 foot (30 cm) in 1 minute.

9. If an engine compresses eight parts of air into a one part volume, what is its compression ratio?

10. Which type of engine normally has a higher compression ratio — a gasoline engine or a diesel engine?

11. Why is the compression ratio higher on a diesel engine?

12. Name the four parts of an engine combustion cycle in the proper sequence.

13. (True or false?) In a four-stroke cycle engine, the crankshaft turns one complete revolution during a complete cycle.

14. Match each item on the left with the correct one on the right.

   a. Diesel engine   1. Fuel is mixed with air in the cylinder
   b. Gasoline engine   2. Fuel is vaporized before mixing with air
   c. LP-gas and natural gas   3. Fuel is mixed with air before it goes
      engines   into the cylinder

15. What are the two major types of engine cooling systems?

OUO1010,0000E89 -19-14JUL09-1/1

Safety Rules for Engines — Introduction

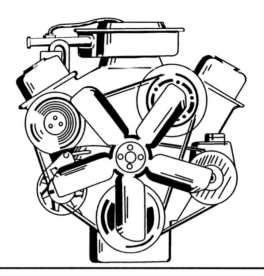

Always Exercise Safety

In the beginning of this book, we explained the many moving parts of the engine and the high forces that can be generated in the engine. These can cause serious injuries and even death, if reasonable care is not exercised.

The engine's cooling system, fuel system, and electrical system also present potential hazards that must be recognized when working on the engine. The following information is designed to alert you to the potential hazards, and to keep you safe if you follow the instructions.

Safety is too expensive to learn by accident. Accidents result in lost time from work and more importantly, may cause a permanent handicap or a loss of health that affects the injured party's family and earning ability. Unsafe practices can also result in property damage. Always exercise safety.

DP79986,000003F -19-22JUL09-1/1

Shop Practices and Work Habits

The best way to minimize hazards associated with engines and the equipment they power is to follow proven shop practices and employ good work habits. Safety must be thought of as a normal part of the shop management process just as supplies, personnel, and overhead costs.

Keep all tools and service equipment in good condition, clean, and organized. A clean, well-equipped shop promotes safety and is indicative of someone who takes pride in doing a good job. You'll be able to do the job quicker and easier if you have the appropriate tools for the job easily accessible and in good working condition.

Wear eye protection. Plastic goggles protect eyes from impact from the front and sides. Unvented or chemical splash goggles also offer protection against chemical vapors and liquids.

Rest regularly to avoid the effects of fatigue. Be alert. Never use drugs or alcohol when working on equipment.

Avoid horseplay. Never work on dangerous equipment when you are ill, angry, or anxious.

Make sure someone knows that you are working in the shop and will check on you and render aid if you are injured.

Understand service procedures before doing work. Always read, understand, and follow the instructions and manuals provided with the equipment. Don't guess. Keep manuals handy in a clean, dry, readily accessible place.

Never lubricate, service, or adjust a machine while it is operating. Keep hands, feet, and clothing from power-driven parts. Stop the engine and remove the key before you begin your service procedure.

Be aware of common machine hazards. Don't take shortcuts. Shortcuts can be hazardous, and because of the problems that they can cause, cost more in time and money than they save.

Securely support any machine elements that must be raised for service work.

Use tools appropriate to the work. Makeshift tools and procedures can create safety hazards. For loosening and tightening hardware, use the correct size tools. DO NOT use U.S. measurement tools on metric fasteners. Avoid bodily injury caused by slipping wrenches.

Use Proper Tools

DXP00148 —UN—23DEC08

Continued on next page OUO1010,0000E99 -19-05AUG09-1/3

Keep floors and benches clean to reduce fire and tripping hazards. Clean up as you go while doing a job, and clean the area completely after the job is done.

Keep lighting, wiring, heating, and ventilation systems in good condition.

Don't let others use tools or service equipment unless they have had adequate instruction.

Keep guards and other safety devices in place and functioning.

Practice Safe Maintenance

OUO1010,0000E99 -19-05AUG09-2/3

Be prepared for emergencies. Keep the fire extinguishers serviced, and the first aid kit replenished with supplies.

Keep emergency numbers for doctors, ambulance service, hospital, and fire department near your telephone.

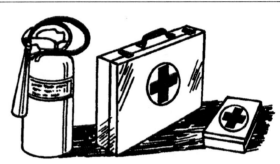

Prepare for Emergencies

OUO1010,0000E99 -19-05AUG09-3/3

Service Cooling Systems Safely

Explosive release of fluids from a pressurized cooling system can cause serious burns.

Before removing the radiator filler cap, shut off the engine and allow to cool. Only remove the filler cap when cool enough to touch with bare hands. Slowly loosen the cap to the first stop to relieve pressure before removing completely.

Use leak-proof containers when draining coolant. Do not use food or beverage containers that may mislead someone into drinking from them.

Properly dispose of used engine coolant. Heavy metals, such as lead, copper, and zinc, accumulate in the ethylene glycol base as a result of corrosion that occurs in a cooling system. Do not pour on ground, down a drain, or into any water source.

Avoid Explosive Release of Fluids

Inquire on the proper way to recycle or dispose of waste from your local environmental or recycling center.

OUO1010,0000E9A -19-15JUL09-1/1

Handle Chemical Products Safely

Direct exposure to hazardous chemicals can cause serious injury and even death. Many chemicals are routinely used in the shop, often without any thought given as to their hazardous nature. Potentially hazardous chemicals include such items as lubricants, coolants, paints, adhesives, and solvents.

A Material Safety Data Sheet (MSDS), supplied by the chemical or equipment manufacturer, provides specific details on chemical products: physical and health hazards, safety procedures, and emergency response techniques.

Check the MSDS before you start any job using a hazardous chemical. That way you will know exactly what the risks are and how to do the job safely. Then follow the procedures and recommended equipment.

Read and Understand the Material Safety Data Sheet

OUO1010,0000E9B -19-12JUN09-1/1

Chemicals and Cleaning Equipment

In service operations, cleaning is needed in order to:

• **Keep dirt from entering the machine when parts are disassembled**
• **Inspect parts for wear and damage**
• **Install and adjust parts properly during assembly**

Let's discuss the safe use of solvents, steam cleaners, and high-pressure washers.

Solvents

Most solvents are toxic, caustic, and flammable. Be careful, therefore, to keep them from being taken internally, from contacting the skin and eyes, and from catching on fire. Read the manufacturer's instructions and precautions before using any commercial cleaner or solvent.

Whenever possible, use a commercially available solvent. Many types are available for general cleaning and for specific cleaning jobs. Always read and follow the Material

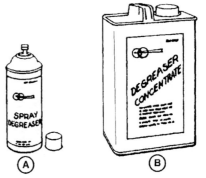

Use Commercial Cleaning Solvents, Not Gasoline

A—Small Pressure Spray Can **B—Bulk Concentrate (Dilute Before Application)**

Safety Data Sheet (MSDS) to get the best results and to be able to handle each product safely.

OUO1010,0000E9C -19-15JUL09-1/7

Never clean with gasoline. It vaporizes at a rate sufficient to form a flammable mixture with air at temperatures as low as –45°C (–50°F). It is always unsafe to use. If you don't use a commercial solvent, use diesel fuel or kerosene. These will burn, but not as easily or explosively as gasoline.

A—1 Gallon of Gasoline **B—83 Pounds of Dynamite**

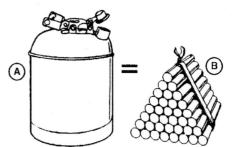

One Gallon of Gasoline Mixed with the Right Amount of Air Equals the Explosive Force of 83 Pounds of Dynamite

Continued on next page
OUO1010,0000E9C -19-15JUL09-2/7

Follow these general safety practices when using chemical cleaners:

1. **Follow the manufacturer's instructions**. Cleaning agents intended for the same purpose may be quite different in chemical composition. Read the label on each container you buy and follow the instructions carefully.

2. **Protect your skin and eyes**. Wear a face shield and rubber gloves when working with strong, concentrated cleaning solutions. Some are caustic, and they can destroy the natural oils of the skin and burn it severely. Always read and follow the manufacturer's instructions carefully. Even when gloves are not called for, avoid long exposures to solvents, and wash your hands when the job is done. Wear an apron if it's needed to keep your clothing dry.

3. **Work in a well-ventilated area**. Do the cleaning outdoors if possible. If you can't, provide ventilation by opening doors and windows.

4. **Keep solvents away from sparks and flame**. Don't let anyone smoke in the immediate area. Don't use solvents near heaters, sparks, or open flames. Some solvent tanks and tubs should be "grounded" to prevent electrical discharges. Contact your electrical supplier for instructions on how to ground your solvent tank or tub.

5. **Never heat solvents unless instructed to do so**. And don't mix them, because one might vaporize more readily and act as a fuse to ignite the other.

Wear Face Shield, Apron, and Rubber Gloves When Cleaning with Cleaning Solutions

A—Concentrated Solvent D—Hard Hat
B—Rubber Gloves E—Apron
C—Face Shield

6. **Wipe up spills promptly**. Keep soaked rags in closed metal containers and dispose of them promptly.

Continued on next page OUO1010,0000E9C -19-15JUL09-3/7

7. **Store solvents in their original containers or in sealed metal cans properly labeled**. In a sealed container, the solvent will be kept clean from further contamination, toxic fumes will be controlled, the fire hazard will be reduced, and there will be less likelihood of spillage. Never use open pans.

8. **If a commeracially made parts washer is used, close the lid when you're finished cleaning**. Never destroy the fusible link that closes the lid automatically in case of fire.

9. **Avoid accidental poisoning**. Wash your hands and arms before eating or smoking. Keep all solutions in labeled containers. Never use empty solvent containers, no matter how thoroughly cleaned, for carrying food or beverages. Keep poison containers sealed even when they are empty. Keep them out of the reach of children.

10. **Be prepared for emergencies**. Keep fire-fighting equipment near your cleaning area. Save the original containers for all solvents until they are completely used. In case of accidental poisoning, follow the instructions given in the MSDS provided by the manufacturer immediately. And when going for medical aid, take the container with you so the chemical can be quickly identified if you are unsure of the contents.

NOTE: If a solvent is splashed into your eyes, flush them thoroughly with water, keeping your eyelids open. Then get medical attention immediately. Some

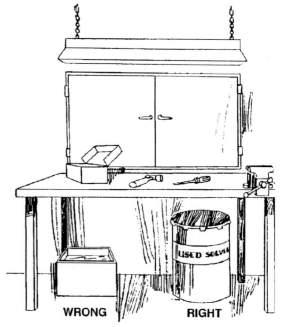

Store Solvents in Sealed and Labeled Cans, Never in Open Pans

solvents may cause skin irritation or dizziness. If that happens, stop. Wash thoroughly with mild soap and water, and get some fresh air. If possible, use solvents in a location where a water source is available and ready for immediate use if needed.

Continued on next page

OUO1010,0000E9C -19-15JUL09-4/7

Steam Cleaners

A number of cleaners are on the market. Operating instructions vary with the type and model. Before using any steam cleaner, become thoroughly familiar with its operation, and read its instruction manual carefully.

⚠ **CAUTION: The high-temperature steam produced by the steam cleaner can cause severe burns to the skin and eyes. Wear a face shield and gloves when using the steam cleaner. Watch for unexpected bystanders when swinging the cleaning wand around.**

NOTE: Do not steam electrical components, carburetors, diesel injection pumps, or air-conditioning lines as damage to the components may result. Avoid excessive steaming of hydraulic lines, wiring, and bearing seals. And don't use cleaning compounds on painted surfaces where appearance is important. Many cleaning compounds will dull the paint.

Certain safe practices apply to the operation of all steam cleaning machines:

1. **Work in a ventilated area.** Clean equipment outdoors when possible. Otherwise, provide adequate ventilation for your indoor cleaning area. Steam must be vented away for good vision, and to prevent damage and rust to your shop facilities and tools caused by condensing moisture.

2. **Protect yourself and others from burns.** Wear a face shield, gloves, and an apron if it's needed to keep your clothing dry. When cleaning, warn others to stay away. Watch for unexpected bystanders when swinging the spray wand around.

Clean Only in a Ventilated Area, Outdoors if Possible

3. **Always hold the wand securely and never lay it down until the cleaner has been shut down.** When shutting it down, turn off the burner and keep the water running until all steam disappears from the nozzle. Then you can be sure that the equipment will not burn or scald anyone.

4. **Keep the cleaner in good condition.** Avoid damage to the steam wand and hose. Remember that tape does not make safe repairs. Keep fuel connections tight. If your cleaner does not ignite readily, burn cleanly, or maintain steam pressure within safe limits, have a qualified technician service it.

OUO1010,0000E9C -19-15JUL09-5/7

Avoid High-Pressure Fluids

Escaping fluid under pressure can penetrate the skin causing serious injury. Protect your hands and body from high-pressure fluids.

Avoid the hazard by relieving pressure before disconnecting hydraulic or other lines. Tighten all connections before applying pressure. Search for leaks with a piece of cardboard, not your hands.

If an accident occurs, see a doctor immediately. Any fluid injected into the skin must be surgically removed within a few hours or gangrene may result. Doctors unfamiliar with this type of injury should consult with medical sources to obtain pertinent information.

Avoid High-Pressure Fluids

Continued on next page

OUO1010,0000E9C -19-15JUL09-6/7

High-Pressure Washers

Cleaning with high-pressure washers is becoming common in many shops. Take these precautions when using this equipment:

1. **Wear eye protection.** The high-pressure stream may cause dirt to be flung into your eyes. In addition, your eyes need protection in case the high-pressure stream should become directed toward your face.

2. **Read and follow the installation and operating instructions.** In addition to an adequate water supply, you'll need a grounding-type electrical circuit or one protected with a ground-fault interrupter. The electrical circuit must have sufficient capacity to handle the load of the motor, and it should be equipped with a safety switch that can be locked in the off position. If you need a new circuit for your washer, have a qualified electrician install one for you.

   Read your operator's manual carefully. Use only the chemicals recommended by the high-pressure washer manufacturer.

3. **Use extension cords of adequate capacity.** Conductors must be large enough to prevent excessive voltage drop. And three-wire cords must be used if the circuit is not protected with a ground-fault interrupter. Refer to the operator's manual for specific recommendations.

4. **Always direct the spray away from the body.** The pressurized stream can penetrate skin. If the water is heated, it can burn. When cleaning, warn others to stay away, and watch for unexpected onlookers when moving the spray wand around.

5. **Hold the wand securely and don't put it down unless the pressure washer is turned off.** Even if

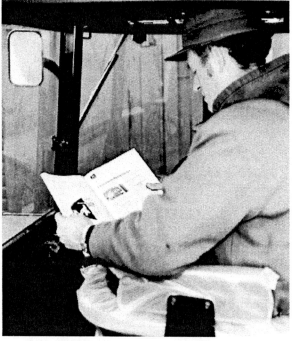

Follow All Safety Instructions for Your High-Pressure Washer

the wand is equipped with a shutoff valve, the valve could fail. Then, the high-pressure discharge would violently whip the wand around and possibly cause a serious injury. If injured by spray, see a doctor at once.

6. **Protect yourself from splatter and throwback.** Always keep yourself out of line of the sludge and water thrown back from the surface you're cleaning. Hold the spray wand at an angle to the surface, and stand to one side when cleaning corners from which the sludge is thrown straight backward.

OUO1010,0000E9C -19-15JUL09-7/7

Electrical System Safety

Avoid these hazards when servicing an electrical system:

- **Battery Explosions**
- **Acid Burns**
- **Bypass Start Hazard**
- **Electric Shock**

NOTE: When servicing the electrical system, always follow the steps outlined in your operator's or service manual to avoid damaging the electrical system. Lock out or disconnect the electrical power source from the electrical system before you begin your service procedure.

For a thorough explanation of electrical system servicing and safety instructions, see the FOS manual on "Electronic and Electrical Systems."

Battery Explosions

When charging and discharging a storage battery, a lead-acid battery generates hydrogen and oxygen gases. Hydrogen will burn and is highly explosive in the presence of oxygen. A spark or flame near the battery could ignite these gases, rupturing the battery.

The following basic safety rules must be observed to avoid battery explosions.

Prevent Battery Explosions

Use a flashlight to check the battery electrolyte level. Never use a match or lighter. These could set off an explosion.

Do not short across the battery terminals by placing a metal object between them.

Do not charge a frozen battery; it may explode. Warm the battery to 60°F (16°C).

OUO1010,0000E9D -19-23JUL09-1/7

When removing the battery, disconnect the grounded battery cable first. When installing the battery, connect the grounded cable last. The ground post lead will be connected to the engine block, machine frame, or other metal surface.

A—Grounded Battery Cable　　**B—Battery**

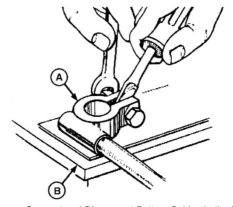

Connect and Disconnect Battery Cables in the Proper Sequence

Continued on next page　　OUO1010,0000E9D -19-23JUL09-2/7

Prevent sparks from battery charger leads. Turn the battery charger off or pull the power cord before connecting or disconnecting charger leads to battery posts.

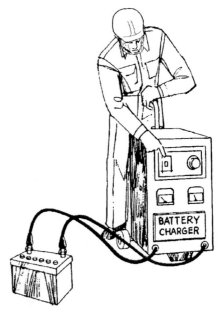

Turn Charger Off Before Disconnecting Clamps from Battery Posts

OUO1010,0000E9D -19-23JUL09-3/7

Acid Burns

Battery electrolyte is approximately 36% full-strength sulfuric acid and 64% water. Even though it's diluted, it is strong enough to burn skin, eat holes in clothing, and cause blindness if splashed into eyes.

Fill new batteries with electrolyte in a well-ventilated area, wear eye protection and rubber gloves, and avoid breathing any fumes from the battery when the electrolyte is added. Avoid spilling or dripping electrolyte when using a hydrometer to check specific gravity readings.

If you spill acid on yourself, flush your skin with water for several minutes.

Apply baking soda or lime to help neutralize the acid.

If acid gets into your eyes, force the lids open and flood the eyes with running water for 15 to 30 minutes. Get medical attention immediately.

If acid is swallowed, drink large amounts of water or milk, but do not exceed 2 quarts (2 liters). Get medical attention immediately.

Avoid Electrolyte Hazards When Using a Hydrometer

Continued on next page

OUO1010,0000E9D -19-23JUL09-4/7

Bypass Start Hazard

Bypass starting of tractors or other equipment is a very serious safety concern.

Never short across the starter terminals with a screwdriver or other device to start a self-propelled machine. If the neutral start switch is bypassed and the machine is in gear when the engine starts, the machine could suddenly lurch forward and crush you. Many people have died doing it. DON'T TRY IT!

⚠ **CAUTION: Never bypass start any tractor or other self-propelled machines, and never start it while standing on the ground. Start self-propelled equipment only from the operator's station and with the transmission in neutral or park.**

Avoid the Bypass Starting Hazard

OUO1010,0000E9D -19-23JUL09-5/7

Electrical Shock

The voltage in the secondary circuit of a spark-ignition system may exceed 40,000 volts. For this reason, don't touch spark plug terminals, spark plug cables, or the coil high tension cable when the ignition switch is turned on or when the engine is running. The cable insulation should protect you, but it could be defective.

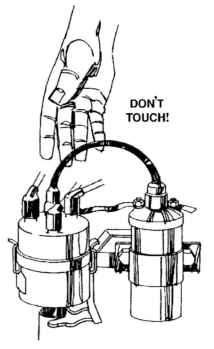

DON'T TOUCH!

Never Touch Wires in the Secondary Circuit or Spark Plug Terminals While the Ignition Switch Is Turned On

Continued on next page   OUO1010,0000E9D -19-23JUL09-6/7

Never run an engine when the wire connected to the output terminal of an alternator or generator is disconnected. If you do, and if you touch the terminal, you could receive a severe shock.

When the battery wire is disconnected, the voltage can go dangerously high, and it may also damage the generator, alternator, regulator, or wiring harness.

A—Output Terminal

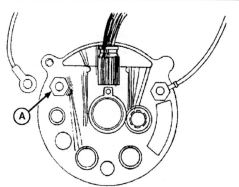

Don't Run the Engine When the Wire Has Been Disconnected from the Alternator or Generator Output Terminal

OUO1010,0000E9D -19-23JUL09-7/7

Safety Rules for LP-Gas

1. Remember first of all that **LP-gas is always under pressure**. It wants to get out of the tank, lines, and other components. They MUST be leak-proof.

2. Never loosen a line or fitting without first reducing the pressure.

3. Natural LP-gas cannot be smelled or seen. It is odorless and colorless — and therefore **dangerous**. However, commercial gases are generally odorized to help in detecting them.

4. LP-gas is heavier than air and will settle in low spots. These pockets are explosive until the leak is stopped and the area is ventilated.

5. Stay outdoors away from buildings when transferring fuel from storage to fuel tank. Some fuel always escapes when hoses are disconnected.

6. Never smoke or use any flame near the fuel system while filling or servicing it.

7. Never fill an LP-gas fuel tank more than 80% full.

8. LP-gas exhaust smoke can fool you — it doesn't smell bad, but it still contains some carbon monoxide. Be sure to ventilate the area before you start or operate an LP-gas engine.

9. When stopping the engine, always close both the vapor and liquid withdrawal valves. Leave the valves closed until the engine is to be started again.

10. Before taking LP-gas equipment indoors, check the state or local regulations. For your own safety, follow the rules.

11. LP-gas expands when heated. If the machine must be taken indoors for service in cold weather, be sure the tank is as near empty as possible. Otherwise, the fuel in the tank may expand until the safety valve opens.

12. Ventilation is a necessity when LP-gas equipment is indoors. Open doors and windows as wide as possible.Or use blowers or fans placed near the floor where vapors might collect if a leak developed. Protect against sparks or fires by using explosion-proof motors, switches, and controls.

OUO1010,0000E9E -19-22JUL09-1/1

Safety Rules for Compressed Natural Gas

1. **Handle compressed natural gas (CNG) safely**. CNG is methane (natural gas) stored at high pressure. It spreads into the air quickly.

   Natural gas fumes can cause sickness or death. Always work in a well-ventilated area.

   Natural gas is colorless, odorless, and tasteless. As a safety measure for leak detection, an odorant is normally added to commercial natural gas. The odorant gives the gas its familiar scent.

   Natural gas is lighter than air and will rise to the ceiling in an enclosed space. Therefore, spark-free vent fans and no ignition source ceiling space heaters must be used where natural gas can collect.

   Handle natural gas with care; it is highly flammable. DO NOT smoke when refueling or working on or around natural gas vehicles or equipment.

   Keep natural gas vehicles away from sparks, flames, and electrical devices in operation, especially if you suspect a natural gas leak.

2. **Service CNG system safely**. Improper installation, service, or operation of CNG storage and delivery components can result in fire, explosion, and serious injury.

Natural Gas Is Highly Flammable, Avoid Fires

3. **Protect against high pressure**. CNG fuel systems operate at high pressures. DO NOT disassemble or remove any CNG fuel system components under pressure. Explosive separation of components, and the escaping natural gas can cause serious injury.

   Relieve CNG fuel system pressure before disconnecting any fuel system component. Properly tighten connections and check for leaks before pressurizing the CNG fuel system.

OUO1010,0000E9F -19-23JUL09-1/2

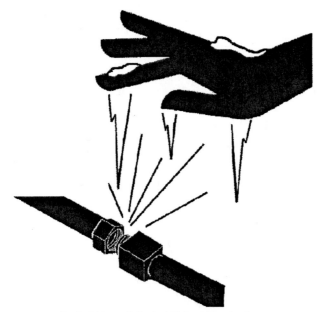

Protect Yourself Against Natural Gas Leakage

4. **Protect against extremely cold CNG leakage**. Gas escaping from the CNG fuel system is very cold. Frostbite and skin damage can occur from contact with cold escaping gas or surrounding components.

Inspect for leaks by spraying a soap and water solution on joints, fittings, and other areas. Look for bubbles that indicate leakage from the system.

OUO1010,0000E9F -19-23JUL09-2/2

Safety Rules for Liquefied Natural Gas (LNG)

The safety rules listed for compressed natural gas (CNG) also apply to liquefied natural gas (LNG). However, some special safety precautions are necessary with cryogenic (extremely cold) equipment.

1. **Protect against extremely cold LNG leakage.** Always wear goggles and a face shield, and insulated gloves when working with or around LNG.

   Accidental contact to LNG or cold issuing gas with the eyes or skin may cause a freezing injury similar to a burn.

2. **Keep the equipment area well ventilate**d. Without adequate ventilation, natural gas will displace the air and give no warning that a non-life-supporting atmosphere is present.

   Natural gas is colorless, odorless, and tasteless. The odorant normally added to natural gas will solidify

at the temperature required to liquefy natural gas. Thus, gas from LNG has no odor as is normal with compressed natural gas.

3. **Protect against high pressure**. The LNG fuel system is a pressurized system. Always empty the system of LNG before working on the LNG fuel system.

   Liquid natural gas will boil as its temperature rises and cause pressure to build in a system. Very small amounts of LNG are converted into large amounts of natural gas. Do not allow liquefied gas to become trapped in piping, as between two closed valves. The liquid will vaporize and rapidly increase pressure, bursting the pipe if a means of relief is not provided.

4. **Keep away from flame or spark**. Natural gas is flammable. Smoking, open flames, and general purpose electrical equipment must be prohibited where liquefied natural gas is stored or handled.

OUO1010,0000EA0 -19-23JUL09-1/1

Use Proper Lifting Equipment

Lifting heavy components incorrectly can cause severe injury or machine damage. Follow recommended procedures for removal and installation of components in the technical manual.

An engine is a heavy piece of equipment that can be awkward to handle when being lifted or suspended. Use extreme caution and NEVER permit any part of the body to be positioned under an engine being lifted or suspended.

Use a safety-approved balance beam-type lifting sling and lifting straps attached to an overhead hoist to lift the engine as shown. The lifting sling provides various attachment points for the lifting brackets and overhead hoist so that the engine can be safely suspended in a balanced and level position.

Follow the recommendations in the engine technical manual for the approved method of lifting an engine.

Use an Approved Sling When Lifting an Engine

A—Lifting Sling

Continued on next page

OUO1010,0000EA1 -19-23JUL09-1/2

100709
PN=59

Mount the engine on a repair stand that has adequate capacity to support the weight of the engine. To avoid structural or personal injury, do not exceed the maximum capacity rating of the repair stand. Refer to the engine Technical Manual or the repair stand manufacturer's instructions for safely mounting the engine.

⚠ CAUTION: Never remove the overhead lifting equipment until the engine is securely mounted onto the repair stand and mounting hardware is tightened to specified torque. Always release the overhead lifting equipment slowly.

To maintain shear strength specifications, alloy steel SAE Grade 8 or higher fasteners must be used to mount the adapters or engine to the repair stand. For full thread engagement, be certain that tapped holes in the adapters and engine block are clean and not damaged. A thread length engagement equal to 1-1/2 screw diameters minimum is required to maintain strength requirements.

To avoid an unsafe off-balance load condition, the center of balance of an engine must be located within 2 in. (50 mm) of the repair stand's rotating shaft. The engine center of balance is generally located in the middle of the engine slightly above the crankshaft centerline.

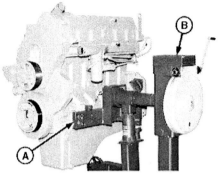

Engine Mounted on a Repair Stand

A—Adapter   B—Repair Stand

If you must work on a lifted machine, securely support the machine. Do not support the machine on cinder blocks, hollow tiles, or props that may crumble under continuous load.

Do not work under a machine that is supported solely by a jack. Follow recommended procedures in the technical manual.

OUO1010,0000EA1 -19-23JUL09-2/2

Dynamometer Safety

A dynamometer is used to apply a load to the engine for testing purposes while it is running. Always follow the safety rules and operating instructions provided by the dynamometer manufacturer to prevent possible damage to the equipment or injury to yourself when operating the dynamometer. Observe the following safety rules:

All driveline shields, chain shields, and sheet metal covers must be in place prior to performing the test and during operation.

Visually inspect the dynamometer driveline and adapters for damage or defects prior to each test. Replace as required. Once a year all driveline components must be magnafluxed for stress cracks and replaced if found defective.

When doing any type of dynamometer testing, make sure that the engine or vehicle is in a safe position to be tested. Place the transmission shift lever in park and block the rear wheels to prevent unexpected movement. Lower the jack stand (if used) at the rear of the dynamometer for added stability.

Check engine oil and coolant levels prior to testing. Make sure there is an adequate supply of fuel to the engine. Engine damage could result if the engine is operated

A PTO Dynamometer

without proper lubrication or cooling or the engine runs out of fuel while under load.

Stay clear of the rotating driveline and adapter assembly during operation. Do not stand near or straddle the driveline while it is in use.

Continued on next page   OUO1010,0000EA2 -19-15JUL09-1/2

Tie long hair behind your head. Do not wear loose fitting clothing, a necktie, scarf, or necklace when you work near moving parts. If these items were to become entangled, severe injury could result.

Prolonged exposure to loud noise can cause impairment or loss of hearing. Wear a suitable hearing protective device such as ear muffs or ear plugs to protect against objectionable or uncomfortable noise levels when testing.

Engine exhaust fumes can cause sickness or death. Perform the dynamometer test in a well-ventilated area, outdoors if possible. If it is necessary to run an engine in an enclosed area, remove the exhaust fumes from the area with an exhaust pipe extension.

A tremendous amount of heat is created as the dynamometer converts the engine's mechanical energy into heat energy. It is the purpose of the dynamometer's cooling system to remove this heat. An insufficient supply of water to the cooling system can result in damage to the equipment or injury to personnel. Read and understand

Service Machines Safely

the cooling requirement instructions in the operator's manual before performing any tests.

Any spillage or leakage of water from the dynamometer cooling system creates a hazardous condition to the operator. Any water leakage in the test area must be corrected before any test is conducted.

OUO1010,0000EA2 -19-15JUL09-2/2

General Safety Instructions

On the previous pages, you have been given safety information that is specific to the engine system. But when working on the engine, it is often necessary to remove and replace non-engine components. Therefore, total care has to be exercised in all service activities.

The following safety instructions, when followed, are designed to help keep you safe. Make it a practice to work safely. It's the thing to do.

This is the safety-alert symbol. When you see this symbol on a machine, on labels, or in instructional material, be alert to the potential for personal injury.

Follow recommended precautions and safe operating practices.

This Safety-Alert Symbol Could Save Your Life

OUO1010,0000EA3 -19-12JUN09-1/1

Understand Signal Words

Signal words like DANGER, WARNING, and CAUTION are used with the safety-alert symbol to draw attention to potentially unsafe areas.

DANGER means that one of the most serious potential hazards is present. Exposure to these hazards would result in a high probability of death or a severe injury if proper precautions are not taken.

WARNING means the hazard presents a lesser degree of risk of injury or death than that associated with Danger.

CAUTION is used to remind of safety instructions that must be followed and to identify property damage hazards and hazards involving minor injuries.

⚠ DANGER

⚠ WARNING

⚠ CAUTION

Signal Words

OUO1010,0000EA4 -19-12JUN09-1/1

Follow Safety Instructions

Carefully read all safety messages in this manual and on the machine's safety signs. No two machines are exactly alike, so don't take any chances.

Become familiar with the safety instructions and follow them. Also insist that those working with you follow the instructions. It will keep you and them safe.

Follow Safety Instructions

OUO1010,0000EA5 -19-12JUN09-1/1

Work in Ventilated Area

Avoid potentially toxic fumes and dust.

Engine exhaust fumes can cause sickness or death. If it is necessary to run an engine in an enclosed area, remove the exhaust fumes from the area with an exhaust pipe extension. If you do not have an exhaust pipe extension, open the doors and get outside air into the area.

Hazardous fumes can be generated when paint is heated by welding or using a torch. Remove paint before welding or heating.

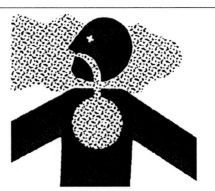

Danger of Hazardous Fumes or Dust

OUO1010,0000EA6 -19-22JUL09-1/1

Handle Fluids Safely

When you work around fuel, do not smoke or work near heaters or other fire hazards.

Store flammable fluids away from fire hazards. Do not incinerate or puncture pressurized containers.

Make sure the machine is clean of trash, grease, and debris.

Do not store oily rags; they can ignite and burn spontaneously.

Handle Fluids Safely, Avoid Fires

OUO1010,0000EA7 -19-12JUN09-1/1

Avoid Heating Near Pressurized Fluid Lines

Flammable spray can be generated by heating near pressurized fluid lines, resulting in severe burns to you and bystanders. Do not heat by welding, soldering, or using a torch near pressurized fluid lines or other flammable materials. Pressurized lines can be accidentally cut when heat goes beyond the immediate flame area.

Avoid Heating Near Pressurized Fluid Lines

OUO1010,0000EA8 -19-12JUN09-1/1

Illuminate Work Area Safely

Illuminate your work area adequately but safely. Use a portable safety light for working inside or under the machine. Make sure the bulb is enclosed by a wire cage. The hot filament of an accidentally broken bulb can ignite spilled fuel and oil.

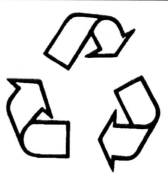

Illuminate Work Area Safely

OUO1010,0000EA9 -19-12JUN09-1/1

Dispose Waste Properly and Recycle Whenever Possible

Improperly disposing of waste can threaten the environment and ecology. Potentially harmful waste includes such items as oil, fuel, coolant, solvent, brake fluid, filters, and batteries.

Use leak-proof containers when draining fluids. Do not use food or beverage containers that may mislead someone into drinking from the container.

Do not pour waste onto the ground, down a drain, or into any water source. Recycle all waste whenever possible.

Air conditioning refrigerants escaping into the air can damage the Earth's atmosphere. Government regulations may require a certified air conditioning service center to recover and recycle used air conditioning refrigerants.

Recycle to Protect the Environment

Inquire on the proper way to recycle or dispose of waste from your local environmental or recycling center.

OUO1010,0000EAA -19-12JUN09-1/1

Live with Safety

Before returning the machine to the customer, make sure the machine is functioning properly, especially the safety systems. Install all guards and shields.

Remember that you and your coworkers contribute to each other's safety. Would you want to work around someone with unsafe working habits? Would you take your machinery to a shop known for unsafe working conditions and practices?

When it comes to safety, be a leader, not a follower. Safety is everybody's business.

Safety: Your Life Depends on It

OUO1010,0000EAB -19-12JUN09-1/1

Test Yourself

Questions

1. (True or false?) The best way to minimize hazards associated with engines and the equipment that they power is to follow proven shop practices and employ good work habits.

2. Why is it important to keep tools and service equipment in good condition, clean, and well organized?

3. (True or false?) A Material Safety Data Sheet (MSDS) provides specific details on chemical products.

4. (True or false?) If a commercial cleaning solvent is not available, gasoline is a suitable substitute.

5. (Fill in the blanks) Always wear _____ and _____ when working with chemicals, solvents, or batteries.

6. What is the proper sequence for connecting and disconnecting the battery cables.

7. Why should you never short across the starter terminals with a screwdriver or other devices to start a self-propelled machine?

8. (True or false?) Both LP-gas and natural gas are heavier than air and will settle in low spots.

9. (Fill in the blanks) Think safety when you see signal words on machinery like _____, _____, and _____.

10. Improperly disposing of waste can threaten the environment and ecology. Name three or more potentially harmful waste products found in the shop.

DP79986,000001E -19-15JUL09-1/1

Basic Engine — Introduction

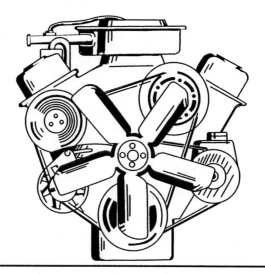

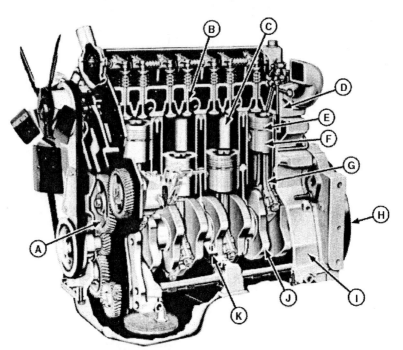

Cutaway of Basic Engine

A—Timing Valves
B—Valves
C—Cylinders

D—Cylinder Head
E—Pistion Rings
F—Pistons

G—Connecting Rods
H—Flywheel
I— Cylinder Block

J—Crankshaft
K—Main Bearings

This chapter covers the basic engine — parts common to all internal combustion engines. Chapter 1 describes how the engine works, while later chapters cover the fuel systems and other accessories. General testing and diagnosis of the engine is outlined in Chapter 13.

In this chapter you will learn about these basic parts of the engine:

- **Cylinder Head (D)** is at the top of the engine and houses the valves (B) and the intake and exhaust passages.
- **Valves (B)** open and close to let fuel and air in and exhaust gases out of each cylinder (C).
- **Camshaft** rotating in the engine block opens the valves (B) by cam action.
- **Cylinder Block (I)** is the main housing of the engine and supports the other main parts.

Continued on next page

SS40167,000009C -19-15JUL09-1/4

- **Cylinders (C)** are hollow tubes in which the piston (F) works. They may be cast into the cylinder block (I) or made of replaceable liners or sleeves.
- **Pistons (F)** move up and down in the cylinders (C) by the force of combustion.
- **Piston Rings (E)** seal the compression in the combustion chamber and also help to transfer heat.
- **Connecting Rods (G)** transmit the motion of the pistons (F) to the crankshaft (J).
- **Crankshaft (J)** receives the force from the pistons (F) and transmits it as rotary driving power.

- **Main Bearings (K)** support the crankshaft (J) in the cylinder block (I).
- **Flywheel (H)** attaches to the crankshaft (J) and gives it momentum to return the pistons (F) to the top of the cylinders (C) after each downward thrust.
- **Balancers** such as shafts or dampers, if used, balance the vibrations in the engine.
- **Timing Drives (A)** link the crankshaft (J), camshaft, and other key parts together to ensure that each is doing its job at the right time.

SS40167,000009C -19-15JUL09-2/4

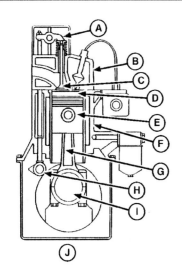
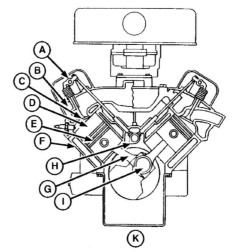

Parts of a Basic Engine

A—Rocker Arms	D—Cylinder	G—Connecting Rod	J— In-Line Block
B—Cylinder Head	E—Piston	H—Camshaft	K—V-Block
C—Valve	F—Cylinder Block	I— Crankshaft	

Engine parts are most commonly arranged in one of two ways:

1. In-Line Block

2. V-Block

SS40167,000009C -19-15JUL09-3/4

A typical V-block engine is shown. Now let us look at each part of the engine in detail — how it works and how to service it.

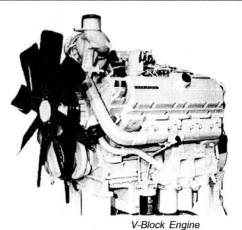

V-Block Engine

SS40167,000009C -19-15JUL09-4/4

Cylinder Heads

Most cylinder heads today are separate one-piece castings that may cover one or more cylinders. The cylinder head includes the valve guides that guide the valves as they open and close the passages (ports) that lead from the intake manifold to the combustion chamber and from the combustion chamber to the exhaust manifold. It also contains the valve seating surfaces, coolant passages (liquid-cooled engines), oil passages, and a number of threaded holes and machined surfaces.

The cylinder head is sealed to the engine block with a gasket. Careful tightening of the cylinder head attaching screws is critical in preventing distortion of the head or gas leaks that lose power from the cylinders.

All cylinder heads have intake ports designed for swirling the incoming air to create a high degree of turbulence in the combustion chamber. This turbulence mixes the air and fuel thoroughly to form a highly volatile mixture that burns evenly and cleanly.

Some engines have specially shaped combustion chambers in the cylinder head or in the top of the piston designed to swirl the air in the cylinder as the piston moves upward on its compression stroke. This forces the trapped air to swirl very rapidly by the time the piston reaches the end of its compression stroke. Combustion then occurs when there is the greatest turbulence and the least space between the piston crown.

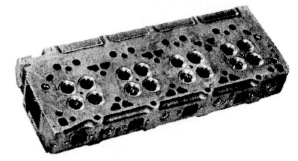

Engine Cylinder Head (V-Block Engine Requires Two)

DXP00185 —UN—23DEC08

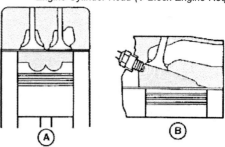

Turbulence Chambers in Cylinder Head

DXP00186 —UN—01JUN09

A—Turbulence Chamber in Top of Piston   **B—Turbulence Chamber in Cylinder Head**

Continued on next page   SS40167,000009D -19-15JUL09-1/10

Valve Arrangements

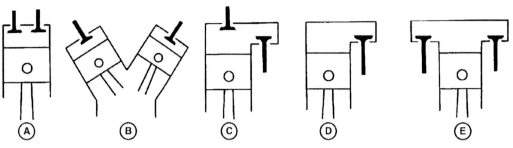

Valve Arrangement (Five Types)

A—I-Head
B—H-Head (2 Cycle)

C—F-Head
D—L-Head

E—T-Head

Valves are located in five major ways:

- **I-Head (A) — Both valves above the cylinder (parallel)**
- **H-Head (two-cycle) (B) — Both valves above angled cylinders (V-block)**
- **F-Head (C) — One valve above and the other to side of cylinder**
- **L-Head (D) — Both valves to one side of cylinder (parallel)**
- **T-Head (E) — One valve on each side of cylinder (parallel)**

I-head valves are widely used on modern engines.

H-head valves are found on V-block engines. Only one rocker arm is needed to operate two intake or two exhaust valves.

F-head valves are sometimes used to allow room for larger valves.

L-head valves are not used much because of the trend toward valve-in-head engines.

T-head was used on early engines but required two camshafts.

Cylinder Head Gaskets

The head gasket must provide an air-tight seal between the cylinders and the cylinder head that can withstand the temperatures and pressures of the combustion process. The gasket must also form a liquid-tight seal between the cylinder head and cylinder block to retain engine coolant and oil in their respective passages.

In the past, asbestos was used in the construction of most head gaskets due to its ability to withstand the high temperatures of the combustion process. However, asbestos is no longer used in the manufacturing process due to health concerns. Modern head gaskets are commonly constructed of plain copper or a special non-asbestos composition such as graphite with a steel core. The surface of the gasket may be treated to improve its liquid-sealing and anti-stick characteristics.

Most head gaskets include a fire ring, located at each cylinder bore, to provide a positive seal for combustion. A U-shaped stainless steel flange holds the fire ring in place.

The head gasket achieves a firm seal by "crushing" when the cylinder head is tightened down on the cylinder block. Because of this, always replace the head gasket after removing the cylinder head, as it is impossible to again obtain the correct "crush" after the initial installation.

Continued on next page

SS40167,000009D -19-15JUL09-2/10

Servicing Cylinder Heads

Removing the Head

Check for oil or coolant leaks around the base of the cylinder head (B) at the gasket line.

Steam clean the engine to prevent dirt from getting on the internal parts.

⚠ **CAUTION: DO NOT drain coolant until the coolant is below operating temperature. Only remove the radiator filler cap when cool enough to touch with bare hands. Slowly loosen the cap to the first stop to relieve pressure before removing completely.**

Drain coolant from the engine before removing the cylinder head (B).

When removing a cylinder head (B), do not use screwdrivers or pry bars between the cylinder block and head to loosen the gasket seal. Damage to the gasket surfaces could result. If necessary, tap the head lightly with a soft hammer to loosen it.

NOTE: If the engine is equipped with cylinder sleeves, do not rotate the crankshaft with the cylinder head (B) removed unless all sleeves are secured with cap screws and flat washers.

Clean and Inspect the Head

Check the combustion face for evidence of physical damage, oil or coolant leakage, or head gasket failure prior to cleaning the cylinder head. Repair or replace the cylinder head if there is evidence of physical damage such as cracking, abrasion, distortion, or valve seat "torching".

Engines with I-head valves have an oil gallery leading from the engine block to the top of the cylinder head to carry oil to the rocker arms. When removing the old head gasket, check for leaks around the area.

IMPORTANT: DO NOT USE scouring pads or steel wire brush to clean the gasket-sealing surface (combustion face). Doing so may affect the sealing ability of the gasket joint.

Scrape gasket material, carbon deposits, oil, and rust from the head. Use a powered brass or copper (soft) wire brush to clean the sealing surface.

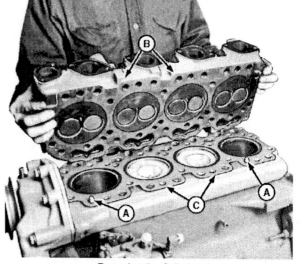

Removing the Cylinder Head

A—Dowel   C—Head Gasket
B—Cylinder Head

Remove expansion plugs from the head and thoroughly clean all passages. Dry with compressed air and blow out all passages.

Diesel engines with unit injectors have fuel rail passages in the head. Think of the fuel rails as internal passages of an injection pump, therefore, cleanliness of the fuel supply rails is extremely important.

IMPORTANT: DO NOT "hot tank" clean the cylinder head unless all plugs and valve guides are removed for replacement. Hot tank solution will destroy the lubricating properties of the valve guides.

Check for lime deposits in the water passages. Use a recommended solution and dip the head to remove scale and lime. (To prevent lime deposits, condition the cooling system as explained in Chapter 10).

Continued on next page   SS40167,000009D -19-15JUL09-3/10

Checking Head for Flatness

After long operation, the cylinder head may contour to match the block — this is normal.

However, if the engine has overheated or had compression leaks, the extra heat may have warped the head.

Check the machined surface of the head for flatness as follows:

1. Clean the machined surface of the head.

2. Check the head for flatness using a heavy, accurate straight edge (A) and feeler gauge (B). Check for flatness lengthwise, crosswise, and diagonally in several places. Compare the measurement to the maximum acceptable out-of-flat specification listed in the engine technical annual.

   The out-of-flat specification is normally given as a maximum dimension for the entire length or width of the head. The specification may also be given as a maximum dimension for a designated linear length. Consult the engine technical manual for details.

3. If the head is warped beyond the allowable limits, decide whether to replace the head or resurface it. This will of course be determined by the allowable amount of metal that can be removed from the head surface. Consult the engine technical manual for resurfacing limits.

Checking the Head for Flatness

Check Head Along Length and Width in Several Places

A—Straight Edge   **B—Feeler Gauge**

SS40167,000009D -19-15JUL09-4/10

Checking Head for Cracks or Leaks

Three methods can be used to check a head for cracks (A) or leaks:

- Water and air pressure method
- Magnetic crack detector method
- Fluorescent chemical method

Using the water and air pressure method, the cylinder head coolant passages are sealed and connected to an air hose. The cylinder head is pressurized with air and immersed in hot water (180–200°F [82–93°C]) for 15 minutes. Leaks are detected by air bubbles which appear in the water.

The magnetic crack detector is placed over the suspected area, setting up a magnetic field as shown. Fine white metallic powder is then sprinkled over the area and the tool is rotated 90 degrees. After the excess powder is blown off, any cracks (A) are clearly shown in white.

The fluorescent chemical method may be used to detect hairline cracks that may not be visible or detected by other methods. A special liquid penetrant that is fluorescent and glows under ultraviolet (black) light is applied to the

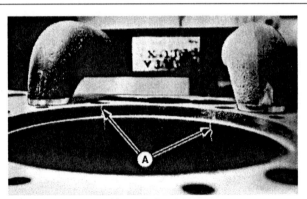

Magnetic Crack Detector

A—Cracks

surface of the cylinder head. Then, the excess penetrant is wiped off and the head is dried. A special developing powder is then applied which helps to draw the penetrant out of the cracks by capillary action. Inspection is carried out under an ultraviolet light.

Continued on next page   SS40167,000009D -19-15JUL09-5/10

Installing the Cylinder Head

Install the cylinder head as follows to prevent leaks and blow-by:

1. Check for defects such as scratches or nicks on the sealing surfaces of the head and block.

2. Clean the cylinder block and head contact surfaces. The surfaces must be dry and free of any oil.

NOTE: Make sure that all threaded holes in the cylinder block are free of oil, water, and dirt by blowing them out with compressed air. Fluid in the threaded holes can cause a hydraulic lock when the cylinder head cap screws are installed. This can cause inaccurate tightening of the cap screws, resulting in insufficient clamping force being applied to the cylinder head, or possible damage to the cylinder block.

3. Clean the threaded holes for the cylinder head mounting cap screws in the top deck of the cylinder block using the correct size tap (A) as shown. Use compressed air to remove any debris or fluid that may be present in the cap screw holes.

4. Install a new cylinder head gasket. Most engine manufacturers DO NOT recommend applying sealing compounds to the head gasket. Consult the engine technical manual for the manufacturer's recommendations.

5. Carefully set the cylinder head on the engine block without disturbing the head gasket. Most engines use dowels to correctly align the cylinder head on the block.

NOTE: Some engines use special cylinder head cap screws that cannot be reused — new cap screws must be installed. Consult the engine technical manual for details.

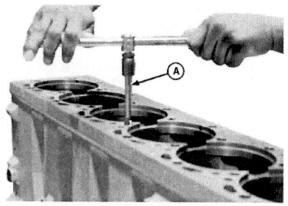

Cleaning the Threaded Holes in the Cylinder Block

A—Special Tap

6. Be sure the cylinder head cap screws are clean. Cap screws that are rusted or pitted from corrosion should not be reused. Lightly lubricate the cap screw threads with SAE 30 engine oil. If cap screws of more than one length are used, make sure they are correctly installed.

NOTE: Draw the cylinder head down gradually and uniformly to be sure of a good seal between the cylinder head and block. Tightening the cylinder head cap screws in one step may distort the head or cylinder liners and result in compression leaks.

7. Tighten the cylinder head cap screws or nuts finger tight. Following the sequence recommended by the engine manufacturer, tighten each cap screw or nut about one-half turn at a time until the specified torque value is reached. Consult the engine technical manual for the specific tightening torque values.

SS40167,000009D -19-15JUL09-6/10

8. The illustration shows one sequence for tightening cylinder head cap screws or nuts. Some manufacturers use this method for older engines. Start at the center of the head and work out toward both ends of the head until the specified torque is reached.

A—Start at the Center

B—Tightening Sequence (Alternate from Side to Side, Working Outward)

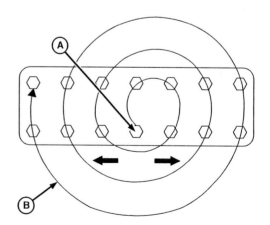

Sequence Guide to Properly Tightening Cylinder Head Cap Screws (Older Engines)

Continued on next page
SS40167,000009D -19-15JUL09-7/10

9. The illustration shows another sequence for tightening the cylinder head cap screws. Tighten cap screw No. 17 first. This prevents the cylinder head from tipping during the tightening sequence. Tighten the remaining screws in sequence, beginning with No. 1. Consult the engine technical manual for the specific torque values.

A—Front of Engine

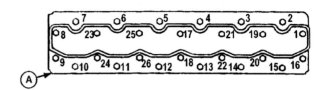

Typical Sequence for Tightening Cylinder Head Cap Screws (Newer Engines)

SS40167,000009D -19-15JUL09-8/10

10. The illustration shows the Torque-Turn tightening procedure recommended for some modern engines. In this procedure, the cylinder head cap screws are turned a specific number of degrees instead of being tightened to a specific numerical torque value. An example of this tightening procedure is as follows:

a. Sequentially (start at cap screw No. 1) turn each cap screw 90 degrees as shown. Line on top of cap screw will be perpendicular to the crankshaft. This 90-degree turning sequence may be repeated one or more times in order to establish the proper tension on the cap screws. Consult the engine technical manual for specific Torque-Turn tightening instructions.

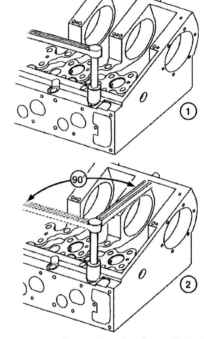

Torque-Turn Cap Screw Tightening Procedure

Continued on next page   SS40167,000009D -19-15JUL09-9/10

b. Sequentially (start at cap screw No. 1 and proceed through remainder of cap screws) tighten all cap screws to the initial torque value specified in the engine technical manual. Next, draw a line parallel to the crankshaft across the entire top of each cap screw. This line is used as a reference.

11. Most engines require a "break-in" or "run-in" after an engine overhaul. The cylinder head cap screws should generally be retightened to the specified torque after this run-in. Retightening cylinder head cap screws is not recommended for some newer engines. Follow the engine manufacturer's specifications.

12. Start the engine and run until normal operating temperature is reached. Check for fuel, oil, and coolant leaks. If the engine manufacturer recommends retightening the cylinder head cap screws after a

DXP00193 —UN—23DEC08

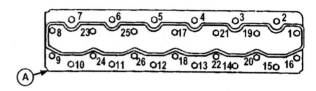

Typical Sequence for Tightening Cylinder Head Cap Screws (Newer Engines)

A—Front of Engine

warm-up, stop the engine and allow it to cool before tightening the cap screws to the specified torque.

SS40167,000009D -19-15JUL09-10/10

Valves

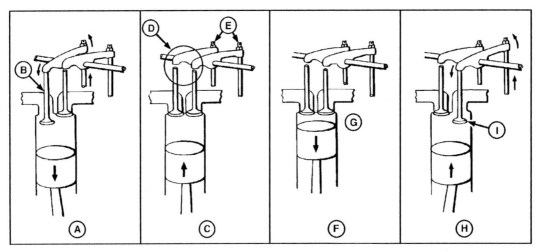

Valve Operation in a Four-Cycle Engine

The engine must take in air-fuel and exhaust spent gases at precise intervals. The valves do this job by opening and closing the intake and exhaust ports to the cylinder.

The illustration shows details the operating sequence of the valves in a typical four-cycle engine. Each cylinder may have two or four valves — one intake and one exhaust or two of each.

During the intake stroke (A), the intake valve(s) opens as shown.

During the exhaust stroke (H), the exhaust valve(s) opens. All valves are closed to seal in the combustible mixture during the other two strokes, compression (C) and power (F).

Two Intake and Two Exhaust Valves in One Cylinder

A—Intake Stroke
B—Intake Valve Open
C—Compression Stroke
D—Valve Clearance
E—Clearance Adjustment
F—Power Stroke
G—Both Valves Closed
H—Exhaust Stroke
I— Exhaust Valve Open

Continued on next page

SS40167,000009F -19-15JUL09-1/17

HOW VALVES ARE COOLED

The illustration shows how valves are cooled during the heat of engine operation.

When the valve is closed, heat in the valve head is transferred to the cylinder head, then to the water passages as shown.

At all times, heat is transferred from the valve stem to the guide.

A—Valve Heat Transfers to
    Head, Then to Water
    Passages

B—Heat Transfers from Valve
    to Guide

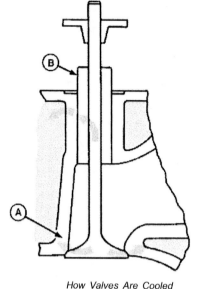

How Valves Are Cooled

SS40167,000009F -19-15JUL09-2/17

TYPES OF VALVES

On valves that are shaped like a mushroom, the valve is made up of a head (D) and a stem (H). The sealing edge is called the valve face.

The valve head is conical in shape so that it centers itself when it closes.

The illustration shows three forms of conical engine valves.

- Standard Valve (A) — Commonly used in American engines.
- Tulip Valve (B) — Used more in aircraft and racing engines for better flow of gases.
- Flat-Top Valve (C) — Combines the standard and tulip valve features. The larger taper under the head helps in better flow of gases and also strengthens the valve.

Valves are usually constructed from one or two pieces of special alloy steel — chrome-nickel for intake valves, silichrome for exhaust valves (because of the greater heat).

SERVICING VALVES

Analyzing Valve Troubles

When rebuilding an engine, some service technicians replace almost every part in the valve train while others replace some of the exhaust valves and replace the others. These are extreme cases. The best method is somewhere between the two.

What is best for one valve overhaul may not be best for another.

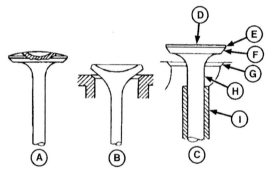

Forms of Engine Valves

A—SAE Standard Valve
B—Tulip Valve
C—Flat-Top Valve
D—Head
E—Margin
F—Face
G—Seat
H—Stem
I— Valve Guide

Major causes of valve failure:

- **Distortion of the Valve Seat**
- **Deposits on Valve**
- **Too Little Tappet Clearance — Burned Valves**
- **Preignition — Burned Valves**
- **Erosion**
- **Heat Failure**
- **Breaks**
- **Worn Valve Guides**

Continued on next page   SS40167,000009F -19-15JUL09-3/17

Distortion of Valve Seat

The valve shown in the illustration is burned because of distortion of the valve seat.

Major causes of seat distortion:

- Failure in the cooling system.
- Out-of-round or loose seat (see Seat Inserts in this chapter). This may stop the transfer of heat between the insert and the head or block.
- Warped sealing surfaces on heads or blocks. This often distorts the seats when the head is tightened. Improper tightening, too much torque, and the wrong sequence can also distort valve seats.
- Failure to grind the valve seat concentric with the valve guide bore.

Valve Face Burned from Distortion of Seat

SS40167,000009F -19-15JUL09-4/17

Deposits on Valve Stems

The illustration shows a valve that failed because face deposits built up and then broke off. This damaged the seat and the resulting blow-by burned the valve.

Other factors that may cause this type of failure:

- Weak valve springs. They cause a poor seal between the seat and the face, allowing deposits to form.
- Too little tappet clearance. This also causes a poor valve-to-seat seal.

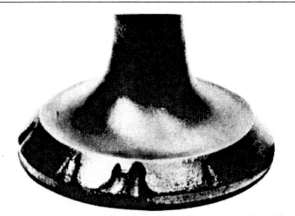

Valve Damaged Because of Face Deposits Breaking Off

SS40167,000009F -19-15JUL09-5/17

- Valves sticking in the valve guide. This allows deposits to build up on the valve face and stem.
- Valve seats that are too wide. They cut down the seating pressure and reduce the crushing of deposits when the valve closes.
- Lack of valve rotation (required to scrub valve).

Valve Stem Deposits

Continued on next page   SS40167,000009F -19-15JUL09-6/17

Too Little Tappet Clearance — Burned Valves

Failure of the valve shown in the illustration was caused by inadequate tappet clearance.

The valve was held off its seat and blow-by caused face burning.

Causes of too little tappet clearance:

- Tappet clearance not set to specifications.
- Failed valve rotators and, as a result, burned valves.
- Cooling system not operating properly or wrong thermostat installed. (Heat affects tappet clearance.)
- Tappet clearance not rechecked after break-in and retorquing of head as recommended by engine manufacturers.

Valve Burned from Inadequate Tappet Clearance

SS40167,000009F -19-15JUL09-7/17

Preignition — Burned Valves

The illustration shows a valve that has burned and failed as a result of preignition. Valve temperatures can reach the point where portions of the valve face melt away. Preignition can also cause excessive cylinder wear, and piston ring and land failures.

General causes of preignition:

- Improper Timing — Timing for maximum power is not always recommended, as preignition cannot always be heard. Follow the recommendations in the engine technical manual on timing.
- Combustion Chamber Deposits — Long idling periods, rich air-fuel mixtures, and cold engine operation can result in excessive deposits. Oil burning can also cause this.
- High Heat Range Spark Plugs or Cracked Porcelains — Replace defective plugs as recommended for the operating conditions.

Valve Burned by Preignition

- Too-High Compression Ratios — When cylinder heads or blocks are resurfaced, the fuel octane requirements also increase.

SS40167,000009F -19-15JUL09-8/17

Erosion of Valves

The valve in the illustration is eroded but has not failed. However, it would have broken after much more service because of the erosion under the head.

Causes of valve erosion:

- Wrong type of fuel.
- Faulty combustion.
- Too-high valve temperatures.
- Lean air-fuel mixtures, which overheat valves and erode them.

Erosion Under Head of Valve

Continued on next page

SS40167,000009F -19-15JUL09-9/17

Valves Cracked from Heat

Overheating of valves can cause cracks in the valve head. This is sometimes called "thermal fatigue." More cracking may cause parts of the valve to break off.

Causes of cracked valves from overheating:

• Worn guides.
• Distorted seats.
• Preignition.
• Lean air-fuel mixtures.

Valve Cracked from Heat

SS40167,000009F -19-15JUL09-10/17

Broken Valves

The two leading causes of broken valves:

• Fatique Break (C) — This is the gradual breakdown of the valve due to high heat and pressure. A fatigue break usually shows lines of progression as shown at the top of the illustration.
• Impact Break (A) — The mechanical breakage of the valve. The cause of this breakage is seating the valve with too much force, often caused by too much valve clearance. An impact break does not show the lines of progression but rather the familiar crow's feet as shown at the bottom of the illustration.

Broken valves are not always clearly one type or the other. Combinations of heat and high seating force can produce failures of varying degree and appearance. The illustration shows the difference between fatigue breaks and impact breaks.

A—Impact Break
B—Starting Point of Break
C—Fatigue Break
D—Starting Point of Break

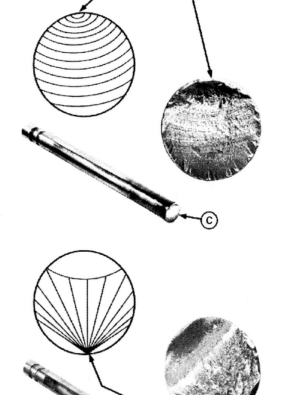

Broken Valves (Two Causes)

Continued on next page   SS40167,000009F -19-15JUL09-11/17

Worn Valve Guides

The valve shown in the illustration has failed because of face burning. Wear on the stem and carbon projecting into the guide shows the valve guide was worn and was the likely cause of failure. **Worn valve guides lead to valve failures:**

- Worn guides prevent concentric grinding of the valve seat. This leads to out-of-square seating, which in turn allows burning gas to leak out and burn the valves.
- Worn guides cause the valves to strike at an angle. This damages the sealing surfaces and leads to blow-by and burning.
- Excessive stem-to-guide clearance allows too much oil to run down the stem, resulting in excessive carbon deposits that cause valve sticking.
- When the inside edges of the valve guides wear, they can no longer act as carbon scrapers.

Installing new guides when new valves are installed does not mean they will be trouble-free.

Other factors leading to premature valve guide wear are:

Effect of Worn Guide on Valve Stem

- Worn Rocker Arms — cause excessive side thrust on the valve stem.
- Poor Lubrication — results in scoring.
- Carbon Deposits — deposits on the valve stem wear the valve guide into a bell-mouthed shape.
- Cocked Valve Springs — place side thrust on the valve stem and result in excessive wear.

SS40167,000009F -19-15JUL09-12/17

Dimensions Needed When Servicing Valves

The illustration shows the dimensions which must be known to recondition valves, seats, and guides.

Refer to the engine technical manual for the correct dimension for each engine.

Reworking or Replacing Valves

When removing valves from the engine, place them in a numbered rack or otherwise mark them. This returns each valve to its valve guide after servicing.

Special tools are a must when reworking engine valves. Valves can't be refaced on a shop-grinding wheel.

1. CLEANING

   Hold each valve firmly against a soft brass or copper wire wheel on a bench grinder. DO NOT use a wire wheel on the plated portion of the valve stem.

   Remove **all** carbon from the valve head, face, and unplated portion of the stem. Any carbon left on the stem will affect alignment in the valve refacer. Polish the valve stems with steel wool or crocus cloth to remove any scratch marks left by the wire brush.

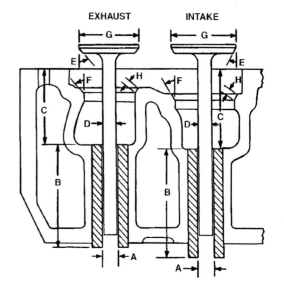

Valve Dimensions Required When Servicing Valves

A—I.D. of Valve Guide
B—Length of Valve Guide
C—Distance of Guide from Head
D—O.D. of Valve Stem

E—Angle of Valve Face
F—Angle of Valve Seat
G—Width of Valve Head
H—Width of Valve Seat

Continued on next page SS40167,000009F -19-15JUL09-13/17

2. INSPECTION

Use a small hole gauge and micrometer to measure
the inside of the valve guide and the outside of the
valve stem. Compare the two measurements to
determine the clearance of the valve stem in the guide.
If the measurements exceed the specified wear limits,
replace either the valve or guide or both.

A—Measure Inside of Guide
  with Gauge
B—Read Measurement with
  Micrometer

C—Measure Outside of Valve
  System

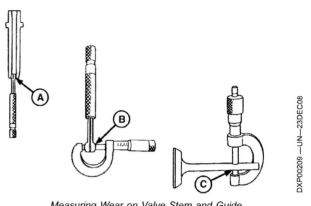

Measuring Wear on Valve Stem and Guide

SS40167,000009F -19-15JUL09-14/17

3. TESTING VALVES FOR BREAKS

Using a valve inspection center, determine if the
valves are out-of-round, bent, or warped. Replace
valves that exceed the runout limits specified in the
engine technical manual.

A—Valve Inspection Center

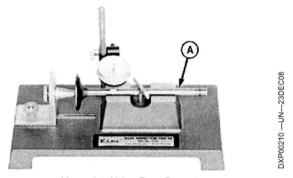

Measuring Valve Face Runout

Continued on next page

SS40167,000009F -19-15JUL09-15/17

4. REFACING THE VALVES

Locate the working head of the valve refacer at the specified angle with the valve head. For quicker seating, an interference angle is normally used. For more information on interference angle, see Valve Seats in this chapter.

First be sure the chamfer on the end of the stem is uniform and then be sure the valve is positioned all the way in the chuck and is seated against the conical centering stop.

IMPORTANT: **When valve faces are ground, it is important not to nick the valve head-to-stem radius with the facing stone. A nick could cause the valve to break. Radius all sharp edges after grinding.**

When grinding, the first cut from the face of the valve must be light. Observe if there is an unevenness of the metal being removed. If only 1/3 or 1/2 of the valve's face has been touched, check to see if the machine is dirty or the valve is warped, worn, or distorted. When the cut is even around the whole valve, continue until the complete face is ground clean. Be sure the correct angle is maintained.

Scrap and replace any valves that cannot be entirely refaced while keeping a good valve margin. The amount of grinding necessary to true a valve tells whether the head is worn or warped.

Avoid a knife-edge (B) around any part of the valve head. Heavy valve heads are required for strength and proper heat dissipation. A knife edge can lead to valve breakage, burning, and preignition because heat localizes on the edge of the valve.

5. GRINDING END OF VALVE STEM

If the end of the valve stem is pitted, worn, or mushroomed, true it and clean it up on the refacer attachment — not the grinding wheel. A very light grind is usually enough to square the stem and remove any pits or burrs.

Renew the chamfer on the end of the valve stem with the chamfering vee attachment. The chamfer need not exceed 0.040 in. (1 mm), but it must be uniform around the circumference of the valve stem.

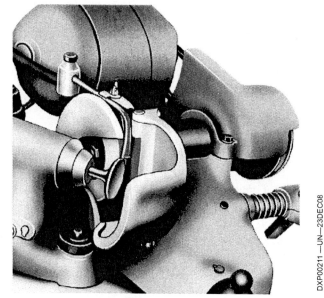

Refacing the Valves

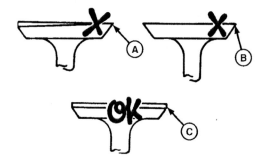

Valve Head Margin

A—Warped Valve with Knife Edge
B—Knife Edge
C—Good Margin

NOTE: *The amount of grinding allowed on a valve can vary between engines. Consult the technical manual for your particular engine for the manufacturer's specifications.*

Continued on next page SS40167,000009F -19-15JUL09-16/17

DXP00211 —UN—23DEC08

DXP00212 —UN—23DEC08

100709
PN=81

6. ASSEMBLING VALVES

Apply oil to the valve stems and return them to the same ports from which they were removed. (Commercial valve stem lubricants are available for this purpose.) Work the valves back and forth to make sure they slip through their guides easily and seat properly. A properly seated valve will bounce when dropped on its seat (without lubrication).

Be sure to seat the valve keepers (locks) (A) and spring retainers properly. Improper seating of these parts may cause valve breakage.

It is a good practice to use **new** valve keepers when reinstalling the valves.

With valve-in-head models, "pop" each spring and valve assembly three or four times by tapping on the end of each valve stem with a soft mallet to ensure good seating of the keepers.

If a failed valve is replaced, also inspect the rocker arm and push rod for that valve. The rocker arm or push rod may be bent or otherwise damaged when the valve failed, even though this may not be noticeable.

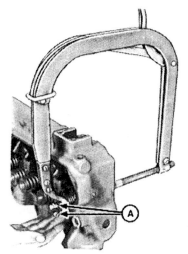

Installing Valves (Valve-in-Head Engine Shown)

A—Valve Keepers

SS40167,000009F -19-15JUL09-17/17

Valve Seats

The machined surface of the block or the cylinder head on which the valve rests when closed is the valve seat (B). This seat normally makes an angle of 20, 30, or 45 degrees with the plane of the valve head or bore.

The valve seat in the block or head is normally of the same material as the block or head and is ground for a tight seal with the face of the valve.

A—Angle of Valve Seat   B—Valve Seat

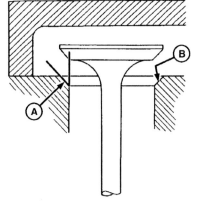

Valve Seat

Continued on next page   SS40167,00000A0 -19-23JUL09-1/8

SEAT INSERTS

Valve seat inserts (A) are used on some engines, often only for exhaust valves. The hardened insert reduces wear, prevents leakage, and reduces valve grinding frequency. It also allows the seat to be replaced.

The insert is a ring of special high-grade alloy metal placed in the block or head to serve as a valve seat.

The seat insert is generally shrink fit, although screw-in types are sometimes used. In most applications the inserts are shrunk in place by using dry ice to get the necessary fit without distorting the insert during installation.

A—Valve Seat Insert

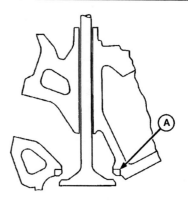

Valve Seat Insert

Continued on next page

SS40167,00000A0 -19-23JUL09-2/8

PN=83

SERVICING VALVE SEATS

1. Cleaning Valve Seats:

   a. Use an electric hand drill with wire cleaning brush and remove all carbon.

   b. Use solvent to help loosen very hard carbon.

2. Cleaning Valve Guides. For cleaning instructions, see Valve Guides later in this chapter.

3. Measuring Valve Guides:

   a. Measure the inside diameter of the guide for wear before machining the seat or counterbore.

   b. Replace the guide or ream the guide hole for oversized valves if necessary.

4. Removing and Installing Valve Seat Inserts:

   a. Check the valve seat insert for wear. Use an eccentrimeter (A) to measure valve seat runout.

   b. Replace valve seat inserts that are not within specifications.

   c. Valve seat inserts may be removed using one of the following methods:

   • Use a valve seat puller (B) as shown in the illustration. The adjusting screw on the puller may need to be retightened during removal of the inserts.
   • Carefully heat the insert at four points around the face of the insert until it becomes red-hot. Allow the seat to cool. Then it can be easily pried out with a screwdriver.

   A—Eccentrimeter

   B—Valve Seat Puller

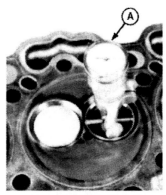

Measuring Valve Seat Runout

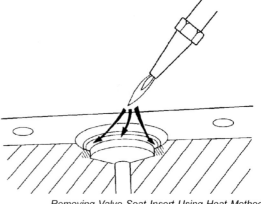

Removing Valve Seat Insert Using a Puller

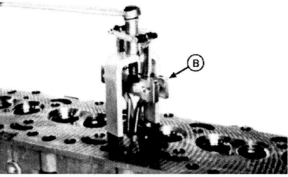

Removing Valve Seat Insert Using Heat Method

Continued on next page   SS40167,00000A0 -19-23JUL09-3/8

IMPORTANT: Be careful not to damage the cylinder head when removing valve seat inserts.

d. After removal of the inserts, thoroughly clean the area around the valve seat bore in the cylinder head and inspect for damage or cracks. Use a vernier caliper or scale to measure the valve seat bores and compare with specifications given in the engine technical manual. If valve seat bores are not within specifications, they can be machined for installation of oversize seat inserts.

e. When installing new valve seat inserts, follow the recommendations for press fits or other methods. Normally, chill each insert and the driver in dry ice. Use a driver that pilots in the guide bore to install valve seat inserts in the cylinder head. Blow out any chips from under the seat before installation. Also check for any nicks or raised edges.

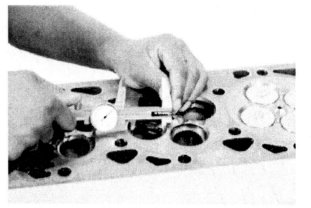

Measuring Valve Seat Bore

SS40167,00000A0 -19-23JUL09-4/8

f. Grind the valve seat insert as required to maintain the correct valve face-to-seat seal.

Grinding Valve Seats

Continued on next page

SS40167,00000A0 -19-23JUL09-5/8

5. Grinding the Valve Seats.

Seat width (C) must be maintained within specifications. Use a vernier caliper or scale to measure the seat width after grinding.

Determine where the seat contacts the valve face using a new or reconditioned valve as follows:

a. Mark the face of the valve with bluing or a series of pencil marks across the face of the valve.

b. Insert the valve in the guide, press down lightly on the valve head, and rotate the valve one-quarter turn.

c. Remove the valve and check the contact area on the valve face. The bluing or pencil marks will remain at the point where the seat contacts the valve. The contact area must be approximately centered on the valve face and even all the way around. There is a tendency to cut seats too wide. This can be corrected using narrowing stones. The objective in narrowing the seat is twofold:

- To make the seat the correct width.
- To make the seat contact the center of the valve face.

If the seat width is too wide after grinding, perform the following:

1. Grind the lower seat surface using a 70-degree seat grinder until seat width is **close** to specifications. *(Using the 70-degree seat grinder narrows the seat from the port side and raises the point of contact on the valve face.)*

2. Grind the upper seat surface using a 15-degree seat grinder until seat width is narrowed to specifications. *(Using the 15-degree seat grinder narrows the seat from the combustion chamber side and lowers the point of contact on the valve face.)*

C—Valve Seat Width
D—Cylinder Head
E—Lower Seat Surface 70°
F—Valve Seat
G—Upper Seat Surface 15°

H—Correct Seat Contact Area
I— Contact Area Too Wide
J— Contact Area Too Low
K—Contact Area Too High

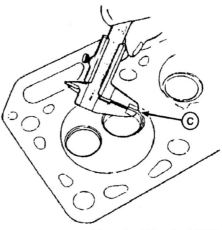

Measuring Valve Seat Width

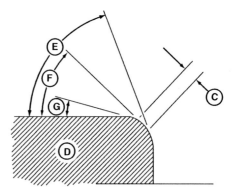

Valve Seat Width and Narrowing Angles

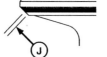
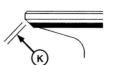

Checking Valve Seat Contact Area on Valve Face

Continued on next page   SS40167,00000A0 -19-23JUL09-6/8

Wide seats (A) improve heat transfer, while narrow seats (B) crush valve deposit better. Usually the exhaust valve seat is wider than the intake valve seat because of the need for better heat dissipation. If the valve seat is too narrow, the valve may burn or erode. A valve seat that is too narrow will also cause the valve and seat to wear more rapidly.

For even better seating forces, an interference angle (C) of the seat to the valve face is used. An interference angle exists when the face angle is 1/2 degree to 1-1/2 degrees less than the seat angle. This interference angle provides a positive seal on the combustion side of the valve and seat.

Be sure to thoroughly clean the cylinder head after servicing the valve seats. If the cylinder head is not cleaned, hard abrasive particles remain in the cylinder head ports, water jackets, and lubrication passages. These abrasives will cause rapid wear of the piston rings, cylinder walls, and the bearing surfaces of all lubricated parts.

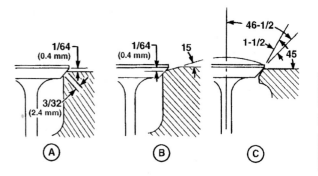

Typical Valve Seating Locations

A—Wide Seat
B—Narrow Seat

C—Interference Angle

SS40167,00000A0 -19-23JUL09-7/8

Precautions for Grinding Valve Seats

1. **Do not grind too long.** Only a few seconds are required to recondition the average valve seat.

2. **Do not use too much pressure.** While grinding, support the weight of the driver to avoid excess pressure on the stone.

3. **Keep the work area clean.** There can be no precision workmanship in the presence of dirt. Always keep tools and work area clean.

4. **Check the seat width and contact pattern with bluing.** If there are any uneven spots, regrind the seat.

NOTE: Lapping the seats with grinding compound may ruin a good grinding job. Always follow the manufacturer's recommendations regarding grinding and lapping.

5. **Check the runout (concentricity) of the valve seat with a dial indicator.** Hold the indicator reading within the specifications shown in the engine technical manual.

   This is perhaps the most important valve check. If the indicator reading is low, the grinding equipment and stone are in good condition, and together with the bluing check, ensures a good seal of the valve to the seat.

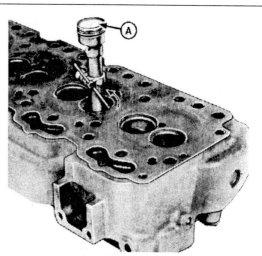

Checking Concentricity of Valve Seat

A—Turn Indicator to Check
   Runout of Valve Seat

Rotate the pilot 90 degrees in the guide and take a second runout reading. If this reading is also low, the pilot is reasonably straight and the guide is within limits.

SS40167,00000A0 -19-23JUL09-8/8

Valve Guides

The valve guide (A) holds the valve centered for proper seating. It also transfers heat from the valve stem to cool the valve.

The guide may be a drilled bore in the head or block or it may be an insert.

The guide is usually as long as space permits to reduce any "cocking" of the valve by the rocker arm.

The valve guide should be lubricated enough to prevent scuffing of the valve stem — although excessive lubrication may lead to heavy deposits at the hot end of the valve stem. In some engines, an oil seal is provided on the valve stem to prevent this.

SERVICING VALVE GUIDES

It is a good policy to replace valve guides, and to knurl and ream the integral type. However, this is not always practical. Refer to the technical manual for your particular engine for specific valve guide servicing information.

IMPORTANT: Never grasp the valve guides when turning over or handling the cylinder head as you may bend or loosen the guides.

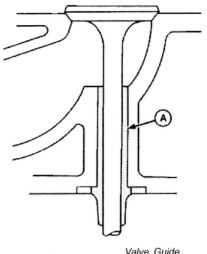

Valve Guide

A—Valve Guide

DXP00226 —UN—23DEC08

SS40167,00000A1 -19-16JUL09-1/5

Cleaning Valve Guides

Use the correct size valve guide cleaning brush (C) to clean valve guides before inspection or repair. Run the brush up and down the full length of the guide.

NOTE: A few drops of light oil or kerosene will help to fully clean the guide.

B—Cleaning Brush

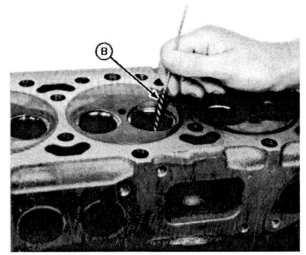

Cleaning Valve Guides

DXP00227 —UN—01JUN09

Continued on next page

SS40167,00000A1 -19-16JUL09-2/5

Measuring Valve Guides

Guides do not wear round or uniformly. Therefore, don't use plugs, valve stems, or pilots to measure them. Measure the inside diameter of the guides at different points within the guide using a small bore gauge (D). Read the bore gauge measurement with a micrometer (C).

C—Micrometer   D—Small Bore Gauge

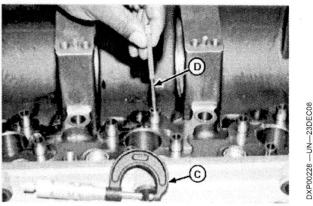

Measuring Valve Guide Inside Diameter

SS40167,00000A1 -19-16JUL09-3/5

Replacing Valve Guide Inserts

To remove a valve guide insert, position the cylinder head with the combustion side of the head facing up. Then using a press with the correct size driver, press the insert out the top of the head.

To install new valve guides (E), position cylinder head with combustion face facing down. Install the valve guides with tapered ends down. The valve guides must be installed to the height (G) specified in the engine technical manual.

Special valve guide installation adapter tools may be used to install the guides to the correct height. The guides are pressed into the head until the adapter bottoms on the machined surface of the head.

Some guides are equipped with a shoulder or snap ring around the outer diameter that determines the correct guide height. These guides should be pressed into the cylinder head until the shoulder or snap ring contacts the surface of the head.

Some new valve guides will compress slightly when installed. They must be precision-reamed to specifications after they are installed.

Insert a valve stem through the valve guide to check for adequate clearance. Valve stem should move freely in the valve guide.

E—Valve Guides   G—Valve Guide Height
F—Pilot Tool

Installing Valve Guides

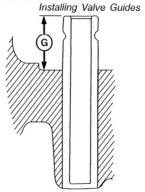

Install Valve Guides to Specified Height

Continued on next page   SS40167,00000A1 -19-16JUL09-4/5

100709
PN=89

Knurling Valve Guides

"Knurling" a valve guide makes a spiral groove on the inside wall of the guide. The groove creates raised metal ridges that reduce the inside diameter of the guide. Some manufacturers recommend knurling as a way of reconditioning valve guides using a special tool designed specifically for this procedure.

NOTE: Use the knurling tool exactly as directed by the manufacturer.

After knurling, ream the valve guide to the finished size to provide the specified stem-to-guide clearance.

H—Reamer
I— Knurler
J— Speed Reducer
K—Lubricant

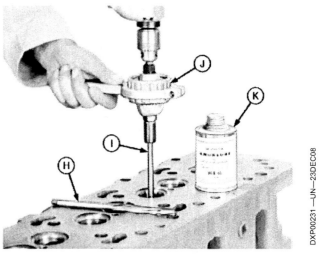

Knurling Valve Guides

SS40167,00000A1 -19-16JUL09-5/5

Valve Rotators

Valve rotators are devices that cause the valve to turn during operation. This removes deposits that form on the valve stem and face. The result: an improved gas seal, cooler valves, and less valve erosion.

Two types of valve rotators are:

• Release Type
• Positive Type

Release-Type Valve Rotators

In the release-type rotator, valve tension is momentarily released, allowing the valve to rotate.

During each valve cycle, the tappet (H) or rocker arm forces the locks away from the shoulder on the valve stem, releasing the spring load from the valve.

The valve then rotates freely for a moment from engine vibrations and moving gases.

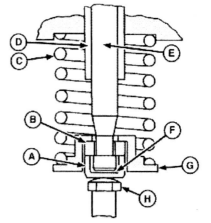

Release-Type Valve Rotators

A—Tip Cup
B—Retainer Lock
C—Valve Spring
D—Valve Guide
E—Valve
F—Built-In Clearance
G—Spring Cap
H—Tappet

Continued on next page
SS40167,00000A2 -19-23JUL09-1/4

Positive-Type Valve Rotators

The positive-type rotator is self-contained and installed in the same way as ordinary valve spring retainers. Rotation occurs when the valve moves and depends on the difference in spring loads between the valve open (D) and closed (A) positions. The illustration shows this rotator in operation.

When the valve is closed, the spring washer (F) is located at point 1 and at point 2.

As the valve starts to open, the extra spring load causes the washer to flatten, transferring the load from point 2 to point 3. This forces the balls (E) to roll down the inclined races.

This rotates the whole assembly, which transmits the movement to the valve.

As the valve closes again, the spring washer is released from the balls. The balls then return to their closed position by action of the return springs (G) (shown at top).

SERVICING VALVE ROTATORS

Positive-Type Rotators (Rotocap)

There is no maintenance for these valve rotators. However, when valves are replaced or reground, replace the valve rotators.

Check rotation of the rotator by looking for carbon buildup on the valve or by inspecting the seat. (Rotation can also be checked by watching the valve with the engine idling.)

The valve rotator may not rotate when held in the hand, but this does not mean it won't operate in the engine.

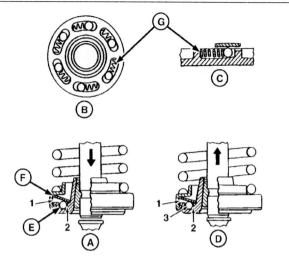

Positive-Type Valve Rotator

A—Valve Closed
B—Top View of Cap
C—Side View
D—Valve Open
E—Balls
F—Spring Washer
G—Return Springs

On valves that use rotators, inspect the valve stem, valve stem cap, and the keepers or locks for wear.

Wear at these points results in false tappet settings and rapid increases in tappet clearance.

Continued on next page

SS40167,00000A2 -19-23JUL09-2/4

Release-Type Rotators

Check the release-type rotator for clearance built into the valve tip cup (B). This built-in clearance should be maintained. Wear is uniform up to the wear limit, after which wear rapidly increases and may cause excessive valve clearance.

To check clearance, use a special micrometer (A) as shown in the illustration and do the following:

1. Remove tip cup from valve end.

2. Position the special micrometer on end of valve stem and set at the "0" mark. Hold the plunger pin firmly on top of the valve stem tip (E) and tighten the clamp screw.

3. Put the tip cup against the plunger pin and check the clearance or lack of clearance that is transferred to the micrometer spindle barrel.

4. Readings to the right of "0" indicate too much clearance, and readings to the left of "0" indicate too little clearance.

Another means of checking valve tip clearance is the Plastigage® method. A plastic strip is inserted between

Plastigage is a registered trademark of AE Clevite, Inc.

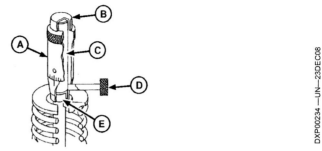

Checking Clearance of Release-Type Rotator Using a Micrometer

A—Special Micrometer
B—Tip Cup
C—Plunger Pin
D—Clamp Screw
E—Valve Stem Tip

the tip cup and the valve stem tip. With the locks held firmly against the shoulder on the valve, the cup is pressed in place. The width of the plastic strip is then checked.

SS40167,00000A2 -19-23JUL09-3/4

To reduce clearance, grind the open end of the tip cup. Grind the valve stem tip to increase clearance.

A—Valve Stem
B—Tip Cup

C—To Increase Clearance, Grind Here
D—To Reduce Clearance, Grind Here

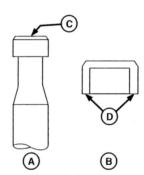

Grinding to Maintain Clearance Between Valve Stem and Cup

SS40167,00000A2 -19-23JUL09-4/4

Camshafts

Camshafts for small and medium size engines are generally made of a one-piece casting or forging. One intake and one exhaust cam lobe is provided for each cylinder, along with bearing journals. The journal diameters are larger than the cam lobes to permit endwise removal of the shaft from its bore.

The camshaft is normally driven by gearing from the crankshaft.

How the cam lobes are arranged on the shaft determines the firing order of the engine. The contour of each cam lobe determines the time and rate of opening of each valve.

Camshafts are made of a low-carbon alloy steel with the cam and journal surfaces carbonized before finish grinding. Some high-speed engines use alloy cast-iron camshafts with hardened cam lobes and journals.

On some large engines, the camshaft may have integral cams or it may have individual cams assembled on a shaft.

DXP00236 —UN—23DEC08

Camshaft, Thrust Plate, and Drive Gear

Overhead camshafts are supported on pedestals mounted on the cylinder head with removable caps, or else in a tunnel formed in the valve housing.

In some dual overhead camshaft engines, one camshaft operates the two intake valves per cylinder while the other operates the two exhaust valves.

Camshafts also drive oil pumps, fuel pumps, fuel injectors, and distributors using extra cam lobes or a gear on the shaft.

Continued on next page   SS40167,00000A3 -19-16JUL09-1/5

CAMSHAFT TIMING

On 4-cycle engines, the camshaft turns at one-half the speed of the crankshaft, so that each valve is opened and closed once during two revolutions of the crankshaft.

The **exhaust valve** should open before the end of the power stroke and close after the completion of the exhaust stroke. It should open before the end of the power stroke because, due to the angularity of the connecting rod and crankshaft, the pressure in the cylinder near the end of the power stroke has very little effect in turning the crankshaft. The valve should close after the end of the exhaust stroke to permit better scavenging due to the inertia of the outflowing gases in the exhaust manifold.

The **intake valve** should open before the end of the exhaust stroke and close after the end of the intake stroke. The intake valve opens before the end of the exhaust stroke to take advantage of the inertia of the out-flowing gases in the exhaust manifold. The valve should close after the end of the intake stroke to take advantage of the ramming effect, due to inertia, of the incoming air or air-fuel mixture, thereby more completely filling the cylinder.

To summarize:

- Exhaust valve opens before end of power stroke, closes after end of exhaust stroke.
- Intake valve opens before end of exhaust stroke, closes after end of intake stroke.

The illustration shows the valve timing for a typical high-speed engine.

The intake valve opens at 10 degrees before top dead center (TDC). It stays open until 50 degrees after bottom dead center (BDC) as shown.

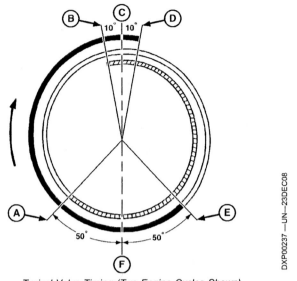

Typical Valve Timing (Two Engine Cycles Shown)

A—Intake Valve Closes
B—Intake Valve Opens
C—Top Dead Center (TDC)
D—Exhaust Valve Closes
E—Exhaust Valve Opens
F—Bottom Dead Center (BDC)

During this time, the engine has taken in its full charge of air-fuel and the piston has started its compression stroke.

The fuel is then compressed and ignited, and the power stroke begins.

When the piston reaches a point 50 degrees before bottom dead center (BDC), the exhaust valve opens as shown and stays open until the piston has traveled to 10 degrees beyond TDC, when it closes.

Continued on next page   SS40167,00000A3 -19-16JUL09-2/5

CAM LOBE PROFILES

The cam lobe profile is the opening ramp on the cam lobe which "takes up the slack" in the valve train. It also includes the nose, which holds the valve open, and the back side, which allows the valve to close.

The slope of the opening ramp determines how fast the valve opens.

Cam No. 1 has curved lobe faces, which cause the valve to open faster at the start and to be held open wide until the end of the closing face is under the valve tappet.

Cam No. 2 provides a rapid opening and closing with a long period at wide open.

Cam No. 3 is used in high-speed engines (racing cars) to give the longest possible opening of the valve.

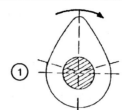

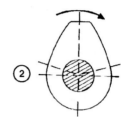

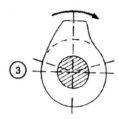

Cam Lobe Profiles (End View of Camshafts)

Continued on next page   SS40167,00000A3 -19-16JUL09-3/5

DXP00238 —UN—23DEC08

CHECKING VALVE LIFT

Measuring valve lift can give an indication of wear on the camshaft lobes and cam followers or bent push rods.

NOTE: For a more accurate measurement, it is recommended that valve lift be measured at 0.00 in. (0.00 mm) valve clearance with the engine cold.

Remove the valve cover and set the valve clearance to 0.00 mm (in.) clearance on the valve being checked. Be sure that the valve is fully closed. Using a dial indicator, put the tip of the indicator on the valve spring retainer. Set the dial indicator pointer at zero.

Manually turn the crankshaft in its running direction and observe the dial indicator reading as the valve is moved to the fully open position. Record the reading and the valve position. Repeat the procedure on all remaining valves.

Compare the readings with the valve lift specifications given in the engine Technical Manual. Any reading that falls below the specified wear tolerance indicates a problem that must be corrected.

SERVICING CAMSHAFTS

Always check the camshaft when you recondition the engine.

In modern engines, the cam lobes have a difficult job to do. They must raise the valves, one by one, at the proper time and under pressure, by rapidly moving the tappets, push rods, and rocker arms at high engine speeds.

When removing the camshaft, identify the rocker arms, push rods, and cam followers so they can be installed in their original locations. The mating surfaces of these components become "seated-in" in much the same way

DXP00239 —UN—23DEC08

Measuring Valve Lift

as the piston rings do in an engine cylinder. If these components are mismatched when reinstalled, the seating-in process must begin again. The result may be premature and excessive wear.

The camshaft bearings can be damaged by the sharp edges on the camshaft lobes. Carefully remove and install the camshaft so that the camshaft lobes do not drag in the bearing bores. Rotate the camshaft during removal and installation to avoid obstruction in any bores.

Continued on next page   SS40167,00000A3 -19-16JUL09-4/5

Injection of Bearing Journals and Bores

Inspect camshaft journals (A) for signs of wear or out-of-round condition.

Measure the camshaft journals with an outside micrometer as shown.

Measure the camshaft bores or bearings with an inside micrometer (or telescoping gauge and outside micrometer) as shown.

Compare the results with specifications.

Checking Cam Lobes

Using a micrometer, check the height of each cam lobe. Compare each intake cam lobe to the other intake cam lobes, and exhaust cam lobes to their counterparts; then check all with specifications. Also inspect the cam lobe faces on which the tappets slide for surface fatigue. The nose of the cam lobe must be polished and flat. If rough or wavy, inspect the cam and its mating follower for possible service.

Regrinding and polishing the camshaft is not satisfactory on modern high-speed engines. Originally the cam surfaces are treated to provide an oil-absorbing surface for better lubrication. Most engine reconditioners cannot duplicate this process. Therefore, the camshaft must be replaced.

Measuring valve lift can give an indication of wear on cam lobes, cam followers, and push rods.

Selecting Bearings

Select the proper bearings to match the camshaft journals. Some manufacturers have undersize bearings for replacement parts as well as special installation tools. Line up oil holes prior to assembly and after assembly. Be sure to check the registry of these holes with those in the cylinder block with a wire.

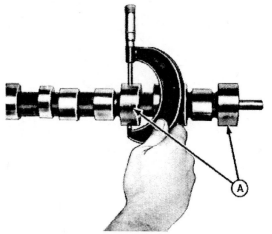

Measuring Camshaft Journals

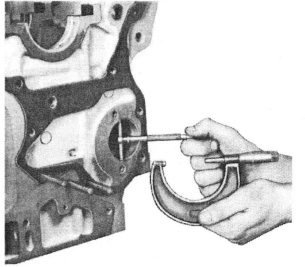

Measuring Camshaft Bores

A—Camshaft Journals

SS40167,00000A3 -19-16JUL09-5/5

Cam Followers and Hydraulic Valve Lifters

CAM FOLLOWERS

Cam followers (D) work off the lobes of the camshaft (C) and drive the push rods (E) to operate the valves (B).

A cam follower is usually a piston-like part working in a vertical guide against the cam lobe. The follower is commonly hardened and ground to provide a long wearing surface face against the cam lobe.

Some cam followers have a roller on the driven end that works against the camshaft lobe. This reduces friction and wear on the follower and the camshaft lobe.

NOTE: Cam followers (D) may also be called "valve tappets" or "solid lifters."

Servicing of Cam Followers

Proper inspection and service of cam followers is vital to good engine performance.

A change in valve clearance or valve lift can usually be detected by excessive noise at idle speeds.

If you suspect a change in clearance, remove the followers and inspect them for excessive wear.

Mark each one so it can be returned to the same bore.

After the followers are removed, clean all parts thoroughly with fuel oil or solvent and dry them with compressed air.

Look for signs of fatigue on each tappet. If the follower is scuffed or "dished," the camshaft may need replacing.

Examine follower bores to make sure they are clean, smooth, and free of score marks. Clean up any marks.

If the camshaft is replaced, also replace the cam followers.

On the roller-type followers, the rollers must turn smoothly and freely on their pins and be free of flat spots or scuffs.

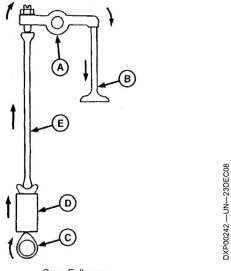

Cam Follower

A—Rocker Arm
B—Valve
C—Camshaft

D—Cam Follower
E—Push Rod

If the rollers do not turn freely or have been scored or worn flat, examine the cam lobes on which they operate. For more information, see Camshafts in this chapter.

Measure the maximum wear on roller-type followers. If clearances do no meet specifications, replace the cam roller and pin, which are usually serviced as a set.

When installing cam followers, lubricate the body and bore with engine oil so that the follower slides in easily. Also apply oil to the camshaft lobes.

Continued on next page

SS40167,00000A4 -19-16JUL09-1/2

HYDRAULIC VALVE LIFTERS

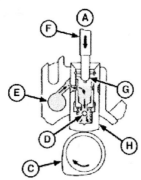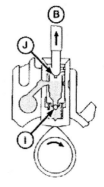

Hydraulic Valve Lifter

A—Valve Closed	D—Ball Check Valve Open	G—Plunger Extended — Zero	J— Push Rod Presses Against
B—Valve Open	E—Oil Under-Pressure	Clearance	Cup
C—Camshaft	F—Push Rod	H—Lifter Body	
		I— Oil Forced Upward Closes	
		Check Valve	

With hydraulic valve lifters, the valve clearances are always zero and valve operation is quiet. Hydraulic lifters eliminate the need for periodic valve adjustments and reduce unnecessary valve train wear.

The lifter body (H) has an opening through which engine oil under pressure is supplied to the pressure chamber in the lower part of the lifter.

The oil unseats the check ball and fills the pressure chamber below the plunger. The small spring below the plunger keeps the lifter in contact with the cam when the valve is closed.

As the camshaft (C) moves to open the valve (B), the pressure of the oil under the plunger increases and closes the check ball. The plunger and valve are lifted by the oil trapped above the check ball.

As the valve is closed (A), the pressure of the oil under the plunger decreases, and the valve ball opens.

Servicing of Hydraulic Valve Lifters

Proper inspection and service of hydraulic valve lifters is vital to good engine performance.

Excessive hydraulic valve lifter noise can be caused by either of the following:

• Worn, damaged, or sticking lifters.
• Low engine oil pressure or restricted oil passages to the lifters.

Check engine oil pressure to make sure it is within specification.

Remove the hydraulic valve lifters and inspect them for excessive wear.

Mark each one so it can be returned to the same bore.

After the lifters are removed, clean all parts thoroughly with fuel oil or solvent and dry them with compressed air.

Look for signs of fatigue or collapse. If the lifter is scuffed or "dished," the camshaft may need replacing.

Examine lifter bores to make sure they are clean, smooth, and free of score marks. Clean up any marks.

Make sure oil passages are clear of obstructions.

If the camshaft is replaced, also replace the hydraulic valve lifters.

When installing hydraulic valve lifters, lubricate the body and bore with engine oil so that the lifter slides in easily. Also apply oil to the camshaft lobes.

Check Lifter Preload

After installing the valve lifters, it is necessary to check the lifter preload. Lifter preload is the distance between the retaining snap ring and the push rod seat in the lifter when the lifter is on the heel of the cam lobe with the valve closed. Lifter preload is affected by a number of things:

• Resurfacing the heads and/or block deck
• Changes in camshaft diameter
• Changes in push rod length
• Changes in valve length
• Changes in rocker arm length or geometry
• Changes in head gasket thickness
• Changes in lifter height
• Valve job
• Different rocker arm stands or shafts

Make sure to check the valve geometry. Use either longer or shorter push rods to correct the geometry. In some cases, a shim under the rocker shaft may be required to maintain correct geometry. When installing new hydraulic valve lifters, follow the original manufacturer's procedures for checking and adjusting lifter preload.

SS40167,00000A4 -19-16JUL09-2/2

Valve Push Rods

The push rod (C), used in valve-in-head engines, is a steel rod that transmits camshaft motion to the rocker arm (A). Push rods are usually made from hollow steel pipe with solid steel ends.

The lower end of the push rod is usually formed into a half-round head to match a spherical seat in the tappet. The top end may have a spherical socket to match the adjusting screw at the rocker arm, or a half-round head similar to the lower end of the push rod.

Servicing Push Rods

1. When removing push rods, identify each rod so it can be assembled with the same mating parts.

2. Inspect push rod ends for wear or damage.

3. Examine push rods for bent condition. Push rods should be replaced if they are bent more than specifications allow. Measure by placing the push rods on a flat plate. Rotate the push rod and measure the deflection with a dial indicator.

4. If a failed valve has been replaced, also inspect its mating rocker arm and push rod. Often the rocker arm or push rod is bent or otherwise damaged when the valve fails, even though this condition may not be noticeable.

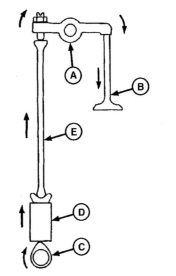

Valve Push Rod for Valve-in-Head Engine

A—Rocker Arm
B—Valve
C—Push Rod
D—Cam Follower
E—Camshaft

SS40167,00000A5 -19-23JUL09-1/1

Valve Springs

The main function of the valve spring is to close the valve and keep it closed and seated until opened by the camshaft. Another important function of the valve spring is to cause the valve-operating mechanism to follow the cam lobe profile for controlled opening and closing of the valve.

Cylindrical springs are used by most engine manufacturers. Valve-spring action is dampened by spring design or by a separate spring damper to reduce spring vibrations.

A variable-rate valve spring (unevenly spaced coils) is one method used to reduce spring vibration and flutter. The valve spring dampening coils are coils that are wound closer together at one end than the other. The end with the closer coils should be installed against the cylinder head.

In some cases, two springs are used for each valve to cut down spring vibration. A reverse-wound secondary spring inside the main valve spring reduces spring vibration by friction as well as different vibration frequencies.

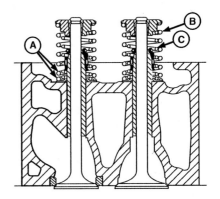

Install Damper Coil End of Spring Toward Cylinder Head

A—Damper Coils
B—Outer Spring
C—Inner Spring

Continued on next page

SS40167,00000A7 -19-16JUL09-1/3

Servicing Valve Springs

When inspecting valve springs, look for the following:

1. Wear on the casting where the springs rotate.

2. Wear on the spring caps. The spring caps can wear almost to a knife-edge.

3. Wear on ends of spring.

4. Rusted or corroded springs. Rust or pitting from corrosion can weaken the spring and lead to spring failure.

5. Burned paint on the springs indicates that the valves have been overheated. Overheating can cause the springs to lose tension.

6. Lack of spring tension. Check against specifications using a spring tester. A tension loss exceeding 10% of the specified load means that the spring needs replacing.

7. Also check for any pressure loss caused by valve grinding. As the valve seat is ground deeper, the spring operating height is longer, decreasing the spring load. Shims can sometimes be added between the spring and the cylinder head to correct this, but be sure to allow for the shims when testing the spring.

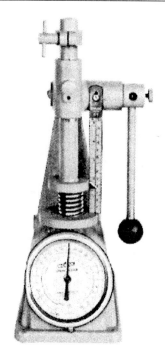

Checking Valve Spring in Tester

SS40167,00000A7 -19-16JUL09-2/3

8. Inspect for "cocked" springs. This places a greater than normal side thrust on the valve stem and results in excessive wear of the valve stem and guide. Measure by placing the spring on a flat surface and using a square. Replace springs if they are cocked.

9. Measure the free length (C) of the springs using a vernier caliper or scale. Compare the results with specifications. Don't be concerned if the valve springs have slightly different lengths but are within the specified limits. The free length can vary and the springs will still have the same length when compressed. Remember, the **force** of the spring is the critical factor.

A—Square Gauge   C—Free Length
B—Spring Inclination

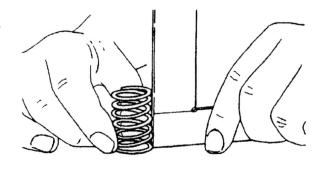

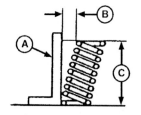

Checking Valve Spring Inclination

SS40167,00000A7 -19-16JUL09-3/3

Valve Rocker Arms

In valve-in-head engines, the rocker arms (A) transmit camshaft motion to the valves.

Most rocker arms are mounted on a single hollow shaft located at the top of the engine.

When the push rods move up, the mating rocker arm is moved down, contacting its valve stem tip and opening the valve.

On engines with two intake and two exhaust valves per cylinder, the rocker arm contacts a crosshead (G) instead of the valve stem tip. The crosshead then contacts both intake or both exhaust valves and causes two valves to open simultaneously.

Lubricating oil is usually supplied through the drilled rocker arm shaft (C) to each rocker arm.

A—Rocker Arms
B—Spring
C—Shaft
D—Plug
E—Brackets

F—Intake Valves
G—Crosshead
H—Crosshead Adjusting Screw
I— Wear Cap
J— Rocker Arm Adjusting Screw

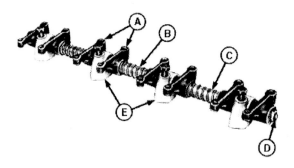

Valve Rocker Arm Assembly

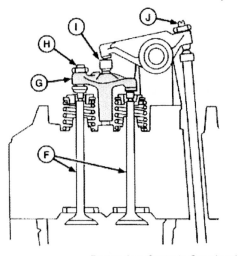

Rocker Arm Contacts Crosshead

SS40167,00000A8 -19-02JUL09-1/2

Servicing Rocker Arms

1. Inspect the rocker arm shaft for scratches, scores, burrs, or excessive wear at points of rocker arm contact. Be sure that all oil holes are open and clean. If a valve has failed, examine the shaft for cracks.

2. Check for cups or concave wear on ends of rocker arms where they contact the valve tips. Concave wear at this point makes it difficult to get the proper tappet adjustment. A worn rocker arm also creates side thrust on the valve stem and causes wear.

3. Replace or recondition rocker arms when noticeable grooving or wear occurs.

4. Check the rocker arm adjusting screw and nut for damage. Examine the spacer springs on the shaft between the rocker arms and be sure they are strong enough to exert a positive pressure on the arms.

5. Thoroughly clean holes in rocker arm mounting brackets. This is very important in some engines, because these holes feed oil to the rocker arm shaft.

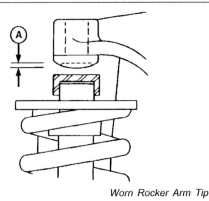

Worn Rocker Arm Tip

A—Worn Section

6. When installing rocker arms, be sure to install the arms and springs in the same sequence they were removed.

SS40167,00000A8 -19-02JUL09-2/2

Valve Clearance

When valves are properly adjusted, there is a small clearance (C) between the valve stem and the end of the rocker arm. (This clearance is sometimes referred to as "valve lash" or "tappet clearance.")

Valve clearance allows for the heat expansion of parts. Without this clearance, expansion of heated parts would cause the valves to stay partly open during operation.

This clearance is small, varying from approximately 0.006 inch (0.15 mm) to 0.030 inch (0.76 mm). Valve clearance can also vary between engine manufacturers and their models.

The valve clearance may vary, depending on the engine model and whether the engine is hot or cold during adjustment (some engines run hotter than others).

Too little valve clearance throws the valves out of time. This causes the valves to open too early and close too late. Also, valve stems may lengthen from heating and prevent the valves from seating completely. Hot combustion gases rushing past the valves cause overheating because the valves seat so briefly or so poorly that normal heat transfer into the cooling system does not have time to take place. This causes burned valves.

Too much valve clearance causes a lag in valve timing, which throws the engine out of balance. The fuel-air mixture is late entering the cylinder during the intake stroke. The exhaust valve closes early and prevents waste gases from being completely removed.

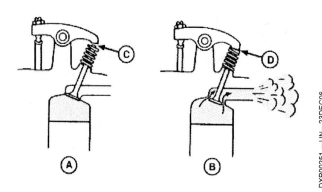

Problem of Inadequate Valve Clearance

A—Valve Closes and Seals Gases in Cylinder
B—Valve Doesn't Seat; Power is Lost and Valve Overheats
C—Valve Clearance
D—No Valve Clearance

The valves themselves also become damaged. When valve clearance is properly adjusted, the camshaft slows the speed of valve movement as it closes. But with too much clearance, the valves close with great impact, cracking or breaking the valve and scuffing the cam and follower.

Continued on next page   SS40167,00000A9 -19-16JUL09-1/3

HOW TO ADJUST VALVE CLEARANCE

Follow the manufacturer's procedure for correct valve adjustment. Here is a typical valve-in-head procedure:

1. Be sure the engine is at the recommended temperature before you begin valve adjustment. Generally, the valves are adjusted with the engine cold.

2. Clean all dirt and oil from around the rocker arm cover. Remove the cover.

3. Turn the engine crankshaft until the piston in the No. 1 or first cylinder is at top dead center (TDC) of its compression stroke. Most engines have timing marks on the flywheel or fan drive pulley to mark the TDC or other timing point. The engine is at No. 1 TDC (compression stroke) if the rocker arms (B) for No. 1 cylinder are loose (valves closed).

   A positive way to find when a piston is at TDC (compression stroke) is to remove the spark plug or injection nozzle and hold your finger over the opening. On the compression stroke, air will be forced out against your finger until the piston reaches the TDC position.

4. With the No. 1 piston at TDC, check the clearance between the rocker arms (B) and valves for the No. 1 cylinder using a feeler gauge. Compare the measured clearance with the clearance specified in the engine technical manual.

NOTE: Be sure to determine which are intake and which are exhaust valves, because the clearances are usually different for the two. To identify the valves, observe the position of the valves in relation to the intake and exhaust ports in the cylinder head.

5. If valve clearance needs to be adjusted, loosen the lock nut on the rocker arm adjusting screw (C). Install proper thickness feeler gauge (A) at the joint between the rocker arm (B) and valve stem tip. Turn the rocker arm adjusting screw up or down until the feeler gauge slips between the rocker arm and valve with a slight drag. Hold the adjusting screw from turning with a screwdriver, and tighten the lock nut.

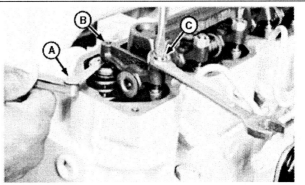

Adjusting Valve Clearance

A—Feeler Gauge
B—Rocker Arm

C—Adjusting Screw

NOTE: On engines with two intake and two exhaust valves per cylinder, there may be adjusting screws on both the crosshead and the rocker arm. The crosshead adjustment is to ensure that both intake or both exhaust valves are actuated at the same time. The rocker arm adjustment sets the valve clearance.

6. If the engine is equipped with unit injectors, the unit injector preload should also be adjusted at this time. Consult the engine technical manual for the specific adjustment procedure.

7. Rotate the engine crankshaft in normal running direction (counterclockwise as viewed from rear) until the next piston in the firing order reaches TDC of its compression stroke. Check and adjust the valve clearances for this cylinder. Repeat this procedure for the remainder of the cylinders.

NOTE: The "firing order" will vary, depending on the design of the engine and the number of cylinders. Consult the engine technical manual for specifications.

Continued on next page   SS40167,00000A9 -19-16JUL09-2/3

8. Often two or three sets of valves may be set at a time with one rotation of the crankshaft, allowing all the valves to be set with only two rotations of the crankshaft. A typical two-position valve clearance adjustment sequence for a six-cylinder, in-line engine is shown in the illustration.

NOTE: *The two-position valve clearance adjustment sequence is determined by the arrangement of the valves in the cylinder head, the number of engine cylinders, and the engine's firing order. Refer to the engine technical manual for the correct valve clearance adjustment method for each engine.*

In the example shown in the illustration, check and adjust valve clearance on Nos. 1, 3, and 5 exhaust valves and Nos. 1, 2, and 4 intake valves when the No. 1 piston is at TDC of its compression stroke. Then, rotate the flywheel 360 degrees until the No. 6 piston is at TDC of its compression stroke. Check and adjust valve clearance on Nos. 2, 4, and 6 exhaust valves and Nos. 3, 5, and 6 intake valves.

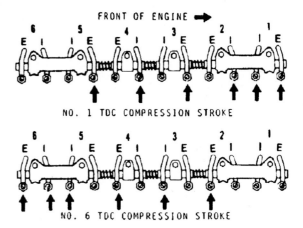

Two-Position Valve Clearance Adjustment Sequence

9. Install the valve cover, using a new gasket.

SS40167,00000A9 -19-16JUL09-3/3

Cylinder Blocks

The cylinder block is the main support for the other basic engine parts. The block is usually a one-piece casting made from gray cast iron.

Blocks are cast with end walls and center webs to support the crankshaft and camshaft and have enlargements in their walls to make room for oil and coolant passages.

The cylinders are often cast into the block and may be bored into the block or machined to receive replaceable liners.

SERVICING CYLINDER BLOCKS

Strip the cylinder block as outlined in the engine technical manual.

After stripping, inspect the block for damage. If the block is still serviceable, prepare it for cleaning as follows:

Scrape all gasket material from the block. Then remove all oil gallery plugs and core hole plugs to allow the

Cylinder Block for V-8 Engine

cleaning solution to contact the inside of the oil and water passages.

Continued on next page   SS40167,00000AA -19-16JUL09-1/4

Cleaning Cast-Iron Cylinder Blocks

1. Remove grease by agitating the cylinder block in a hot bath of commercial grade heavy-duty alkaline solution.

2. Rinse the block in hot water or steam clean it to remove the alkaline solution.

3. If the water passages are heavily scaled, the block can only be cleaned using special equipment as follows:

   a. Agitate the block in a bath of inhibited commercial pickling acid.

   b. Allow the block to remain in the acid bath until the bubbling action stops (about 30 minutes).

   c. Lift the block, drain it, and immerse it in the same acid solution for 10 minutes.

   d. Repeat step c until all scale is removed.

   e. Rinse the block in clear hot water to remove the acid solution.

   f. Neutralize the acid clinging to the casting by immersing in an alkaline bath.

   g. Rinse the block in clean water or steam clean it.

4. Make sure all water passages, oil galleries, and oil holes have been thoroughly cleaned.

Cleaning Aluminum Cylinder Blocks

1. Make a solution of one part emulsion-type soap and five parts water. Using this solution in a steam cleaner, steam clean the cylinder block.

2. Rinse the block using clean water and then dry the block with compressed air.

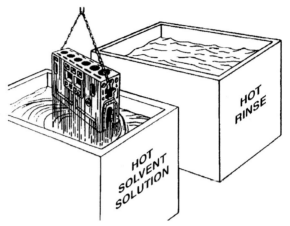

Cleaning a Cylinder Block (Cast-Iron Type)

3. If further cleaning is necessary:

   a. Brush a commercial chlorinated solvent, suitable for aluminum, on the block. Allow the solvent to remain on the block for several hours.

   b. Steam clean the block with soap solution used in step 1.

   c. Rinse the block with clear water and dry it with compressed air.

4. After the cylinder block is thoroughly cleaned and dried, reinstall any core hole plugs. Apply sealer to the threads of the plugs.

SS40167,00000AA -19-16JUL09-2/4

Testing Cylinder Block for Cracks and Leaks

Overheating of the engine may result in cracks between the water jackets and the oil passages.

The block can be pressure tested for cracks and leaks by the following method:

1. Seal the water openings in the block by using plates with gaskets. Drill and tap one of the plates to provide a connection for an air line.

2. Immerse the block for 15 minutes in a tank of hot water [180–200°F (82–93°C)].

⚠ **CAUTION: DO NOT exceed maximum cooling system pressure recommendations specifications.**

3. Apply air pressure to the block and observe the water in the tank for bubbles that would indicate cracks or

leaks. If a large tank is not available, fill the block with water and apply air pressure, then check for water leakage.

4. Always replace a cracked cylinder block.

Continued on next page   SS40167,00000AA -19-16JUL09-3/4

Inspecting the Cylinder Block

After cleaning and pressure testing the cylinder block, check or inspect the block for the following:

1. Check all dowel pins, pipe plugs, expansion plugs, and studs for looseness, wear, or damage and replace as necessary. Apply joint sealing compound to all parts before installing.

2. Inspect all machined surfaces and threaded holes in the block. Carefully remove any nicks or burrs from the machine surfaces with a file. Clean out tapped holes and clean up damaged threads.

3. Check the top of the block for flatness with an accurate straight edge and a feeler gauge. This is the critical area for sealing oil, water, and compression. If warped, replace or resurface the block as recommended in the engine technical manual.

4. Inspect the cylinder bores. For information, see Cylinders and Cylinder Liners in this chapter.

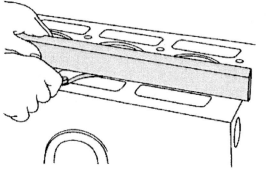

Checking Top of Cylinder Block for Flatness

5. After inspection, spray the machined surfaces with engine oil. If the cylinder block is to be stored for a long period, spray or dip the block in a rust-preventive solution. Cast iron will rust when exposed to the atmosphere unless it is greased or oiled.

SS40167,00000AA -19-16JUL09-4/4

Cylinders

A cylinder is simply a hollow tube in which a piston works.

There are two basic types of cylinders:

- Cast-in-Block (Enbloc)
- Individual Castings (Liners or Sleeves)

Let's see how they differ.

Cast-in-Block (Enbloc)

Automotive engines generally feature the enbloc design. The cylinders are cast into the cylinder block so that the cylinders and the block form a single unit.

Individual Cylinder Castings

The chief advantage of individual castings is that replacement is less expensive. In larger engines, where each separate casting is bolted to a base, it is possible to build up an engine of almost any number of cylinders.

In high-speed engines, these individual castings are called liners or sleeves.

Cylinder liners are of two types:

- Dry Liners (A)
- Wet Liners (C)

DRY LINERS are sleeves that fit inside an already completed cylinder. This liner is simply a wearing surface for the piston. They are not exposed to the engine coolant — so they are called "dry."

WET LINERS not only form the cylinder wall, but also the inside of the water jacket. When using wet liners, seals must also be used to prevent coolant leakage.

A wet liner usually has a flange at the top to allow the liner to seat in a counterbore at the top of the block. With the cylinder head installed, the liner is held firmly in place.

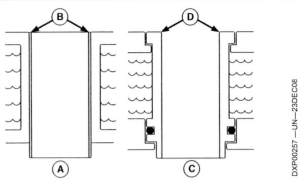

Types of Cylinder Liners

A—Dry Cylinder Liner
B—Liner Is a Sleeve Inside the Cylinder
C—Wet Cylinder Liner
D—Liner Forms the Cylinder Itself

SERVICING CYLINDERS

Because the servicing for integral cylinders and removable cylinder liners is identical except for removal and installation of the liners, the information in this chapter will serve for both types of cylinders.

Preliminary Inspection

As soon as the engine head or heads have been removed, inspect the condition of the cylinders in the area of ring travel. When the appearance is good and there are no scuff marks deep enough to require cylinder reconditioning, check the cylinder for wear or out-of-roundness, described later in this chapter. These measurements help you decide whether to rebore or resleeve the engine.

SS40167,00000AC -19-23JUL09-1/13

Removing Ridge from Cylinders

As the cylinder wears, a ridge is formed at the top of the piston ring travel zone. If this ridge gets too high, pistons can be damaged when they are removed.

Remove any ridges from cylinders using a ridge reamer. Refer to Chapter 12 for more details on the use of the ridge reamer.

Do not cut down into the ring travel zone when removing the cylinder ridge.

It is possible to cut so deeply into the cylinder wall or so far down into the ring travel that reboring or replacement of the engine block or liner is necessary.

Most worn cylinders are out-of-round when cold. Regardless of how uneven the wear, blend the cut made with the ridge reamer so that the area where the worn

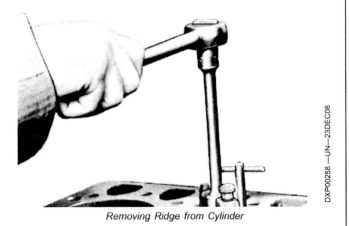

Removing Ridge from Cylinder

surface meets the reamed surface is as smooth as possible.

Continued on next page   SS40167,00000AC -19-23JUL09-2/13

Removing Cylinder Liners

After long service, liners collect carbon and rust or "freeze" in place. Details of removal will vary with the size and type of engine, but use these general tips:

1. Use pullers to remove liners. They may be the conventional screw-threaded type or impact pullers.

2. In stubborn cases, other aids must be used. Removing scales from the water jacket, with inhibited acid compounds, may do the trick. Solvents and penetrating oil may help. Heat may also be used — putting steam or hot water into the water jackets while filling the liner with cold water.

NOTE: Measure cylinder liners for wear before removing them from the block. See Measuring Cylinders for Taper and Out-of-Roundness in this chapter.

Cleaning the Cylinder Area

Clean cylinders as instructed in Cylinder Blocks earlier in this chapter. Use a wire brush and solvents to clean out hardened carbon and gum deposits.

Removing Cylinder Liner

Thoroughly clean the counterbores in the cylinder block with a scraper. Deposits can unseat the liner, distort it, and possibly break the flange.

For wet liners, thoroughly clean the lower sealing ring surface in the block to prevent coolant leakage past the sealing rings.

Continued on next page   SS40167,00000AC -19-23JUL09-3/13

MEASURING CYLINDERS FOR TAPER AND OUT-OF-ROUND

Use a cylinder dial gauge, an inside micrometer, or a telescope gauge with an outside micrometer to measure cylinder bores for taper and out-of round.

Measure cylinder bores as follows:

1. Measure the bore parallel to the crankshaft at the top end of the ring travel zone (A).

2. Measure the bore in the same position at the bottom of the ring travel zone (B).

3. Measure the bore perpendicular to the crankshaft at the top end of the ring travel zone (C).

4. Measure the bore in the same position at the bottom end of the ring travel zone (D).

Compare the measurements (A and C) to find the out-of-round wear at the top of the cylinder.

Compare the measurements (B and D) to find the out-of-round wear at the bottom of the cylinder.

Compare the results of measurements (A, B, C, and D) to find out whether or not the bore is tapered.

How Much Cylinder Wear?

The amount of cylinder taper and out-of-roundness that can be tolerated without reboring or resleeving the engine depends on the design, the condition, and the type of service for which the engine is used.

Generally, whenever cylinder taper or out-of-roundness exceeds 0.005 inch (0.127 mm), the engine should be rebored or resleeved. However, follow the manufacturer's rules for the exact wear limits.

When a cylinder bore is worn beyond the wear limit, the piston generally is also worn beyond its limits. Therefore,

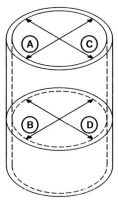

Measure Cylinder Bore at Top and Bottom Ends of Ring Travel

Measuring Cylinder Bore

normally replace the piston when the cylinder is replaced or reconditioned. For more information, see Pistons in this chapter.

Continued on next page   SS40167,00000AC -19-23JUL09-4/13

WET CYLINDER LINER EROSION

The outer surface of wet cylinder liners, which is in contact with the engine coolant, is susceptible to erosion (D) or pitting (A). This erosion or pitting occurs when tiny vapor bubbles in the coolant collapse against the surface of the liner.

Over a period of time, this pitting will progress completely through the metal. Such action is accelerated by impurities and the lack of proper additives in the coolant.

If coolant is allowed to enter the combustion chamber, engine failure or other serious damage will result.

Inspect the exterior of the liner and the liner-packing step (C) for pitting or erosion. If pitting or erosion is observed, measure the depth of the pits and erosion with a wire or needle.

Generally, the liner should be replaced if the pitting depth is one-half the liner thickness (B) or more, or if the erosion depth is one-half the packing step depth or more. Follow the recommendations specified in the engine manufacturer's technical manual.

It is an acceptable practice to reuse liners with erosion or pits provided that the depth of the erosion or pits is less than the amount specified in the engine technical manual. When installing these liners, rotate 90 degrees from original position. The liners should also be deglazed and new ring sets installed on the pistons.

Liners should also be replaced if any of the following conditions are observed:

1. Crosshatch honing pattern not visible immediately below the top ring turnaround area.

2. Deep vertical scratches that can be detected by the fingernail.

3. Cracks in flange area or ring travel area.

Reboring the Cylinder

Cylinder liners for smaller engines are generally priced for replacement when they have reached their wear limits.

Liners for larger engines and all integral cylinders are commonly rebored.

They are sized to the smallest standard oversize diameter at which they will clean up. Oversize pistons must then be

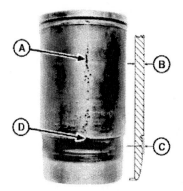

Cylinder Liner Erosion and Pitting

A—Liner Pitting
B—Liner Thickness
C—Packing Step
D—Liner Erosion

fitted to provide the correct piston-to-liner clearance. The final finish should then be obtained by honing.

A number of reboring bars are available which can produce a good finish (ready for honing) in the cylinders. Reboring should be done only by qualified service personnel.

NOTE: Use the reboring bar exactly as directed by the manufacturer. For more information, see Chapter 12.

When reboring cylinders, take these precautions:

1. Grind the cutting tool properly before using it.

2. Be sure the top of the engine block is free of all deposits and irregularities.

3. Clean the base of the boring bar before the bar is set up. Otherwise, the boring bar will tilt and the cylinder will be bored crooked.

4. Make an initial rough cut, followed by a finish cut. Then hone the cylinder to the exact size.

When the boring bar is operated incorrectly, it will produce a rough cylinder surface that may not clean up even when honed. The results: a noisy engine at least during break-in, and faster ring wear.

Continued on next page

SS40167,00000AC -19-23JUL09-5/13

Honing Cylinders After Reboring

Use the recommended grit size for the honing stone to produce the specified finish. Too smooth a finish can retard piston ring seating, while a rough finish will wear out the rings.

Hone the cylinders as follows:

1. When reboring a cylinder, use a rigid hone (A) with a stone grit size equal to the one recommended by the manufacturer of your particular engine. A ridged hone with 120 to 200 grit stones is generally used for removing large amounts of material.

   When honing a cylinder to final size, a flexible hone with a stone grit size of 220 to 280 grit is commonly used. But always follow the recommendation of the engine manufacturer for the proper size of stone grit to use to produce the final finish.

2. Check the stones for wear or damage. Clean them frequently with a wire brush to prevent loading.

3. Follow the manufacturer's instructions for your particular hone for the type and weight of oil to use on the stones. If recommended, apply honing oil to the cylinder. **Do not use cutting agents with a dry hone.**

4. Align the center of the bore to the drill press center.

5. Insert the hone into the bore. Then adjust the hone so the lower end of the stones and cylinder are flush.

6. Adjust the stones until they contact the narrowest part of the cylinder snugly. When correctly adjusted, the hone will not shake in the bore, but it will drag freely up and down when the hone is not running.

7. Start the drill press and run the hone at about 250 rpm. Move the hone up and down the bore with short, overlapping strokes of about 1 in. (25 mm) to obtain a 45-degree crosshatch pattern (B).

   During the first cut, concentrate on any high spots that may cause an increased drag on the stones. As they're removed, the drag will become lighter and smoother.

   NOTE: Measure the bore when the cylinder is cool.

8. Periodically stop the hone and check the bore diameter. Be careful so as not to oversize the cylinder

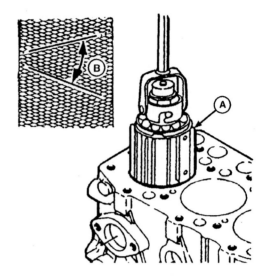

Honing Cylinder Bore with a Rigid Hone

A—Rigid Hone
B—45-Degree Crosshatch Pattern

bore. Increase the diameter of the hone in small increments.

Some stones cut more rapidly than others, even under low pressure. Do not remain in one spot too long or the bore will become irregular.

NOTE: Moving the hone from the top to the bottom of the bore will not correct an out-of-round condition.

Where and how much to hone can be judged by feel. A heavy cut in a distorted bore produces a steady drag on the hone and makes it difficult to feel the high spots. Therefore, use a light cut with frequent adjustments.

9. Continue moving the hone up and down to maintain the 45-degree crosshatch pattern.

10. Remove the rigid hone when the cylinder is within 0.001 in. (0.03 mm) of the desired size.

Continued on next page

SS40167,00000AC -19-23JUL09-6/13

100709
PN=112

11. Then use a flexible hone with stones of the recommended grit size (normally 220 to 280 grit) for honing to the final size.

NOTE: The finish should not be smooth. It should have a 45-degree crosshatch pattern. The crosshatch surface provides small cavities for holding oil during the piston ring break-in period.

12. It is extremely important to thoroughly clean the cylinder block to remove all the abrasive material produced during the honing process. For more information, see Cleaning the Cylinders later in this chapter.

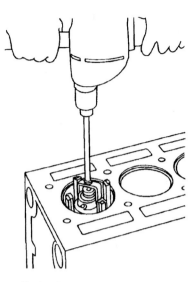

Honing Cylinder Bore with a Flex Hone

SS40167,00000AC -19-23JUL09-7/13

Deglazing Cylinders

Cylinders that are not worn in excess of the specified wear limits are usually deglazed, but not rebored.

Deglazing gives a fine finish by removing any scuffs or scratch marks, but does not enlarge the cylinder diameter. Therefore, standard-size pistons may still be used.

The purpose for deglazing a cylinder is to produce a crosshatch surface in the cylinder that provides cavities for holding oil during piston ring break-in.

The reason for the angled crosshatch pattern is to prevent the rings from catching in the grooves.

Deglaze the cylinders as follows:

1. Remove the pistons and rods. Cover the crankshaft with clean rags or damp paper to prevent the abrasives and dirt caused by deglazing from falling on the crankshaft. The cylinder liner (C) may also be removed and placed in a fixture (D) or scrap cylinder block if desired.

2. Swab the cylinder walls with a clean cloth or cotton string mop that has been dipped in clean, light engine oil.

3. Use the recommended deglazing tool. A flexible, brush-type hone (B) with coated bristle tips often gives the best job of deglazing.

NOTE: Use honing oil along with the flex hone when deglazing the cylinder.

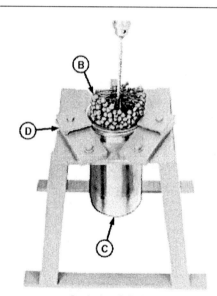

Deglazing Cylinder Liners

B—Brush-Type Hone
C—Cylinder Liner

D—Holding Fixture

4. Use a 1/2-inch variable speed drill to power the hone. Operate the drill at slow speed (approximately 250 rpm) to obtain the recommended finish on the cylinder walls.

Continued on next page

SS40167,00000AC -19-23JUL09-8/13

5. Surface hone each cylinder for 10 to 12 complete strokes. Move the hone up and down in the cylinder rapidly enough to obtain a crosshatch pattern as shown.

NOTE: The finish should not be smooth. It should have a 45-degree crosshatch pattern.

6. If rags or damp paper was used to cover the crankshaft while deglazing, remove the rags or paper as carefully as possible to prevent the abrasives collected during the deglazing process from falling onto the crankshaft.

7. Be sure to thoroughly clean the cylinder bores after honing. For information, see Cleaning the Cylinders in this chapter.

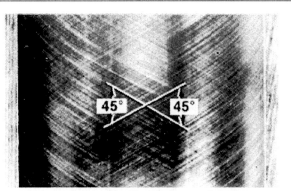

Deglazing Hone Pattern on Cylinder

SS40167,00000AC -19-23JUL09-9/13

Cleaning the Cylinders

If the cylinders are not thoroughly cleaned, harsh abrasives will remain in the engine. These abrasives will circulate with the lubricating oil and become embedded in the engine's metal surfaces, disrupting the thin lubricating oil film that protects the moving parts from friction and wear. This rapidly wears rings, cylinder walls, and the bearing surfaces of all lubricated parts.

If the cylinders are deglazed while in the cylinder block, the entire cylinder block must be thoroughly cleaned.

Use a stiff bristle brush to remove all debris, rust, and scale from the outside diameter of liners, under the liner flange, and in O-ring packing areas. Make certain that there are no nicks or burrs in the areas where the packing will seat.

IMPORTANT: DO NOT use gasoline, kerosene, or commercial solvent to clean the cylinders. Solvents will not remove all the abrasives from the cylinder walls.

Wipe as much of the abrasive deposits from the cylinder wall as possible. Then thoroughly clean the cylinder walls with a 50% solution of hot water and liquid detergent. Rinse thoroughly and wipe dry with a clean cloth.

One cleaning and wiping is not enough. Three operations are usually required — and more may be necessary. Keep on cleaning until a clean white rag shows no discoloration when wiped through the cylinder bore.

Swab the cylinder with clean SAE 10 engine oil immediately after cleaning.

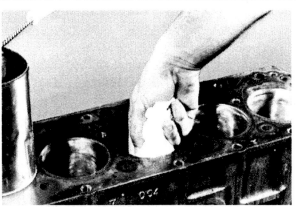

Thoroughly Clean Cylinders After Honing or Deglazing

Installing Dry Cylinder Liners

Before installing dry-type cylinder sleeves, inspect and measure the cylinder bores.

If the bores are distorted, the block must be rebored to take oversize sleeves. Otherwise the cylinder sleeves may conform to the distorted bores, or air pockets will form between sleeves and block, causing hot spots that often result in a breakdown of the rubbing surfaces.

Carefully clean all carbon and gum deposits from the cylinder bores as previously described.

The recommended fit of the sleeve in the bore may be either a loose fit or an interference fit. Always follow the manufacturer's recommendations when installing sleeves.

Continued on next page   SS40167,00000AC -19-23JUL09-10/13

Installing Wet Cylinder Liners

1. Clean all deposits from under the flange of the liner (B) and mating bore in the block (D). The liner must rest flat to prevent distortion.

2. Check the liner bore in the block. If the flange surface is worn unevenly, remachine it.

3. Clean the lower sealing surfaces in the block and liner to prevent coolant leaks when the liner is installed.

4. If a new liner is being installed, install it without seals and check its height (C) in relation to the top of the block as follows: Secure the liners in the block using cap screws and flat washers. Using a liner height gauge with a dial indicator, measure the liner height. Check the height in several places to be sure the liner is seated squarely.

5. Compare the measured height to the engine specifications. If necessary, machine the block to square up the liner, or add shims (if recommended).

NOTE: Always remove the liner and install the seals before final assembly.

6. If liners are being reused, they should be installed into the same cylinder block bore as removed.

A—Water Passage   C—Height of Liner
B—Cylinder Liner   D—Engine Block

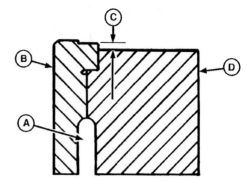

Checking Position of Cylinder Liner in Block

Measuring Liner Height

SS40167,00000AC -19-23JUL09-11/13

7. Put new packing (D) and O-rings on the liner (B) and in the cylinder block (C). Lubricate both the seals and mating surfaces with a liquid soap solution. **Check to be sure the seals are not twisted or crimped.**

NOTE: Do not use petroleum-based lubricant on the cylinder liner seals unless specifically recommended by the engine manufacturer. Oil can cause the packing to swell, which squeezes the liner and could possibly cause a scored piston.

Some engines have a coolant passage (A) located around the top of the cylinder liner to provide increased cooling at the top of the cylinder. The cylinder head gasket seals this upper coolant passage.

8. Work the liner gently into place as far as possible by hand. Finish seating the liner by placing a hard wooden block over it and tapping lightly with a hammer.

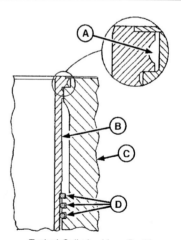

Typical Cylinder Liner Packing

A—Coolant Passage   C—Cylinder Block
B—Cylinder Liner   D—Packing

Continued on next page   SS40167,00000AC -19-23JUL09-12/13

9. The liner (B) may protrude above the top of the cylinder block (C) more than normal due to uncompressed packing and O-rings. But if in doubt, remove the liner and check for twisted seals or carbon deposits (A). When the cylinder head (D) is installed, the liners must seat evenly. The cylinder head may become distorted or head gasket (E) leaks may develop if liners are too high or too low in the cylinder block.

10. Further check for correct liner installation by measuring for out-of-roundness in the liner bore. If a liner is distorted, check for twisted seals or deposits on the liner bore seats.

A—Carbon
B—Cylinder Liner
C—Cylinder Block

D—Cylinder Head
E—Head Gasket

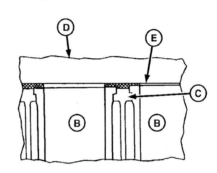

Problems with Cylinder Liners of Different Heights

DXP00271 —UN—23DEC08

SS40167,00000AC -19-23JUL09-13/13

Pistons

Now let's turn to the parts within the cylinder — the piston (B) with its rings (A).

The piston (B) does four jobs:

- Compresses the air/fuel mixture
- Receives the force of combustion
- Transmits this force to the crankshaft
- Carries the piston rings, which seal and wipe the cylinder

DESIGN OF THE PISTON

The piston must be:

1. **Precision-made** — to fit snugly in the cylinder, yet slide freely up and down.

2. **Rugged in Construction** — to withstand the forces of combustion and the rapid stop and start at the end of each stroke.

3. **Carefully Balanced and Weighed** — to overcome inertia and momentum at high speeds.

To be strong yet light, pistons are constructed of cast iron or aluminum alloys. Reinforcing ribs are used to keep the piston as light as possible.

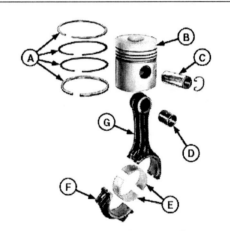

Complete Piston and Connecting Rod Assembly

DXP00272 —UN—23DEC08

A—Piston Rings
B—Piston
C—Piston Pin
D—Pin Bushing

E—Bearing Shells
F—Connecting Rod Cap
G—Connecting Rod

Continued on next page

SS40167,00000AD -19-16JUL09-1/27

PARTS OF THE PISTON

The main parts of the piston are:

- Head
- Skirt
- Ring Grooves
- Lands

The illustration show the parts of a typical piston.

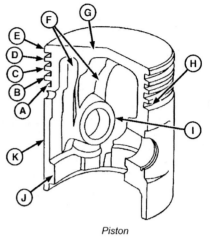

A—Ring Groove
B—4th Land
C—3rd Land
D—2nd Land
E—Top Land
F—Piston Pin Boss
    Reinforcement

G—Head Rib
H—Oil Drain Holes (Behind
    Ring)
I— Piston Pin Boss
J— Skirt Reinforcement
K—Skirt

Piston

DXP00273 —UN—23DEC08

SS40167,00000AD -19-16JUL09-2/27

The head of the piston is the top surface where the combustion gases exert pressure. The piston head may be concave (A), convex or dished (B), or flat (C).

The different shapes of the piston head allow for more or less compression and swirling as needed for the different engines and fuels. Diesel pistons may also have a combustion chamber recessed in the head.

Ribs inside the piston head reinforce it and help to carry heat away from the head to the rings.

A—Concave Head
B—Convex or Dished Head

C—Flat Head

Pistons with Three Types of Heads

DXP00274 —UN—23DEC08

Continued on next page

SS40167,00000AD -19-16JUL09-3/27

The skirt of the piston is the outside part below the ring grooves. The piston is kept in alignment by the skirt.

The piston skirt is usually cam-ground, tapered and elliptical in shape.

Fitting the piston to the cylinder is very critical because:

• Metals expand when heated.
• Space must be provided for lubricants between the piston and cylinder wall.

The piston clearance depends upon the size and thickness of the piston and whether it is made of cast iron or aluminum (which expands more when heated). Also, the piston skirt runs much cooler than the top of the piston, so it needs less clearance.

By making the piston skirt elliptical in shape, it will fit the cylinder during operation when it is hot. The narrower part is at the pin mounts (A), where the metal is thickest, while the wider part is where the metal is thinnest (B).

This fitting prevents noise or slap of the piston during warm-up. Then as the piston gets hot, it expands and becomes round.

Skirts of different shapes are used to allow for the heat and stress on the piston. However, most engines use the trunk-type skirt shown in illustration.

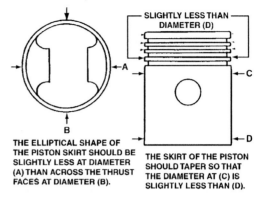

THE ELLIPTICAL SHAPE OF THE PISTON SKIRT SHOULD BE SLIGHTLY LESS AT DIAMETER (A) THAN ACROSS THE THRUST FACES AT DIAMETER (B).

THE SKIRT OF THE PISTON SHOULD TAPER SO THAT THE DIAMETER AT (C) IS SLIGHTLY LESS THAN (D).

Cam-Ground Piston

The ring grooves are cut around the piston to accept the piston rings. They are shaped to the proper rings for good control of oil and blow-by. The lower groove has openings for oil collected by the oil control ring to flow back to the crankcase.

The piston lands are the areas between the ring grooves, which hold and support the piston rings.

SS40167,00000AD -19-16JUL09-4/27

PISTON RINGS

Piston rings do three jobs:

• Form a gas-tight seal between the piston and cylinder.
• Help cool the piston by transferring heat.
• Control lubrication between piston and cylinder wall.

Piston rings are of two types:

• **Compression Rings (A)**
• **Oil Control Rings (B)**

The compression rings prevent gases from leaking by the piston during the compression and power strokes. They seal by expanding out against the cylinder wall. The rings expand by their own tension and also by combustion pressure behind the rings during the power stroke.

Compression rings are split for easy assembly on the piston. The ends that are split do not form a perfect seal, so more than one ring is used.

If the cylinders are worn, ring expanders are sometimes used inside the rings for tighter sealing.

The oil control ring is the bottom ring on a piston. Its job is to wipe excess oil from the cylinder wall. This oil is fed through slots in the rings to holes in the piston groove, where it returns to the crankcase. For better oil control, spring expanders are often used under the oil control ring.

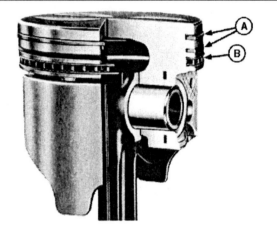

Piston Rings for a Typical Piston

A—Compression Rings   B—Oil Control Ring

In some engines, oil rings are used both above and below the piston pin.

Piston rings are usually made of hardened cast iron and are often plated with metals such as chrome on their contact faces. The hard plating reduces the wear on the ring.

Continued on next page   SS40167,00000AD -19-16JUL09-5/27

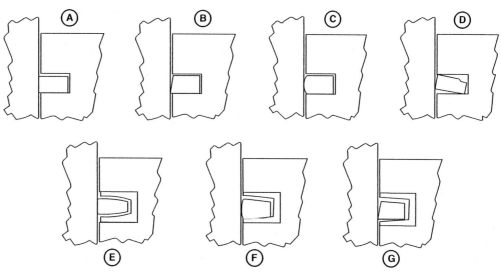

Compression Rings (Seven Major Types)

A—Rectangular Ring
B—Taper-Faced Ring

C—Barrel-Faced Ring
D—Inside Bevel Ring

E—Keystone Ring
F—Keystone Barrel-Faced Ring

G—Keystone Tapered-Faced Ring

DXP00277 —UN—01JUN09

Types of Piston Compression Rings

Various shapes of compression rings are used for pistons. Many pistons use a combination of these rings.

Rectangular Ring (A) — The common ring bears evenly against the cylinder wall along its whole face.

Taper-Faced Ring (B) — The taper ensures that the lower outside edge will have positive contact with the cylinder. This gives quick seating and good wiping of oil on the downstroke, but only fair control of blow-by. Seats very quickly.

Barrel-Faced Ring (C) — This is used as a top ring on some pistons for better control of blow-by.

Inside Bevel Ring (D) — Cutout along the edge of this ring allows it to twist in its ring groove and form a tighter seal during combustion. Used often as an intermediate ring.

Keystone Ring (E) — This shape helps prevent ring sticking by reducing deposits that fill ring grooves and freeze rings. An insert may be used in the groove as a seat for the ring. Keystone rings are used mostly in heavy-duty diesel and aircraft engines.

Keystone rings may also be barrel-faced or taper-faced.

SS40167,00000AD -19-16JUL09-6/27

Joints for Compression Rings

Compression ring joints must be wide enough to prevent the ends from touching when the ring heats up and expands in operation.

The three common ring joints are shown in illustration. The step joint (H) gives better sealing, but the straight joint (J) or angled joint (I) is used most where joint leakage is not a great problem.

H—Step
I— Angle

J— Straight

DXP00278 —UN—23DEC08

Joints for Compression Rings

Continued on next page   SS40167,00000AD -19-16JUL09-7/27

Types of Oil Control Rings

Today's engines throw more oil onto the cylinder walls than is needed for lubrication. (This extra oil absorbs heat.)

To control oil consumption and smoky exhaust, the oil control rings must wipe the excess oil from the cylinder walls and allow it to return to the crankcase.

Oil control rings have a groove in their outer face where oil is collected, fed on through the piston, and returned to the crankcase.

The expander springs that are sometimes used under the oil control ring are made of a steel strip or coil.

Slotted Oil Control Ring

SS40167,00000AD -19-16JUL09-8/27

DIAGNOSING PISTON FAILURES

Why do pistons fail? What do you look for? These are the questions that will be answered on the following pages.

Three Big Engine Problems

Three problems cause trouble in the piston area:

- Oil Consumption and Blow-By
- Combustion Knock (Detonation)
- Preignition (Gasoline Engines)

Let's look at each of these problems and see what the causes and signs are.

Oil Consumption and Blow-By

The piston and rings are supposed to seal the cylinder against two things:

1. Excess oil passing up into the combustion chamber.
2. Gases or blow-by passing down to the crankcase.

OIL CONSUMPTION is the result of excess oil passing into the combustion chamber. The piston rings must leave a film of oil on the cylinder walls or the parts will wear out too fast. Yet if the rings pass too much oil, the engine will consume too much oil.

Slight oil consumption can be expected. The piston rings must pass a small amount of oil for lubrication as the piston works.

One possible cause of excessive oil consumption results from the pumping action of worn piston rings as shown in illustration. The condition has been exaggerated in the drawings for clarity.

However, the more the piston rings and cylinder wall are worn, the greater the oil consumption. (A lighter oil also causes more oil to be used.)

INTAKE

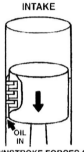

1. DOWNSTROKE FORCES OIL INTO RING GROOVES.

COMPRESSION

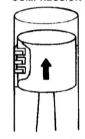

2. UPSTROKE FORCES RINGS TO THE BOTTOM OF THE RING GROOVES, TRAPPING THE OIL.

POWER

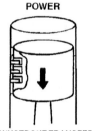

3. DOWNSTROKE TRANSFERS OIL TO CYLINDER WALL. PISTON IS HELD DOWN AGAINST RING LAND BY FORCE OF EXPANSION.

EXHAUST

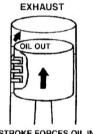

4. UPSTROKE FORCES OIL INTO COMBUSTION CHAMBER, WHERE IT IS BURNED.

Oil Consumption (How the Piston Rings Pump Oil up into the Cylinders, Where it Is Burned)

There are also other ways that oil can reach the combustion chamber from the crankcase: worn valve guides, leaking manifolds or turbochargers. Check all possible causes when diagnosing engine oil consumption and don't forget that external leaks also consume oil.

Continued on next page

SS40167,00000AD -19-16JUL09-9/27

BLOW-BY occurs when combustion gases get past the piston rings and into the crankcase.

A slight amount of blow-by can be expected — this is why crankcases are usually ventilated.

However, severe blow-by can create real problems:

- Pistons are overheated and expand, scoring the piston and cylinder wall.
- Compression is lost, reducing power.
- Crankcase oil is contaminated, causing wear.

Causes of Oil Consumption and Blow-By:

- Piston rings installed wrong
- Stuck rings or plugged oil ring
- Top ring broken or top groove worn
- Overall wear in piston, rings, and cylinder:
  - Abrasive wear
  - Scuffing and scoring
  - Corrosive Wear
- Physical damage to pistons

Lets take a look at each of these causes in detail.

PISTON RINGS INSTALLED WRONG

For a good seal between the piston and cylinder, the piston rings must conform to the cylinder wall and have plenty of tension.

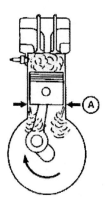

Blow-By of Gases in Cylinder

A—Blow-By

If rings are installed upside down, are the wrong type or size, or are stretched or even broken by improper installation, more oil and gas vapor get by the piston.

For correct procedures, see Installing Rings on Piston in this chapter.

Continued on next page
SS40167,00000AD -19-16JUL09-10/27

STUCK RINGS OR PLUGGED OIL RING

Deposits caused by too much heat, unburned fuel, and excess lubricating oils, collect in the piston ring area. Ring failure usually occurs when these deposits harden and freeze the rings in their grooves.

When the rings are completely stuck, they often break.

Deposits in the top ring groove can cause sticking, scuffing, and scoring because they keep out oil and trap metal particles that wear off the piston.

Sludge deposits in the oil control ring can cause it to plug. This means that oil control has been lost.

Other conditions that lead to stuck or plugged rings are:

- Top groove failure
- Cylinder liner distortion
- Combustion knock
- Preignition
- Overloading
- Cooling system failure
- Improper lube oil
- Cold engine operation (stop and go service)

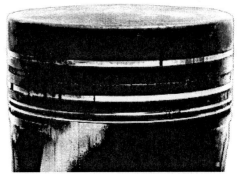

Stuck Rings That Are Broken

Plugged Oil Control Ring

Continued on next page

SS40167,00000AD -19-16JUL09-11/27

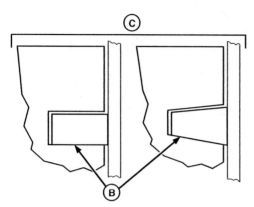 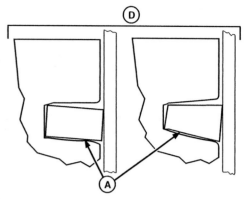

How a Worn Ring Groove Affects Oil Control

A—New Rings in New Grooves   B—New Rings in Worn Grooves   C—Good Oil Control   D—Poor Oil Control

TOP RING BROKEN OR TOP GROOVE WORN

The top piston ring acts as both compression and final oil control ring. It must form a seal between its sides and the ring groove, and between its face and the cylinder wall.

A poor seal will allow oil to bypass the ring seal and be lost.

A poor seal will allow blow-by to contaminate the crankcase oil, forming sludge, which leads to ring sticking, clogging, and possibly scuffing.

The top ring and groove wears most since this area is exposed to the most heat, pressure, and abrasives and gets the least lubrication.

The illustration shows a new compression ring in a new groove at the left. The sides of the groove area are flat, parallel, and smooth. The ring also has the correct side clearance. Combustion gases in the power stroke force the ring down against the lower side of the groove. At the same time, the gases pass behind the ring and force it out against the cylinder wall. The result is a good seal.

At the right in the illustration, we see a new ring installed in a worn groove. The groove permits the ring to sag. This causes the upper corner of the ring face to contact the cylinder wall, resulting in oil being wiped up into the combustion chamber. The flat new ring cannot mate with the worn sides of the ring groove. The result is a bad seal at both the face and sides of the ring.

Top ring groove wear results in broken top rings or damaged pistons.

The following conditions can cause top groove failure:

• Abrasives entering the engine through the intake system
• Combustion knock
• Installation of new ring in worn groove
• Assembly of new ring without ridge reaming the cylinder
• Use of wrong-size piston or ring, resulting in too much or too little side or end clearance

Continued on next page

SS40167,00000AD -19-16JUL09-12/27

OVERALL WEAR ON RINGS, PISTON, AND CYLINDER

1. Abrasive Wear

   When the ring faces are covered with dull gray, vertical scratches and have excessive ring gap (preferably checked in a new cylinder), the rings have been worn by abrasives.

   Other indications of abrasives present in the engine are dull vertical scratches on the piston skirts, scratched cylinder bore, high ridge at the top of the cylinder, loose piston fit, or badly scratched rod and main bearings.

   Badly worn compression rings can be identified by their reduced radial wall. The illustration compares the reduced radial wall of the worn ring (B) on the left with that of a new ring (A) on the right.

   We know what causes abrasive wear, but how do the abrasives enter the engine? Possibly from dirt left in the engine during the last engine overhaul. Through the air intake, crankcase breather, or fuel systems. Or from scuffing and scoring of parts as they wear.

Abrasive Wear Causes Vertical Scratches on Rings

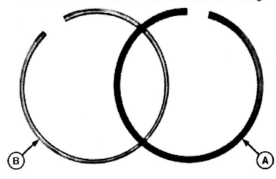

Radial Wall Wear on Worn Compression Ring

A—New Ring   B—Worn Ring

Continued on next page   SS40167,00000AD -19-16JUL09-13/27

2. Scuffing and Scoring

Scuffing is caused by too much heat. When two metal parts rub and the heat builds up to the melting point, a small deposit or "hot spot" (C) of melted metal is pulled out and deposited on the cooler surface.

Scuffing leaves discolored areas on the surface of rings, pistons, and cylinder walls. Under a magnifying glass, the metal in the center of these areas is burnished and smeared in the direction of motion of the part.

Scuffing starts as a very small surface disturbance that may be difficult to see and identify. If not removed, scuffing spreads to adjacent areas and becomes more noticeable and severe, at which time it is called scoring.

Any engine condition which heats rubbing parts to the melting point or which the transfer of heat from these surfaces, has an influence on scuffing.

Possible causes of scuffing and scoring:

- Improper warm-up (fast speeds or big loads too soon)
- Lubricating system not functioning
- Cooling system plugged
- Mineral deposits on wet cylinder liners
- Combustion knock and preignition
- Lugging and overloading

C—Localized Hot Spot

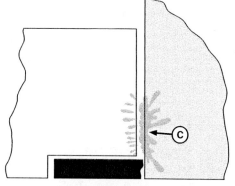

How Scuffing and Scoring Occur

Severe Piston Scoring

SS40167,00000AD -19-16JUL09 14/27

---

3. Corrosive Wear

The third cause of overall wear on pistons and cylinders is corrosion.

Leaking coolant can corrode pistons. Cold engine operation or the wrong lubricating oils can also deposit chemicals in the crankcase, which will corrode parts.

Severe corrosion will show up as a mottled, grayish pitted surface on pistons or cylinder walls. The corrosion is caused by acids from the products of combustion.

Other corrosion may be harder to find. If excessive wear is found, and scuffing and scoring are eliminated as possible causes, suspect corrosive wear.

Corroded Piston with Mottled, Grayish, Pitted Appearance

Continued on next page   SS40167,00000AD -19-16JUL09-15/27

PHYSICAL DAMAGE TO PISTONS

An example of physical damage to the piston is shown in illustration. This piston has lost its pin lock.

Causes of physical damage to pistons:

- Connecting rod out of alignment
- Crankshaft has too much end play
- Crankshaft journal has too much taper
- Cylinder bore is out of alignment
- Piston pin locks installed wrong
- Piston pin locks are faulty
- Ring groove scratched while trying to clean out carbon
- Piston handled carelessly or dropped

Piston Damaged by Loss of Pin Lock

SS40167,00000AD -19-16JUL09-16/27

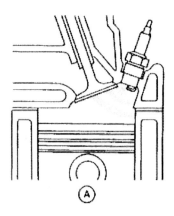

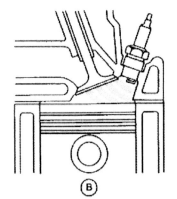

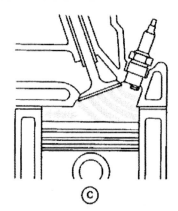

How Combustion Knock (Detonation) Occurs

Combustion Knock (Detonation)

Knocking occurs when combustion of fuel in the cylinder is too early, too rapid, or uneven.

The result is a "knock," which can burn the piston, wear out the top groove, or cause the ring to break or stick.

Causes of combustion knock:

- Lean fuel mixtures
- Fuel octane too low
- Ignition timing has too much advance
- Lugging the engine or overfueling
- Cooling system not working

Combustion Knock Damaged This Piston and Broke the Top Ring

A—Spark Begins Fuel-Air
  Mixture Burning
B—Flame Advances Smoothly,
  Compressing and Heating
  End-Charge

C—End-Charge Suddenly
  Ignites with Violence,
  Producing a Knock

Continued on next page   SS40167,00000AD -19-16JUL09-17/27

Preignition (Gasoline Engines)

Preignition occurs when the fuel ignites before the spark or before compression is completed.

As a result, part of the fuel burns while the piston is still coming up on its compression stroke.

The burning fuel is compressed and overheated by the piston and by further combustion. The heat can get so high that engine parts are melted.

The piston in the illustration was damaged by the heat of preignition.

Causes of preignition:

- Carbon deposits remain incandescent and ignite fuel early.
- Valve operating too hot because of excessive guide clearance or bad seat.
- Hot spots caused by damaged rings.
- Spark plugs with wrong heat range.
- Spark plug loose (results in "hotter" plug).

Preignition Burned a Hole Through the Head of this Piston

Summary: Diagnosing Piston Failures

A premature failure of the engine must have a cause.

That cause must be found and corrected if the engine is to run properly again.

SS40167,00000AD -19-16JUL09-18/27

SERVICING PISTONS

Removing Piston Rings

Clamp the connecting rod in a vise to prevent damaging or burring of the piston. Remove the old rings using a ring expander.

Never reinstall piston rings once they have been removed. Discard the old rings and replace with new ones.

If the pistons are to be removed from the rods, remove the piston pin locks (if used). Piston pins may be floating or press fit. Use the proper procedure for removing piston and pin. Consult engine technical manual for details.

Cleaning Pistons

Careful cleaning of pistons is most important, especially the ring grooves.

Do not scrape with a broken piston ring. Hard scraping will scratch the fine surfaces of the piston.

The chemical-soaking method is recommended for cleaning combustion deposits from the piston.

CHEMICAL SOAKING OF PISTONS

A recommended piston cleaner will soften the carbon on pistons without a lot of effort or any damage. The carbon and residue can then be easily removed with a pressure spray rinse. Follow these steps:

1. Mix the cleaner solution, and heat it as recommended.

2. Before soaking, use a solvent to remove oil film from the pistons (otherwise the cleaning solution will be weakened).

Removing Piston Rings

3. Soak the pistons in the cleaning solution for the specified period.

4. Soak for a second period if needed. A very light scraping may be required in some cases.

5. After soaking, drain and spray rinse the pistons with water and air.

⚠ **CAUTION: Be careful with the cleaning solution. Store and handle with the recommended safety.**

Continued on next page   SS40167,00000AD -19-16JUL09-19/27

CLEANING PISTON RING GROOVES

This method for cleaning piston ring grooves has proved successful if the proper care is taken not to nick or scratch the piston.

IMPORTANT: Always follow the manufacturer's instructions, and safety steps exactly. DO NOT bead blast ring groove area.

1. Carefully remove old rings from the pistons.

2. Clean the pistons by any one of the following methods:

   • Immersion-Solvent "D-Part"
   • Hydra-Jet Rinse Gun
   • Hot water with liquid detergent soap

3. Use a stiff bristle brush — **not a wire brush** — to loosen carbon residue.

4. Clean piston ring grooves using a piston ring groove cleaning tool.

PRECAUTIONS WHEN CLEANING PISTONS

1. Be careful to avoid scratching the sides of the piston ring groove.

Piston Ring Groove Cleaning Tool

2. Never use a wire brush to clean pistons.

3. Be sure the piston ring grooves are thoroughly cleaned. Excessive deposits in ring grooves can force the rings out, causing scuffing and scoring.

SS40167,00000AD -19-16JUL09-20/27

Inspecting Pistons

After cleaning, examine the piston for score marks, damaged ring grooves, or signs of overheating. Always replace a piston that has been severely scored, overheated, or burned.

Carefully inspect the pistons for cracks in the head or skirt areas and for bent or broken lands. Fatigue failures will often show up as cracks in the area of the pin boss.

The magnetic particle inspection methods outlined under "Crankshaft Inspection" may be used in locating cracks in cast-iron pistons.

Damaged pistons must be replaced.

CHECKING RING GROOVES FOR WEAR

Pistons should be replaced or ring grooves reconditioned whenever the ring grooves are worn excessively. Suspect that wear is excessive when ring side clearance is over 0.008 inch (0.20 mm). (Different wear limits apply to many two-cycle and large-bore diesels. Refer to the engine manufacturer's recommendations for each application.)

Measuring Ring Side Clearance with a Feeler Gauge

Check the piston grooves at several points because the grooves wear unevenly. The top groove receives the most wear, but check all of them.

Check rectangular ring groove wear by installing a new ring in the groove. Then insert a feeler gauge between the upper surface of the ring and the land to check the clearance.

Continued on next page   SS40167,00000AD -19-16JUL09-21/27

Keystone groove wear may also be checked using a special wear gauge (A).

Insert the correct size wear gauge (A) in the ring groove. If there is clearance between the gauge shoulder and piston ring land, piston ring groove wear is acceptable. If the gauge shoulder contacts the piston ring groove land, the ring groove is worn. Replace the piston.

Many piston ring manufacturers provide special gauges for checking ring groove wear. These gauges are available for checking both rectangular and keystone grooves.

Excessive ring groove wear can cause the rings to "flutter" in the ring grooves when the engine is running. This can result in loss of compression, loss of power, oil consumption, and ultimately piston ring breakage.

Replace or regroove each piston having excessively worn grooves. Regroove with a ring groove cutting tool (only if recommended). For information, see Re-Machining of Piston Ring Grooves in this chapter.

Piston Ring Groove Inserts

Many pistons for heavy-duty service have a steel or alloy insert cast into the top ring groove. This retards side wear on the groove and ring, and lengthens the life of the pistons and rings.

The piston ring groove insert cannot be replaced or re-machined.

A—Wear Gauge
B—Clearance (Acceptable Ring Groove Wear)
C—Contact (Worn Ring Groove)

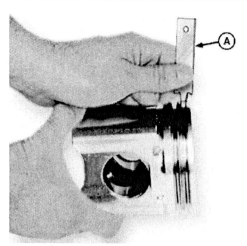

Checking Keystone Ring Groove for Wear

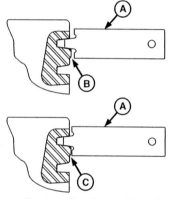

Keystone Ring Groove Wear Gauge

Continued on next page

SS40167,00000AD -19-16JUL09-22/27

Re-Machining of Piston Ring Grooves

Worn piston ring grooves on some pistons can be re-machined with a ring groove cutting tool.

If recommended, machine these grooves for a new standard ring with a flat steel spacer above the ring.

Heat-treated steel spacers must be installed above the ring to compensate for the metal removed by regrooving. The spacers are flat and supply a bearing surface for the full depth of the groove. This ensures a good seal and heat transfer between ring and groove. Top groove spacers (A) are easily installed without special tools.

A—Top Groove Spacer

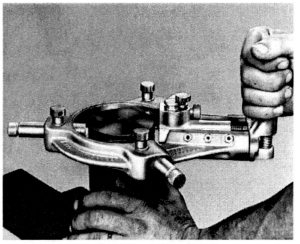

Re-Machining a Piston Ring Groove

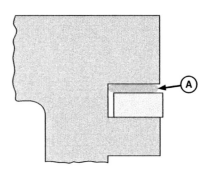

Groove Spacer Above Piston Ring

SS40167,00000AD -19-16JUL09-23/27

CHECK PISTON WEAR

To check for piston wear, measure the diameter of the piston skirt across the thrust faces (at right angles to the piston pin bore). Take a reading at both the top and bottom of the skirt.

Compare these measurements with new dimensions given in the engine technical manual. The difference between these two measurements determines the piston wear.

CHECKING PISTON-TO-CYLINDER CLEARANCE

Since most pistons wear with use and may have excessive clearance, be sure to check the piston clearance at every overhaul.

Pistons should be replaced if their clearance exceeds specifications.

Measure piston clearance as follows:

1. Measure the cylinder diameter at right angles to the crankshaft in the lower or least-worn area of the cylinder. Use a cylinder dial gauge, an inside micrometer, or a telescope gauge with outside micrometer.

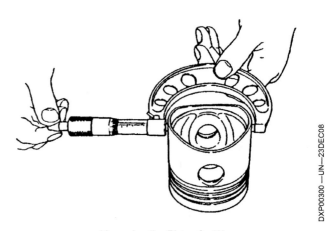

Measuring the Piston for Wear

2. Measure the diameter of the piston across the thrust faces with an outside micrometer.

3. The difference between these two measurements is the piston clearance.

Continued on next page　　　SS40167,00000AD -19-16JUL09-24/27

Piston clearance may also be determined with a feeler ribbon and spring scale as follows:

1. Place the feeler ribbon in the cylinder along one of the thrust sides. (Several thicknesses of feeler ribbon are available.)

2. Turn the piston over and insert it in the cylinder with a thrust face centered against the ribbon.

3. Push the piston down into the lower part of the cylinder. The feeler ribbon must be centered between a thrust side of the cylinder and thrust face of the piston, or an incorrect measurement will result.

4. Pull out the feeler ribbon from between the cylinder and piston, using a 5 to 10 pound (10 to 20 kg) pull on the scale.

5. The thickness of feeler ribbon that can be removed with this pull is the piston clearance.

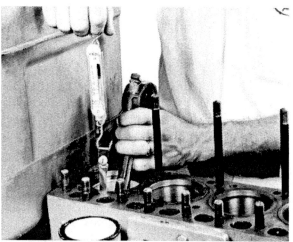

Determining Piston Clearance with Feeler Ribbon and Scale

SS40167,00000AD -19-16JUL09-25/27

INSTALLING RINGS ON PISTONS

First be sure the piston is cleaned thoroughly. Check the ring grooves for any deposits.

Rings must be installed with the top side upward to provide good oil control. Refer to the engine technical manual for directions on installing different ring types properly. Also, each new ring set usually contains an instruction sheet.

Use a ring expander to prevent twisting or stretching the rings during installation.

Be extremely careful not to twist or expand the rings too much as this will permanently distort them and reduce their performance. A ring expanding tool helps avoid this danger.

Normally, ring ends may be located anywhere around the piston, since compression and oil rings are not pinned and usually can rotate around the piston. However, the ring ends should be staggered, and locating of the rings may be specified by the manufacturer.

Installing Piston Rings Using a Ring Expander

Continued on next page

SS40167,00000AD -19-16JUL09-26/27

INSTALLING PISTONS

NOTE: Before installing pistons, complete all the piston pin, connecting rod, and bearing services that follow in this chapter.

Thoroughly clean the piston and cylinder wall before installing the piston. Hard abrasives remaining in the piston ring grooves or on the surface of the cylinder walls will cause rapid wear of the piston rings, pistons, cylinder walls, and the bearing surfaces of all lubricated parts.

IMPORTANT: Many pistons are designed with some form of turbulence chamber in the piston crown. Also, some pistons used in high-compression engines have notches cast or machined in the piston crown to provide operating clearance for the valves. These pistons normally must be installed in the cylinder with the turbulence chamber or notches correctly positioned; otherwise the engine will not operate correctly or engine damage can occur. Always check the top of the piston for identifying markings (an arrow, the word "Front," or other symbols) that indicate the correct position of the piston in the cylinder.

Thoroughly lubricate each piston and its rings with engine oil just prior to installation in the cylinder. This lubricates the pistons, rings, and cylinders during engine cranking and until the "oil throw-off" from the connecting rod journals is adequate.

Several hundred revolutions may be necessary in a dry engine before the lubrication system supplies oil to all the moving parts.

Use a ring compressor to compress the rings while installing the piston. Engine manufacturers usually recommend a specific type of compressor for this purpose.

Apply a gentle downward pressure on the piston to compress the rings into the cylinder. If the piston sticks, compress the rings again and check the cylinder for a partially removed ridge.

Lubricate Piston and Rings Before Installing

Installing Pistons

Do not pound on the piston head, as this may damage the piston or rings.

Refer to the engine manufacturer's instructions on how to position the piston in the cylinder.

SS40167,00000AD -19-16JUL09-27/27

Piston Pins

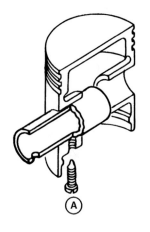 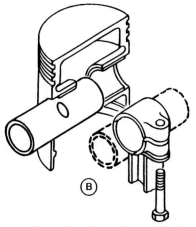 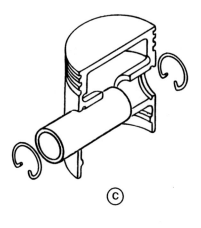

Piston Pins (Three Ways of Fastening)

A—Fixed Pin B—Semi-Floating Pin C—Full-Floating Pin

The piston pin (or wrist pin) connects the piston to the connecting rod.

Piston pins can be fastened in three ways.

- **Fixed Pin (A) — pin fixed to piston, moves in rod**
- **Semi-Floating Pin (B) — pin fixed to rod, moves in piston**
- **Full-Floating Pin (C) — pin moves in both rod and piston**

Most piston pins are "full-floating" with bearings in both the rod and the piston. These pins must be held within the piston bore to prevent contact with the cylinder. Note the pin locks at the right in the illustration.

NOTE: Always install NEW piston pin retaining rings. Make certain the retaining rings have completely expanded in the grooves of the piston. The sharp edge of the retaining ring MUST face toward the outside of the piston.

In high-speed engines, most of the floating pins bear directly on the piston material. This is true with both aluminum-alloy and cast iron pistons.

Servicing of Piston Pins

Piston pins should be checked for looseness, etching, scoring, and excessive damage or wear. Replace them if damaged.

Normally, piston pins need not be replaced except when installing new pistons. This avoids the possibility of noisy pins when the job is done.

Always determine the type of pin and rod assembly from the engine technical manual before attempting to drive the pin from the piston. This will save time and prevent piston damage.

Precision Pin Fits

The piston must take impact loads under all conditions without wearing out or failing. To do this, the pin bearing must have a full bearing surface and still be able to oscillate without any drag.

The requirements for a precision pin fit are:

1. A perfectly **round** pin hole, free from high spots and chatter marks.

2. A **straight** pin hole, free from taper, waviness and bellmouthing.

3. A perfectly **aligned** pin hole between bosses, free from bind and deflection.

4. A **correct surface finish** to sustain and support an adequate oil film.

5. A **good oil clearance**, as recommended for each type of piston and rod.

Continued on next page

SS40167,00000AE -19-16JUL09-1/3

DXP00305 —UN—23DEC08

Checking the Pin Fit

Without a precision pin hole gauge, it is impossible to check a pin fit accurately.

However, these three simple tests will show whether or not you should install new bushings:

1. Check the pin (A) for out-of-roundness or looseness. Clamp the pin in a pin vise and rotate the rod back and forth on the pin several times. Then remove the rod and examine the shiny spots. A good pin fit will show pin contact over the entire surface of the bushing.

2. Check the bore for taper or bellmouthing. Insert the pin from each end of the bushing. If it is free on one end but tight on the opposite end, the pin hole is tapered. If it enters easily from either end but becomes tight in the center, the hole is bellmouthed. A good pin fit must have parallel surfaces.

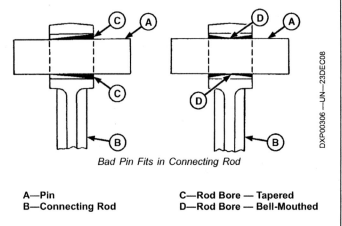

Bad Pin Fits in Connecting Rod

A—Pin
B—Connecting Rod

C—Rod Bore — Tapered
D—Rod Bore — Bell-Mouthed

SS40167,00000AE -19-16JUL09-2/3

3. Check for taper or misalignment between the piston pin holes. Check each pin hole in the piston (B) separately with the pin for equal size and for taper. Both holes should be straight and of equal size.

If pin holes are not tapered, push pin through toward second piston boss. Pin should enter second boss without a "click" and without forcing or binding. A good pin bearing should also have an even drag through both pin holes.

Also make these checks after installing new bushings to be sure the new pin fit is accurate.

Know the exact clearance between pin and pin holes for each piston and pin design to get the precision fit required for today's high-compression engines.

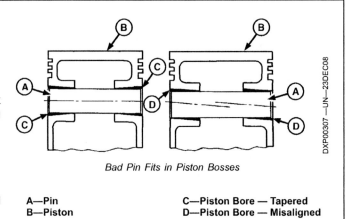

Bad Pin Fits in Piston Bosses

A—Pin
B—Piston

C—Piston Bore — Tapered
D—Piston Bore — Misaligned

SS40167,00000AE -19-16JUL09-3/3

Connecting Rods

Connecting rods must be light and yet strong enough to transmit the thrust of the piston to the crankshaft.

The connecting rod has a shank (C) with a small eye (D) at one end and a large head at the other. The eye (D) forms a bearing for the piston pin, while the head (B) forms a bearing for the crankshaft. The head end of the rod is split and has a bearing cap that is bolted on. Two split bearing halves are inserted into the head (B) and cap (A).

Most connecting rods are forged in one piece, and the cap is then cut off and fitted to the head. On some later connecting rods, a fractured joint is used between the rod and cap. Some connecting rods are designed with a taper at the eye end. This provides for added bearing surface in the heavily loaded areas of the piston and connecting rod during the power stroke (arrows).

A—Cap
B—Head
C—Shank
D—Eye
E—Front Mark (if used)
F—Bearing Inserts

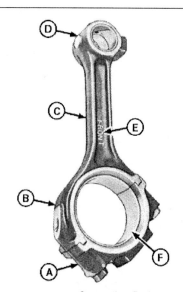

Connecting Rod

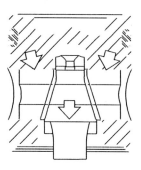

Connecting Rod with Tapered Eye

SS40167,00000AF -19-16JUL09-1/7

SERVICING OF CONNECTING RODS

Bent or Twisted Rods

Generally, connecting rods do not become bent or twisted as the result of normal operation. Both bending and twisting usually result from poor machining or mishandling when reconditioning the engine.

A twisted connecting rod not only places false loads on the connecting rod bearings, but also on the piston, sometimes leading to piston scuffing.

The illustration shows the points of wear from a bent connecting rod.

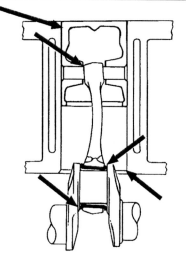

Wear Points from a Bent Rod

Continued on next page   SS40167,00000AF -19-16JUL09-2/7

Rod Alignment

The rod must be aligned within close limits.

The crankshaft bearing bore and the piston pin bushing bore must be parallel to each other (A) and not twisted (B) within 0.001 inch in 6 inches (0.025 mm in 152 mm).

Piston Pin Bushing

New or reconditioned connecting rods usually have enough material in the piston pin bushing end to permit fitting of an oversize piston pin (if one is available) by removal of stock.

IMPORTANT: Boring the connecting rod bushing should be done ONLY by experienced personnel on equipment capable of maintaining bushing finish specification.

If the piston pin bushing is excessively worn, the connecting rods from an engine are normally reused, after checking, by pressing out the old bushing and inserting a new one. Lubricate the connecting rod bore and the outside of the bushing with clean engine oil before pressing the new bushing into the rod. When installing the new bushing, make certain that the oil hole in the bushing is aligned with the hole in the connecting rod.

Precision bore the new bushing to specification to obtain the recommended pin-to-bushing fit. Remove all debris

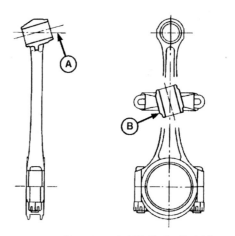

Recommended Limits for Rod Alignment

A—Bores Must Be Parallel to Each Other

B—Bore Must Not Be Twisted

from the boring operation before installing the connecting rod.

Continued on next page

SS40167,00000AF -19-16JUL09-3/7

Piston and Rod Alignment

The piston must work squarely in its bore or the rings won't seal. Misaligned rods not only cause oil consumption and blow-by, they also impose big loads on the rod bearings, the pistons, and the cylinder walls.

To check piston and rod alignment, use a fixture such as the one shown in the illustration.

Corrections are usually made by twisting or bending the rod with a notched bar.

Heavy rods seldom remain aligned after this operation because the rod is not permanently set and the rod soon returns to its warped condition. It is better to replace these rods if bent.

A—Piston and Rod Not Aligned   B—Upper Corner of Rings
                                           Contact Cylinder

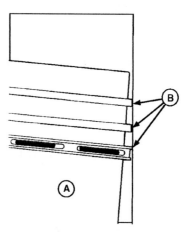

Engine Will Use More Oil if Piston and Rod Are Not Aligned

Checking Piston and Rod Alignment

Continued on next page   SS40167,00000AF -19-16JUL09-4/7

Bearing Clearance

Install connecting rod caps with bearing inserts in place and tighten to the specified torque. Measure the inside diameter of the connecting rod bearings at several places. Also measure the outside diameter of the crankshaft connecting rod journals at several places. Compare the two measurements to find bearing clearance.

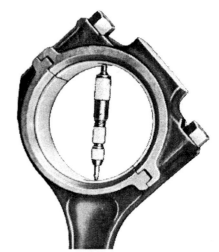

Measuring Connecting Rod Bearing Clearance

Measuring Crankshaft Rod Journals

Continued on next page   SS40167,00000AF -19-16JUL09-5/7

Measuring Bearing Clearance Using Plastigage®

Bearing clearance can also be measured with the crankshaft in place by using Plastigage® — a plastic thread that "crushes" to the exact clearance. While this method will give the bearing clearance, it will not tell you whether the wear is on the bearing or on the crankshaft journal.

1. Place a piece of Plastigage the full length of the bearing insert about 1/4 inch (6 mm) off center.

2. Install the bearing cap and tighten the bolts to the specified torque. Prevent the crankshaft from turning while tightening the bolts.

3. Remove the bearing cap. The flattened Plastigage will be found adhering either to the bearing insert or to the crankshaft.

4. Compare the width of the flattened Plastigage at its widest point with the graduations on the envelope.

The number within the matching graduation on the envelope indicates the total clearance in thousandths of an inch.

Approximate taper may be seen when one end of the flattened Plastigage is wider than the other.

Measure each end of the flattened Plastigage. The approximate taper is the difference between the two readings.

Plastigage is a registered trademark of AE Clevite, Inc.

Placing Plastigage on Bearing Insert

Determining Bearing Clearance with Plastigage

Continued on next page

SS40167,00000AF -19-16JUL09-6/7

Assembly of Rods

Install each connecting rod and piston in the cylinder bore from which it was removed.

Some connecting rods are offset with the center of the bearings to one side from the center line of the shank (C). This makes the engine as compact as possible without skimping on the bearing surface.

Be sure that offset rods face the proper way before you install them. Most offset rods have the word "FRONT" or some other marking on the rod to indicate which way the rod should be installed. See illustration. Refer to the engine manufacturer's instructions on how to position the connecting rod in the cylinder.

Apply clean engine oil to the bearing inserts (F) and crankshaft rod journals. Install the connecting rod caps, making sure that match marks (if present) on the cap and rod are aligned.

Some engine manufacturers specify that the connecting rod cap screws must not be reused. Follow the recommendations in the engine technical manual.

IMPORTANT: DO NOT use a pneumatic wrench to install the connecting rod cap screws — damage will result.

Dip the connecting rod cap screws in clean engine oil. Tighten the cap screws to the torque specified in the engine technical manual.

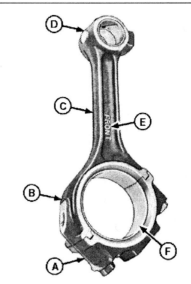

Connecting Rod

A—Cap
B—Head
C—Shank

D—Eye
E—Front Mark (if used)
F—Bearing Inserts

SS40167,00000AF -19-16JUL09-7/7

Crankshaft

The crankshaft converts the up-and-down motion of the pistons into rotary motion. It ties together the reactions of all the pistons into one rotary force that drives the machine.

The crankshaft is forged or cast from a heat-treated steel alloy for extra strength. It is usually made in one piece.

PARTS OF THE CRANKSHAFT

The main parts of the crankshaft are:

- **Journals** — bearing surfaces for support and for connecting rods
- **Throws** — offsets that help provide leverage to rotate the crankshaft
- **Counterweights** — balancing weight opposite the rod journals.

The illustration shows a typical crankshaft.

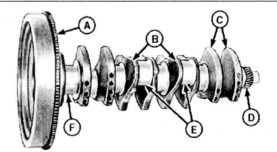

Crankshaft of a Typical Four-Cylinder Engine

A—Flywheel
B—Crankshaft Throws
C—Counterweights

D—Timing Gear
E—Connecting Rod Bearing Journals
F—Main Bearing Journal

Continued on next page   SS40167,00000B2 -19-23JUL09-1/15

ARRANGEMENT OF CRANKSHAFT THROWS

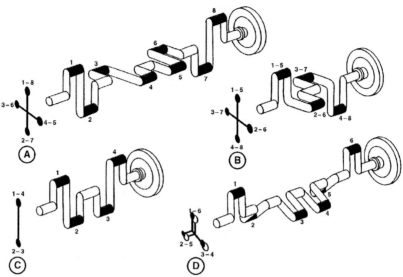

Crankshaft and Throw Arrangements

A—8-Cylinder In-Line Engine **B—8-Cylinder V-Type Engine** **C—4-Cylinder Engine** **D—6-Cylinder Engine**

The arrangement of the crankshaft throws affects:

- The balance of the engine
- Vibration from turning of the shaft
- Loads on the main bearings
- The firing order of the engine

The throws are placed so they counterbalance each other when the crankshaft is driven at great force and speed.

The illustration shows the arrangement of throws on several crankshafts.

Four-Cylinder Engine Crankshafts — have either three or five main support bearings and four throws in one plane. As shown, the throws for No. 1 and No. 4 cylinders are 180 degrees from those for No. 2 and No. 3 cylinders.

Six-Cylinder Engine Crankshafts — have each of three pairs of throws arranged 120 degrees apart. They may have as many as seven main bearings — one at each end and one between each pair of throws.

Eight-Cylinder Engine Crankshafts — have different throw arrangements depending upon whether the engine is a V-8 or in-line model. The crankshafts of V-8 engines are similar to 4-cylinder types or they have the four throws fixed 90 degrees from each other, as shown, for better balance and smoother operation. V-8 engines usually have two connecting rods side-by-side fastened to one throw.

SS40167,00000B2 -19-23JUL09-2/15

BALANCING THE CRANKSHAFT

The power impulses of the engine tend to set up torsional vibrations in the crankshaft.

These vibrations must be controlled, or the crankshaft might break at high speeds.

To balance the force of the pistons and connecting rods, counterweights are placed opposite the rod journals.

The weight of the flywheel and the use of vibration dampers also help to stabilize the turning crankshaft. For more information, see Engine Balancers in this chapter.

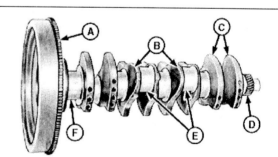

Crankshaft of a Typical Four-Cylinder Engine

A—Flywheel
B—Crankshaft Throws
C—Counterweights
D—Timing Gear
E—Connecting Rod Bearing Journals
F—Main Bearing Journal

Continued on next page

SS40167,00000B2 -19-23JUL09-3/15

LUBRICATING THE CRANKSHAFT

Most engines have pressure oil lubricating the crankshaft.

Oil holes are drilled through the crankshaft journals to match the holes leading in from the block. This oils the main and rod bearings as shown, while the excess oil sprays out to help lubricate the pistons and cylinder.

Bearing Journals

Main and connecting rod journals of most crankshafts are induction hardened. They are also ground and polished to the exact bearing size.

Chrome plating of the journals is not very common, although it is one way to reclaim old journals.

A—Drilled Oil Passage in Crankshaft   B—Oil Spray

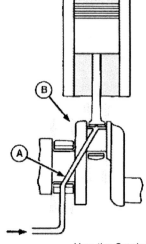

How the Crankshaft Is Lubricated

SS40167,00000B2 -19-23JUL09-4/15

SERVICING OF CRANKSHAFTS

Inspecting the Crankshaft

Bearing trouble is usually related to crankshaft wear. Whenever the main or connecting rod bearings are inspected, the crankshaft should also be inspected. For more information, see Bearing in this chapter.

PRELIMINARY INSPECTION OF BEARINGS

IMPORTANT: Before removing main bearing caps (D), check each cap to make sure it is numbered for reassembly on the same numbered main bearing boss. Keep matched bearings with their respective main bearing caps for comparison with the crankshaft journal (for surface wear) from which removed.

1. Identify the main bearing and connecting rod caps so they can be installed in the order they were removed.

2. Remove the bearing caps (D) one at a time. Examine the crankshaft for scoring, ridging, overheating, cracks, or abnormal wear.

3. Remove the bearing inserts (B) from the main bearing bore in the block and from the main bearing caps. Examine the inserts (B) for evidence of scoring, wear, or "flaking out" of bearing material. Also look for a worn spot on the bearing, indicating a particle has lodged behind the bearing.

4. If one main bearing insert needs replacing, always replace both the bearing inserts.

5. Check clearance and condition of all main bearing inserts at this time. Wear on the damaged insert may be caused by another being out of specifications.

Removing Main Bearing Caps from Crankshaft

A—Thrust Surface	D—Main Bearing Caps
B—Inserts	E—Locking Tangs in Grooves
C—Locking Tang in Groove	F—Locking Tang in Groove

6. If any other inserts are within specifications but show excessive wear, replace them.

7. Install new main bearing inserts at every major overhaul.

Continued on next page

SS40167,00000B2 -19-23JUL09-5/15

100709
PN=142

COMPLETE INSPECTION OF CRANKSHAFT

When a crankshaft has been removed for reconditioning, inspect it as follows:

1. Clean the crankshaft with solvent and dry it with compressed air. Clean and blow out all the oil passages thoroughly.

2. Check the alignment of the crankshaft. Support it on its front and rear journals in V-blocks or a lathe, and check the alignment at the center or intermediate journals using a dial indicator. Protect the crankshaft journals from scratches while turning in the V-blocks by laying a strip of paper in each V.

   The center or intermediate journals should not vary more than the average oil clearance allowable in the total indicator reading. Refer to the engine specifications for the allowable oil clearance.

Checking Crankshaft Alignment with a Dial Indicator

To test either the front or rear main journals, move one of the V-blocks to the center or to an intermediate journal.

SS40167,00000B2 -19-23JUL09-6/15

3. Measure all of the main and connecting rod bearing journals. Measure at several places around the journal to find the smallest diameter, just in case the journals are worn and no longer round. Refer to the engine technical manual for wear limits.

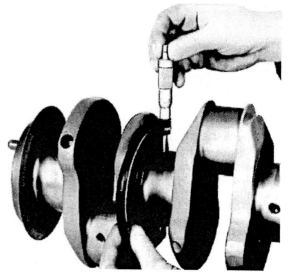

Measuring Crankshaft Journal with a Micrometer

Continued on next page   SS40167,00000B2 -19-23JUL09-7/15

4. Measure the main bearing clearance. With the crankshaft out of the block, install the main bearing caps with bearing inserts in place, and tighten to specifications.

Using an inside micrometer, measure the inside diameter of the main bearings. Compare the reading with the outside diameter of the crankshaft main journals. Then calculate the difference between the two readings to determine the bearing clearance.

Main bearing clearance can also be measured with the crankshaft in place by using Plastigage. For more information, see Connecting Rods earlier in this chapter.

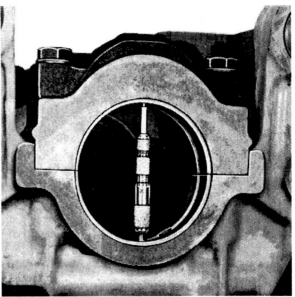

Measuring the Main Bearing with an Inside Micrometer

SS40167,00000B2 -19-23JUL09-8/15

5. Used crankshafts may have some ridging caused by the oil groove in the upper bearing. If this ridge is not removed before new bearings are installed, pressure on the bearings will be too high during operation.

Low ridges can be removed by working crocus cloth (wet with fuel oil) around the journal. Rotate the crankshaft frequently to eliminate an out-of-round condition. If the ridges are greater than 0.0005 inch (0.01 mm), first use 120 grit emery cloth to clean up the ridge, then use 240 grit emery cloth for finishing.

If the ridges are greater than 0.001 inch (0.025 mm), the crankshaft may have to be reground. Finally, use wet crocus cloth for polishing. Consult engine technical manual for details.

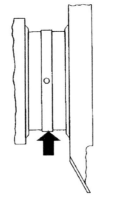

Badly Ridged Crankshaft Journal

Continued on next page

SS40167,00000B2 -19-23JUL09-9/15

6. Check the surfaces of the crankshaft for cracks. Several methods of finding minute cracks not visible to the eye are outlined below:

Magnetic Particle Test — The part is magnetized and then covered with a fine magnetic powder or solution. Flaws such as cracks form a small local magnet, which cause the magnetic particles in the powder or solution to gather, marking the crack. The crankshaft must be demagnetized after this test.

Fluorescent Magnetic Particle Test — This method is similar to the magnetic particle method, but is more sensitive since it employs magnetic particles that are fluorescent and glow under ultraviolet light. Very fine cracks that may be missed under the first test, especially on discolored or dark surfaces, will be disclosed under the light.

Fluorescent Penetrant Test — This is a method that may be used on non-magnetic materials such as stainless steel, aluminum, and plastic. A highly fluorescent liquid penetrant is applied to the part. Then the excess penetrant is wiped off and the part is dried. A developing powder is then applied to help draw the penetrant out of the flaws by way of capillary action. Inspection is carried out under an ultraviolet light.

A majority of the cracks revealed by the above tests are normal and harmless. Few cracks will actually result in damage to the part. Remember that interpreting the results is the most important step. Consult engine technical manual for details.

Crankshaft failures are rare; when one cracks or breaks completely, be sure to make a thorough inspection for the causes. Unless these causes are discovered and corrected, there may be a repeat of the failure.

Two types of loads are imposed on a crankshaft:

- Bending forces
- Twisting, torsional forces

The design of the crankshaft is such that these forces produce almost no stress over most of the surface. Certain critical areas, however, sustain most of the load.

BENDING FATIGUE FAILURE result from crankshaft bending, which takes place once per revolution.

The crankshaft is supported between each of the cylinders by a main bearing, and the force of combustion on the piston is divided between the adjacent bearings. Abnormal bending stress, particularly in the crank fillet, may result from misalignment of the main bearing bores, improperly fitted bearings, failed bearings, a loose or broken bearing cap, or an unbalanced pulley.

Bending failures start at the crank fillet and progress throughout the crank, sometimes extending into the journal fillet.

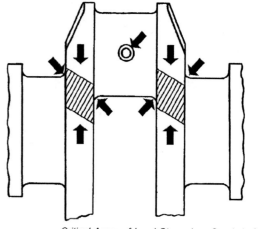

Critical Areas of Load Stress in a Crankshaft

DXP00326 —UN—23DEC08

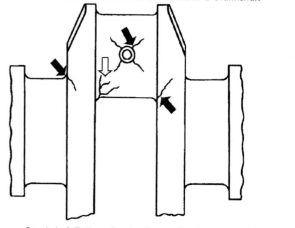

Crankshaft Fatigue Cracks Require Replacement of Crankshaft

DXP00327 —UN—23DEC08

If the main bearings are replaced due to one or more badly damaged bearings, make a careful inspection to determine if any cracks have started in the crankshaft. These cracks are most likely to occur on either side of the damaged bearing.

TORSIONAL FATIGUE FAILURES result from twisting vibrations occurring at high frequency.

A combination of abnormal speed and load conditions may cause the twisting forces to set up a vibration which imposes high stresses at the locations shown in the illustration.

In addition, these stresses occur at the crankshaft journal oil holes near the flywheel end of the shaft.

Torsional stresses may produce a fracture in either the crankshaft rod journal or the main journal. Failures of crankshaft journals are usually at the fillet at a 45-degree angle to the axis of the shaft.

Causes of torsional failures are loose, damaged or defective vibration damper, loose flywheel, or improper or additional fan pulleys or couplings. Other causes may be overspeeding the engine or resetting the governor.

Continued on next page

SS40167,00000B2 -19-23JUL09-10/15

Two types to look for are circumferential fillet cracks at the critical areas and 45-degree cracks (with axis of shaft) starting from either the critical fillet locations or the crankshaft journal holes as shown. These cracks require replacement of the crankshaft.

7. Check the crankshaft thrust surfaces for evidence of excessive wear or roughness. In many instances, only slight grinding or "dressing up" of the thrust surface is necessary. In these cases, new standard thrust washers will probably hold the end thrust clearance within the specified limits (if oversize thrust bearings are used).

8. Inspect the crankshaft keyways for evidence of cracks or wear, and replace the shaft if necessary.

9. Carefully inspect the crankshaft in the area of the rear oil seal contact surface for evidence of rough or grooved conditions. Marring of this surface can result in oil leaks.

SS40167,00000B2 -19-23JUL09-11/15

Reconditioning Crankshaft Journals

Two methods are used to recondition the rod and main journals:

- Grinding the journals by removing material from the surface.
- Rebuilding the journals by adding material to the surface.

REGRINDING THE CRANKSHAFT

1. Before grinding the crankshaft, make a careful check for cracks that start at an oil hole and follow the journal surface at an angle of 45 degrees to axis. Any crankshaft with such cracks must be replaced — regrinding only increases the effect of stress.

   Also, when a shaft is inspected by the magnetic particle method, minute cracks may be found beneath the surface. These are not harmful provided the regrinding does not bring them out onto the surface.

2. Measure the crankshaft journals at dimensions (A and B) and compare with the diameters required for various undersize bearings (obtained from engine or bearing manufacturer). This will determine the size to which the crankshaft journals must be reground.

In addition to the standard main and connecting rod bearings, undersize 0.002, 0.010, 0.020, and 0.030-inch (0.05, 0.25, 0.50, 0.75 mm) bearings are usually available.

It is not advisable to regrind the average crankshaft below 0.030 inch (0.75 mm). This applies to small high-speed engines, but there are some large heavy-duty engines that permit regrinding to sizes below this. As a guide, check to see if undersize bearings are available below 0.030 inch (0.75 mm).

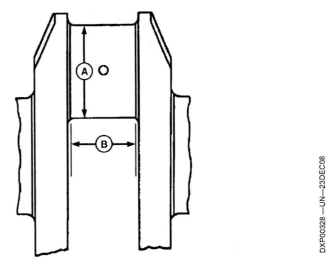

Crankshaft Dimensions to be Measured Before Regrinding

A—Crankshaft Diameter   B—Crankshaft Width

REBUILDING CRANKSHAFTS

Rebuilding crankshafts is adding material to the surface of the crank pins and journals.

This is a specialty job and should not be attempted unless the equipment is available and the job is recommended by the engine manufacturer.

Several methods of rebuilding include:

1. Chromium Plating
2. Electro-Welding
3. Metal Spraying
4. Gas Welding

Continued on next page   SS40167,00000B2 -19-23JUL09-12/15

Installing the Crankshaft

1. Install new bearing inserts (B) in the cylinder block and place the bearing caps (D) in the same location from which they were removed. Usually the inserts and bores have locking devices that align to ensure proper fit and avoid turning. Be sure that these locks align and that the oil holes in the inserts line up with oil passages in the cylinder block.

2. Apply a few drops of clean engine oil to the bearing and spread it over the bearing surface. Carefully position the crankshaft on the bearings. Be sure to hold the crankshaft parallel to the bore and gently lower it into position. This must be done with extreme care because the thrust bearing surface (A) can be easily damaged.

3. Install the remaining bearing halves and caps, making sure the caps are installed on the mains from which they were removed by referring to identification marks made during removal. Loosely install the cap screws in the bearing caps until they are finger tight.

4. Before tightening the caps, align the thrust bearing or washers as follows: Tap the crankshaft to the rear to line up the front flanges. Then tap the crankshaft to the front to line up the rear flanges.

5. Now tighten the bearing cap screws to the specified torque, starting with the center cap and working alternately toward both ends of the block.

6. If the bearings have been installed properly, the crankshaft will turn freely after all the main bearing caps are drawn down to the specified torque.

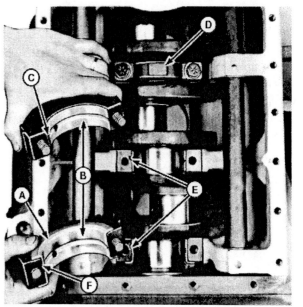

Installing a Typical Crankshaft

A—Thrust Surface
B—Inserts
C—Locking Tang in Groove
D—Main Bearing Caps
E—Locking Tangs in Grooves
F—Locking Tang in Groove

SS40167,00000B2 -19-23JUL09-13/15

Checking Crankshaft End Play

The crankshaft must have a certain amount of end play to get the proper thrust during operation and to avoid excessive wobble and wear.

The main bearing which absorbs the thrust usually has a double flange. Separate thrust washers (A) are sometimes used for the same purpose.

A—Separate Thrust Washers
B—Bearing
C—Thrust Flanges on Bearing
D—Cap
E—Shaft
F—Case

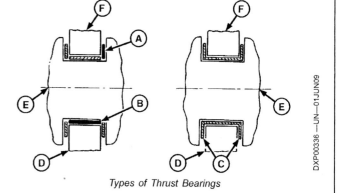

Types of Thrust Bearings

Continued on next page

SS40167,00000B2 -19-23JUL09-14/15

To check the crankshaft end play, force the crankshaft toward the dial indicator as shown. Keep a constant pressure on the tool and set the dial indicator on zero. Then, force the crankshaft in the opposite direction and note the amount of end play shown on the indicator. Refer to the engine specifications for the correct end play.

Too little end play can be the result of a misaligned thrust bearing, or a burr or dirt on the inner face of the bearing flange or thrust washer.

Too much end play means that the thrust surfaces are worn and need replacement.

A—Pry Bar

Checking Crankshaft End Play

SS40167,00000B2 -19-23JUL09-15/15

Main Bearings

All the major wear and load points in an engine use bushings or bearings to reduce friction.

Let's define the two words:

- **Bushing — small full-round sleeve, pressed in, for lighter loads or slower speeds.**
- **Bearing — full-round or halves, for heavier loads and higher speeds.**

BUSHINGS are used at the piston end of the connecting rod, at the rocker arms, the oil pump, etc.

BEARINGS are used at the crankshaft main journals, at the connecting rod journals, at the camshaft, etc.

Actually, the words **bushing** and **bearing** are synonymous when talking in general about a wear or friction point in an engine.

In the remainder of this chapter we will use the word **bearing** for all applications.

WHAT BEARINGS ARE MADE OF

The bearing material depends upon the expected wear and stress. Generally, the bearing has a steel backing with one of three linings:

1. Tin or lead based babbitt

2. Copper or aluminum alloys

3. Multi-layer bearings in copper or aluminum alloys and silver combinations

Normally these materials are applied to the steel backings as a thin layer which is 0.013 to 0.025 inch (0.33 to 0.64 mm) in thickness in small bearings and somewhat thicker in large bearings. Bearings having one material deposited on the backing are referred to as "bimetal" bearings, while those having extra overlay coatings are called "trimetal" bearings.

WHAT THE BEARING MUST DO

Each type of bearing lining must have these qualities:

1. **Conformability** — the ability to creep or flow slightly so that the shaft and bearing will conform to each other.

2. **Embeddability** — the ability to let small dirt particles embed themselves to avoid scratching the shaft.

Bushings and Bearings for a Typical Engine

Bushings and Bearings for a Typical Engine

3. **Seizure Resistance** — smooth surface action, since it is difficult to avoid some metal-to-metal contact during starting or when the oil film becomes thin during operation.

4. **Corrosion Resistance** — the ability of the bearing material to resist chemical corrosion.

5. **Temperature Strength** — the ability of a bearing material to carry its load at high operating temperatures.

Other items such as Wear Rate, Cost, Thermal Conductivity, and the Ability to Form a Good Bond with the backing material must also be considered.

Continued on next page   SS40167,00000B3 -19-16JUL09-1/13

BEARING LOCKS

Unless it is designed for full-floating operation, the bearing insert must be locked in place to prevent rotation with the shaft.

This is usually done by means of a locking lug on each half of the insert, which fits in a notch in both the housing and bearing cap. The lug on one insert prevents rotation in one direction while the other lug works in the opposite direction.

In some thick-walled bearings, a dowel (A) may be used in the cap or housing to hold the bearing.

A—Dowel   C—Lip Slot
B—Dowel Hole   D—Locking Lip

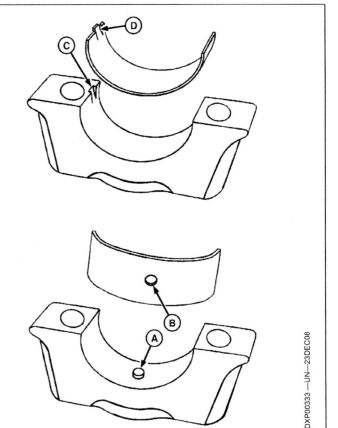

Types of Bearing Locks

DXP00333 —UN—23DEC08

SS40167,00000B3 -19-16JUL09-2/13

BEARING CRUSH

A crush fit provides a close contact between a bearing insert and its housing and cap (D). When the cap is tightened down fully, the bearing (C) is crushed into place.

The illustration shows the allowable crush height of a bearing insert half (A).

This tight fit aids the bearing lock in preventing movement of the bearing in the housing and provides full support.

BEARING SPREAD

Most main and connecting rod bearing halves are purposely manufactured with spread for a tighter fit. This is an extra distance across the parting faces of the bearing half in excess of the actual diameter of the housing bore.

A—Crush Height of Each   C—Bearing
    Bearing Half   D—Cap
B—Rod

Bearing Crush Allowance

DXP00334 —UN—23DEC08

Continued on next page

SS40167,00000B3 -19-16JUL09-3/13

100709
PN=150

BEARING OIL GROOVES

In most engines, the lubricating oil is supplied to the main bearings through the cap or housing. Part of this oil must pass through a hole in the main bearing journals and then through the cranks to the connecting rod journals.

Some of this oil must get to a passage in the connecting rod and on to the piston pin and piston. Oil grooves (B) must be provided for this oil flow. In many cases, either partial or complete circumferential grooves are provided in the bearing surface. In other cases, at least part of the grooving is in the cap or housing outside of the bearing insert.

Distributing grooves are often machined into each bearing half along the parting faces to help distribute the oil the full length of the bearing so that the shaft can pick up the oil and carry it on to the load-supporting area. The term "mudpocket" is sometimes used to describe these grooves because the grooves will temporarily trap wear particles or dirt.

"Thumbnail" grooves (A) are often placed in select locations on flange bearing faces to help distribute the oil evenly over the thrust surfaces.

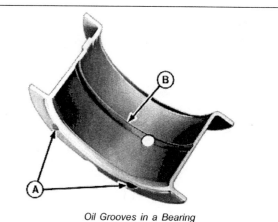

Oil Grooves in a Bearing

A—Thumbnail Grooves   B—Annular Oil Groove

SS40167,00000B3 -19-16JUL09-4/13

CRANKSHAFT THRUST BEARINGS

The crankshaft (E) is usually held in position axially at one of the main bearings (B) near the flywheel end. This may be done by flanges (C), which are part of the bearing inserts, or separate flat thrust washers (A) on each side of one main bearing. Thrust flanges must be on both halves of the bearing.

If new thrust bearing inserts are installed, they must be installed as a matched set. Make certain that the "thumbnail" grooves in the face of the inserts face the crankshaft.

SERVICING OF BEARINGS

Analyzing Bearing Failures

Failure of a bearing in service can be recognized in most cases by the following signs:

1. A drop in lubricating oil pressure
2. Excessive oil consumption
3. Engine noise — rhythmic knock

Wear will vary widely with different engines and operations. For example, continued overloads or frequent shutdowns. Normal wear rate is affected by an abnormal operation, even starting the engine.

Major Causes of Bearing Failure

When bearings fail, find the cause and correct it before installing new inserts. Otherwise, another failure can be expected in a very short time.

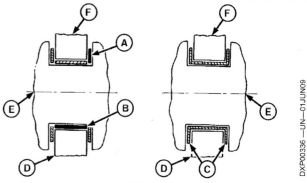

Types of Thrust Bearings

A—Separate Thrust Washers   D—Cap
B—Bearing   E—Shaft
C—Thrust Flanges on Bearing   F—Case

One bearing manufacturer found the probability of bearing failures were as follows:

Dirt	42.9%
Lack of Lubrication	15.3%
Improper Assembly	13.4%
Misalignment	9.8%
Overloading	8.7%
Corrosion	4.5%
Undetermined and Other Causes	5.4%

Continued on next page

SS40167,00000B3 -19-16JUL09-5/13

Diagnosing Bearing Failures

To locate the cause of an engine bearing failure, take the following steps:

1. Remove the oil pan from the engine after draining.

2. While removing the bearings, mark each one for its correct location. Mark it lightly with a soft lead pencil. A code can be used: "1U" for No. 1 cylinder upper half of bearing, "1L" for the lower half, etc.

NOTE: Any bearings showing signs of damage or excessive wear should be discarded and new bearings installed.

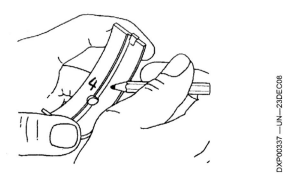

Mark the Bearing to Identify Its Position in the Engine During Removal

3. Clean all of the bearing halves removed in a suitable solvent to remove sludge and oil. Blow or wipe them dry.

4. Observe the sludge or settlings in the engine oil pan. Sometimes sand or metal particles as well as disintegrated bearing linings can be found and identified.

5. Place the cleaned bearings on a flat surface in the order of position. Make two groups: connecting rods farther away, main bearings nearer. The bearing surfaces should face up, with 1U on the upper left, 1L below it, and 2U next to 1U, and so on.

6. Keep an open mind; that is, avoid reaching a conclusion at this point.

7. Examine the connecting rod bearing operating surfaces, noting the distress areas, the rubbing areas, and so on.

8. Examine the main bearings in the same way. Closely inspect the mains that match the connecting rod bearings which are most distressed.

9. Turn all the bearings over to show the bearing backs. Examine the fit of the bearings into their housing by noting the transfer pattern on each half. Inspect the backs for interference by dirt particles, and check for a damaged fit of the locking lips into the recesses.

10. Turn the bearings over once more. After noting the condition of all bearings — overheated condition, poorly fitted, poorly installed, dirty, and so on — then make a conclusion.

11. Repair the engine as necessary to eliminate the cause of bearing failure. Then replace the bearings with new, approved sets.

SS40167,00000B3 -19-16JUL09-6/13

Damage Revealed by Bearings

Here are some examples of bearing damage plus the causes and remedies.

Oil starvation was the cause of damage to the bearing shown in the illustration. Lack of oil can occur immediately after overhaul. This is when priming of the engine's lubricating system is vital to ensure initial lubrication.

After break-in, other things can happen. Both local and general oil starvation can result from external leaks and mechanical supply failures. Blocked oil suction screen, oil pump failure, oil passages plugged or leaking, failed pressure relief valve springs, or badly worn bearings can stop the circulation of lubricating oil.

A mislocated oil hole will also cut off the oil supply to a bearing, causing rapid failure. Always check to be sure that the oil hole in the bearing is in line with the oil supply hole in the connecting rod or crankcase.

Oil Starvation Caused This Damage

Also, the oil supply may become diluted by seepage of fuel into the crankcase from a defective fuel pump. This will reduce the oil's film strength and score the bearings.

Continued on next page

SS40167,00000B3 -19-16JUL09-7/13

Corrosion from acid formation in the oil is seen by a finely pitted surface and large areas of deterioration.

Corrosion occurs when the oil temperature goes above 300°F (150°C) and when excessive blow-by occurs. It is also aggravated by stop-and-go operation, which causes condensation in the crankcase.

Prevent corrosion by changing oil at correct intervals and by selecting oil of the proper quality and classification for the machine and type of service.

Follow the manufacturer's recommendations.

Corrosion from Acid Formation in Oil

SS40167,00000B3 -19-16JUL09-8/13

Dirt can embed in the soft bearing material. This causes wear and decreases the life of the bearing and its journal.

Prevent this by cleaning the engine thoroughly during bearing installation and by proper maintenance of both air and oil filters.

Damage from Dirt Embedded in Bearing

SS40167,00000B3 -19-16JUL09-9/13

Bent connecting rods (A) can cause concentrated wear on the bearings. This angular loading from a bent rod will cause the excessive wear shown on one edge of the upper bearing insert and the opposite edge of the lower bearing insert. When this wear pattern is found, check the connecting rod for poor alignment.

A—Connecting Rod

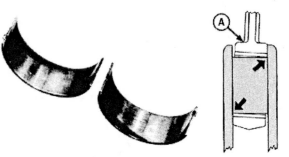

Excessive Wear Caused by a Bent Connecting Rod

SS40167,00000B3 -19-16JUL09-10/13

Tapered journals allow areas of excessive clearance between the journal and bearing, which distributes more wear on one edge of the bearing. This wear is further concentrated by the force of the piston on the insert carrying the greater load.

DXP00342 —UN—23DEC08

Wear on One Edge of Bearing Caused by Tapered Journals

Continued on next page

SS40167,00000B3 -19-16JUL09-11/13

Overheating from overloads on the engine causes a metal failure which breaks away and voids the surface bearing.

Bearing Fatigue Caused by Overloading and Heat

SS40167,00000B3 -19-16JUL09-12/13

To measure bearing wear, assemble the bearing without the crankshaft. Properly tighten the cap screws (B and D). Use an inside micrometer (E) to measure the bearing surface. Also check for wear on the matching crankshaft journals.

A—Bearing Insert
B—Cap Screw
C—Main Bearing Cap
D—Cap Screw
E—Inside Micrometer

Checking Bearing for Wear

SS40167,00000B3 -19-16JUL09-13/13

Engine Balancers

Just as the crankshaft must be balanced, so the whole engine must be balanced. This is a harder job because some parts rotate while others reciprocate — or move up and down.

UNBALANCING FORCES IN THE ENGINE

Two unbalancing forces act on the working engine:

- **Centrifugal Force — rotary outward force at the crankshaft**
- **Inertial Force — up-and-down force at the pistons**

CENTRIFUGAL FORCE (A) at the CRANKSHAFT is created by the revolving weight of the heavy crankshaft throws. Many engines can use counterweights on the crankshaft to balance this force and reduce the bending force on the crankshaft and main bearings. Such a crankshaft is said to be statically and dynamically balanced.

INERTIAL FORCE (B) at the PISTONS is greatest at the top and bottom of each stroke where the piston must reverse itself at high speeds. Other inertial forces are

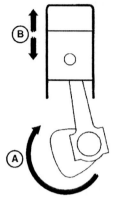

Unbalancing Forces in the Engine

A—Centrifugal Force   B—Inertial Force

created by the motions of the connecting rod and the piston as they "pedal" at high speeds. Counterweight on the crankshaft can also balance these forces in some cases, but not all.

SS40167,00000B4 -19-23JUL09-1/4

BALANCING THE ENGINE

In four-stroke cycle engines with more than four cylinders, inertial forces can be balanced by correct arrangement of the crankshaft throws.

With four cylinders or less, however, the crankshaft (E) can be balanced as a whole but the individual cranks are difficult to balance. To reduce the vibrations on high-speed engines, balancer weights (D) or balancing shafts are used.

The balancer shown is driven by the crankshaft and has two rotating gears (C) with counterweights.

The balancer weights are timed to the crankshaft so that their position counteracts the forces on the cranks. These forces are caused by the different speeds of the pistons at the tops and bottoms of their strokes.

For example, when the crank is pulled up, the weights are down. Result: the upward force of the piston is cancelled by the downward centrifugal force of the weights.

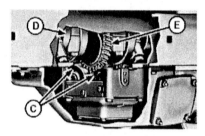

Engine Balancer on a Four-Cylinder/Four-Stroke Cycle Engine

C—Balancer Drive Gear   E—Crankshaft
D—Balancer Weights

Other positions of the crankshaft and weights are also shown in the illustration.

Continued on next page   SS40167,00000B4 -19-23JUL09-2/4

ANOTHER FORCE — TORSION VIBRATION

As the cylinders fire, the pressure "twists" the crankshaft, producing torsional vibrations in the crankshaft (C). We saw earlier that the crankshaft throws and the flywheel are weighted and arranged to reduce these forces. However, some engines are operated under extra stress — sudden loads, varying speeds, and the like.

A **vibration damper** can be mounted on the free end of the crankshaft to control the vibrations.

Most vibration dampers resemble a miniature flywheel. A friction facing is mounted between the hub face and a small damper flywheel (G). The damper flywheel is mounted on the hub face with bolts that go through rubber cones (A) in the flywheel. These cones permit the damper to move slightly on the end of the crankshaft. This reduces the effects of torsional vibration in the crankshaft.

Several other types of vibration dampers are used which employ springs, rubber bonding, "floating" action of fluids, and loose pins. However, the principle is the same as described above.

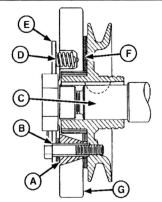

A Typical Vibration Damper

A—Rubber Cone
B—Sleeve
C—Crankshaft
D—Tension Spring
E—Damper Plate
F—Friction Facing
G—Damper Flywheel

SS40167,00000B4 -19-23JUL09-3/4

Servicing Vibration Damper

IMPORTANT: Do not use a jaw-type puller to remove the vibration damper. Damage could result to the damper. Never apply thrust on the outer ring of the damper. The damper is sensitive to impact damage from being dropped or struck with a hammer.

Most vibration dampers have threaded holes in the hub, allowing a puller to be used for removal. Use the correct size hub puller (A) to remove the damper. Do not strike the puller screw or the damper with a hammer as the damper or crankshaft or both could be damaged.

IMPORTANT: Do not immerse the damper assembly in petroleum products (such as gasoline, oil, or solvents). Doing so can damage the rubber portion of the assembly.

Most vibration dampers are not repairable. Check the damper for torn or split rubber protruding from the front and back of the assembly. If there is evidence that the outer ring of the damper has rotated on the hub or if rotation can be felt when turning the outer ring by hand, the damper is defective and should be replaced.

NOTE: Some engine manufacturers recommend that the damper be replaced after a certain number

Removing Vibration Damper

A—Hub Puller

of years or hours of operation. Also, some manufacturers recommend that the damper be replaced whenever the crankshaft is replaced or at engine major overhaul. For more details, refer to the engine technical manual.

SS40167,00000B4 -19-23JUL09-4/4

Crankshaft Oil Seals

Oil seals are located at the front and rear ends of the crankshaft. The purpose of the seals is twofold:

- To prevent oil leakage
- To stop dirt from entering the engine

The front oil seal (E), since it is normally exposed to the environment, usually has a dust seal (F) that faces outward and an oil seal that faces inward. The dust seal stops dirt from entering the engine, and the oil seal prevents oil leakage.

The rear oil seal (D), because of its more protected location, typically isn't exposed to the same type of contaminants as the front oil seal. The rear seal has an oil seal that faces toward the engine to prevent oil leakage. If the engine is used with a wet flywheel housing, it may also have an oil seal that faces away from the engine to seal the engine from the oil in the flywheel housing.

A two-piece, rope-type oil seal is used on many early engines. This seal consists of a flexible material that fits into the bore of a two-piece housing. The housing halves are then assembled around the crankshaft flange and fastened together. This type seal is not commonly used on modern high-speed engines.

Most modern engines use a circular seal with a spring-loaded lip that locates around the crankshaft

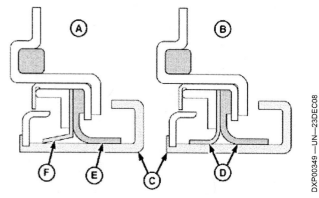

Crankshaft Front and Rear Oil Seals

A—Front Seal
B—Rear Seal
C—Wear Sleeve
D—Rear Oil Seal
E—Front Oil Seal
F—Dust Seal

flange. The seal may ride directly on the crankshaft flange, or it may contact a wear sleeve (C) fitted to the crankshaft flange. The sleeve provides a sacrificial wear surface for the seal, thus preventing wear to the crankshaft surface.

Continued on next page   SS40167,00000B5 -19-16JUL09-1/3

Servicing Crankshaft Oil Seals

The crankshaft oil seals can usually be replaced without removing the crankshaft from the engine. A front seal (B) like the one shown in the illustration is usually mounted in the engine front cover. The rear seal is usually mounted in a seal housing that is bolted to the engine block. If the front cover or rear seal housing is to be removed, the oil seal can be removed after the cover or housing is removed.

If the oil seal housing is not going to be removed, the following method may be used to remove the seal. Drill a small hole in the seal casing. Install a sheet metal screw in the drilled hole with a slide handle puller (A) attached. Carefully pull the seal from the housing.

The preferred method of removing the wear sleeve is with a wear sleeve puller. If a puller is not available, the sleeve can be removed using one of the following procedures.

1. Use the ball side of a ballpeen hammer and tap the wear sleeve across its width in a straight line (to deform and stretch the sleeve). Carefully pry the sleeve from the crankshaft flange.

2. Score the wear sleeve in several places around the outside diameter using a chisel (do not cut through the sleeve). Then carefully pry the sleeve from the crankshaft flange.

Use caution not to nick or scratch the crankshaft when removing the seal and wear ring. Nicks or burrs should be removed with a medium-grit stone. A polishing cloth (180 grit or finer) may also be used when a stone is not available.

NOTE: The seal housing that is bolted to the engine block must be centered carefully on the crankshaft flange. Many engine manufacturers use dowel pins to center the housing; others use no dowel pins and the housing must be centered using a dial indicator or other centering device.

The oil seal housing runout must be checked before installing the seal in the housing. To check housing runout, position a magnetic base dial indicator (D) on the end of the crankshaft flange as shown. Preset the dial

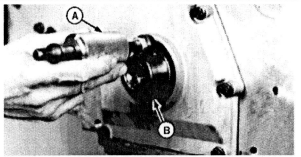

Removing Crankshaft Front Oil Seal

Checking Oil Seal Housing Runout

A—Seal Puller   C—Oil Seal Housing Bore
B—Old Oil Seal   D—Dial Indicator

indicator tip on the inside diameter of the seal housing bore. Rotate the crankshaft one full revolution, and observe full indicator movement.

Refer to the engine technical manual for the maximum oil seal housing bore runout specification. If runout exceeds the specification, loosen the cap screws and adjust the seal housing to obtain an acceptable runout.

Continued on next page   SS40167,00000B5 -19-16JUL09-2/3

It is important to use the correct tools when replacing the crankshaft oil seals. Special seal alignment and installation tools are usually available from the engine manufacturer or other sources.

A seal housing alignment tool (F) is used to center the seal housing bore with the crankshaft flange. A seal installing tool (G) must be used to install the seal (H), otherwise serious damage to the seal may result. If the seal is cocked, installed to the wrong dimension, or otherwise damaged, it will not seal properly.

In some applications, the replacement seal and wear ring are assembled. If the parts become separated, they cannot be reassembled without causing damage to the seal.

Some seals should be lubricated with engine oil prior to installation, and some seals should be installed dry. Refer to the engine technical manual for the recommended installation procedure.

The oil seal must be installed with the open side toward the engine. The seal is designed to function correctly with the direction of rotation of the crankshaft. If the seal is reversed, oil leakage will occur.

NOTE: *If the seal rides directly on the crankshaft flange, a groove may be worn into the crankshaft at the old seal's point of contact. When a new seal is fitted to a worn crankshaft, it may be permissible to position the seal so that it contacts an unworn portion of the crankshaft flange. Refer to the engine technical manual for installation instructions.*

Centering Rear Seal Housing with Alignment Tool

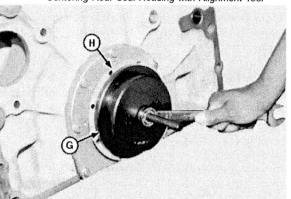

Installing Crankshaft Rear Oil Seal

E—Rear Seal Housing
F—Seal Housing Aligner
G—Seal Installer
H—New Oil Seal

SS40167,00000B5 -19-16JUL09-3/3

Flywheel

The flywheel (A) does three things for the engine:

- **Stores energy for momentum between power strokes**
- **Smoothes out speed of crankshaft**
- **Transmits power to the machine**

Mounted on the rear of the crankshaft, the heavy flywheel is a stabilizer for the whole engine.

In a 4-stroke cycle engine, the flywheel must be heavy enough to turn the engine during the exhaust, intake, and compression strokes. At the same time, it must transmit power to the driven machine.

The more cylinders in the engine, the less need for a flywheel. This is because the power impulses are closer together and the engine has more momentum of its own.

A lighter flywheel is needed for an engine operated at variable speeds where faster acceleration is required.

Flywheels are generally made of heavy cast iron or steel and are fastened to the crankshaft by dowel pins and bolts.

Flywheels may also have two other jobs:

1. Provide a drive from the starting motor via the ring gear (B).
2. Serve as a facing for the engine clutch.

SERVICING THE FLYWHEEL

The flywheel puts a heavy load on the engine rear main bearing, which should be checked carefully at each overhaul. The flywheel must be removed when servicing many parts of the engine.

 CAUTION: The flywheel is heavy. Use proper lifting procedures to avoid personal injury. If the flywheel does not have dowel pins, it may fall on the floor when the last attaching cap screw is removed, possibly causing injury to you or a fellow worker.

Inspecting the Flywheel

After removal, check the clutch contact face of the flywheel for scoring, overheating, or cracks. If scored, most flywheels can be refaced. However, do not remove too much material from the flywheel and keep the surface perfectly flat all around.

The ring gear for the starting motor is generally shrunk in place on the flywheel rim. If it becomes damaged, remove and replace it.

Removing Ring Gear from Flywheel

Note whether the ring gear teeth are chamfered. The replacement gear must be installed so that the chamfer

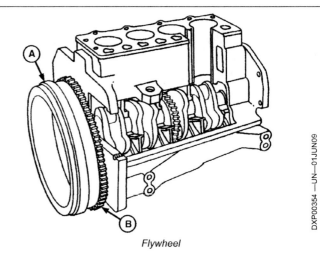

Flywheel

A—Flywheel   B—Ring Gear for Starting Motor

on the teeth faces the same direction. Then remove the ring gear as follows:

1. Support the flywheel, crankshaft side down, on a solid flat surface or hardwood block that is slightly smaller than the inside diameter of the ring gear.

2. Drive the ring gear off the flywheel, with a suitable drift and hammer. Work around the circumference of the ring gear to avoid binding the gear on the flywheel.

Installing Ring Gear on Flywheel

1. Support the flywheel, ring gear side up, on a solid flat surface.

2. Rest the ring gear on a flat metal surface and heat the ring gear uniformly with an acetylene torch, keeping the torch moving around the gear to avoid hot spots.

IMPORTANT: Never overheat the gear because this may destroy the original heat treatment.

If the engine manufacturer specifies a certain heat range, heat crayons that give the temperature may be obtained from most tool vendors.

3. Use a pair of tongs to place the gear on the flywheel with the chamfer, if any, facing the same direction as the gear just removed.

4. Tap the gear into place against the shoulder on the flywheel. If the gear cannot be tapped into place readily, remove it and apply additional heat, but DO NOT OVERHEAT.

Engine Clutches

For information on engine clutches, refer to the FOS manual on "Powertrains."

SS40167,00000B6 -19-16JUL09-1/1

Timing Drives

The crankshaft is the "hub" around which other parts of the engine can be timed and driven. This is done by the meshing of gears as shown.

The camshaft runs at one-half engine speed, so a 2-to-1 reduction of gears is used. An idler gear transmits the rotation to the large camshaft gear (B), which turns half as fast as the smaller crankshaft gear (E). A chain drive can also be used to turn the camshaft.

Other accessories that can be driven by the crankshaft are fuel pumps, oil pumps, injection pumps (diesel), ventilator pumps, and water pumps.

A—Lower Idler Gear
B—Camshaft Gear
C—Upper Idler Gear
D—Injection Pump Gear
    (Diesel)

E—Crankshaft Gear
F—Oil Pump Gear
G—Vent Pump Gear

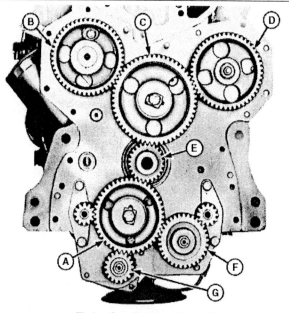

Timing Gear Train on Typical Engine

SS40167,00000B7 -19-20AUG09-1/2

Timing the Gear Train

Many gears in the engine gear train must be timed to each other.

For example, the camshaft operates the valves and must be synchronized with the pistons and crankshaft.

NOTE: On some engines, the crankshaft must make several revolutions before all the timing marks will be aligned.

Timing of these gears is done by matching timing marks (C) when the gears are installed. A timing tool may also be required for timing the gears on some engines. Check the engine technical manual for timing instructions.

Gear Train Backlash

Backlash is the amount of "play" between two gears in mesh. As gear teeth wear, more backlash occurs.

Check the engine technical manual for the allowable backlash between the engine timing gears.

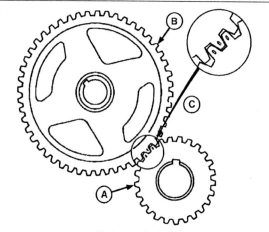

Timing of Engine Gears

A—Crankshaft Gear
B—Camshaft Gear

C—Align Timing Marks When
    Installing Gears

SS40167,00000B7 -19-20AUG09-2/2

Engine Break-In

Modern engines need less break-in than the old ones did. The reason is that design, workmanship, fuels, and lubricants are all better today.

Short but careful break-in periods are the rule on new or overhauled engines.

It is a mistake to "baby" the engine and then suddenly put it under full load. This breaks in the engine for light loads only.

In all cases, follow the recommendations in the engine technical manual.

The steps below are widely recommended.

Break-In Steps

1. Adjust valve tappets, carburetor or injection pump, and engine timing as accurately as you can before starting the engine.

2. Run the engine at half-throttle for a short period until the engine coolant is warmed up to normal.

3. Check for good oil pressure in the engine. Look for oil or coolant leaks.

4. Operate the engine at the specified speed and load for the short time recommended to help seat the head gasket, etc.

5. Stop the engine, then perform the following:

   a. Retighten the cylinder head (if recommended by the engine manufacturer).

   b. Recheck valve tappet clearances (not necessary if equipped with hydraulic lifters).

   c. Recheck engine timing.

6. Operate the engine at normal loads for the first 100 hours (or as recommended). Avoid light loads and excessive idling. Never "lug down" the engine. Check the crankcase oil level more often during this time (special break-in oil may be recommended).

7. At the end of the break-in period, service the engine as specified. This may include changing the oil and replacing the filter.

SS40167,00000B8 -19-16JUL09-1/1

Test Yourself

Questions

1. What is one sequence for tightening cylinder head bolts?

2. What does "knurling" a valve guide do?

3. Why are valve rotators used on some valves?

4. (Choose the correct answer.) The camshaft turns at (1/4 - 1/2 - 2/3) the speed of the engine crankshaft.

5. (Fill in the blanks.) Valve clearance is normally adjusted with the piston at _____ of its _____ stroke.

6. Explain the terms "wet" and "dry" cylinder liners.

7. Where is a cylinder normally worn the most by its piston?

   a. Top inch of the ring travel.

   b. Bottom inch of ring travel.

   c. Center of ring travel.

8. (True or false?) Measure the cylinder liner as soon as it is removed from the engine block.

9. Why are piston skirts not straight up and down?

10. Match each item at the left with the correct item on the right.

a. Blow-By	1. Gasoline igniting before spark occurs
b. Knock (Detonation)	2. Leaking of gases past the pistons
c. Preignition	3. Too-rapid combustion of fuel

11. (True or false?) When removing piston rings, replace only the damaged one.

12. Why mark the position of each connecting rod and piston when removing them from the engine?

13. Why are crankshafts so heavy?

14. What engine part gives momentum between power impulses?

15. (True or false?) For an engine operated at variable speeds, a lighter flywheel is needed.

16. When breaking in an engine, what three adjustments should be rechecked after a brief run-in?

SS40167,00000B9 -19-16JUL09-1/1

Gasoline Fuel Systems

Gasoline Fuel Systems — Introduction

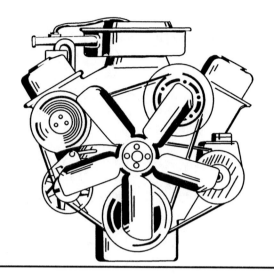

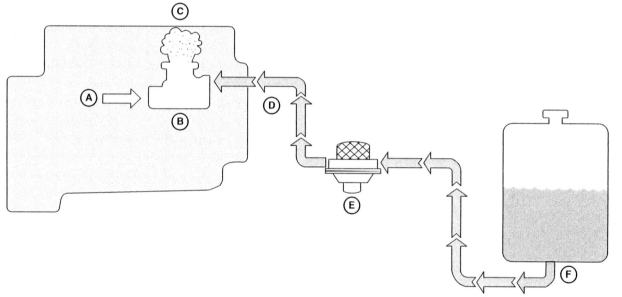

Gasoline Fuel System

A—Air Intake
B—Carburetor
C—Air-Fuel Mixture
D—Fuel Intake
E—Fuel Pump
F—Fuel Tank

The gasoline fuel system supplies a combustible mixture of air and fuel (C) to power the engine.

The gasoline fuel system has three basic parts:

- **Fuel Tank (F)**
- **Fuel Pump (E)**
- **Carburetor (B)**

The **fuel tank** stores the gasoline for the engine.

The **fuel pump** moves the fuel from the tank to the **carburetor**. This is an optional feature needed with pressure-feed supply systems. The fuel pump draws the gasoline through a fuel line from the tank and forces it to the float chamber of the carburetor, where it is stopped.

Continued on next page

SS40167,00000BA -19-17JUL09-1/2

100709
PN=163

The **carburetor** atomizes the fuel, and mixes the fuel with air in the proper ratio. Filtered air (D) is drawn in at one end, and an air-fuel mixture (M) flows out at the other end.

A pressure differential is created when air flows through the narrow neck of the carburetor, called the venturi (B). Air flow moves faster through a restriction, and this lowers the air pressure.

At the same time, the engine (L) creates a partial vacuum on the intake stroke and this causes the air-fuel mixture to flow into the combustion chamber (I) of the cylinder as shown.

The fuel is forced into the air stream from the nozzle (E), which projects into the tube at the venturi. As low-pressure air rushes by, small drops of fuel are forced out and mixed with the air.

The air-fuel mixture must pass the throttle valve (A), which opens or closes to let the correct volume of air-fuel mixture into the engine. The throttle valve controls the engine speed and is usually connected to a foot pedal or hand throttle, which is controlled by the operator.

The choke valve (C) also controls the supply of fuel to the engine. When starting the engine in cold weather, for example, it can be partly closed, to form a restriction. This restriction causes more fuel and less air to be drawn into the combustion chambers. This results in a richer mixture in the cylinders to assist the harder job of starting in cold weather.

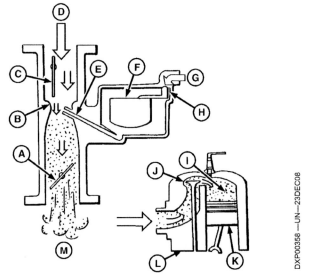

Basic Carburetor at Partial Throttle or Low Power

A—Throttle Valve (Partly Closed)
B—Venturi
C—Choke Valve (Open)
D—Filtered Air
E—Nozzle
F—Float
G—Fuel Inlet
H—Fuel Intake Valve
I— Combustion Chamber
J— Intake Valve
K—Piston
L—Engine
M—Air-Fuel Mixture

DXP00358—UN—23DEC08

SS40167,00000BA -19-17JUL09-2/2

Fuel Supply Systems

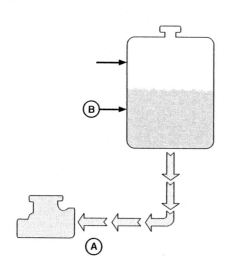

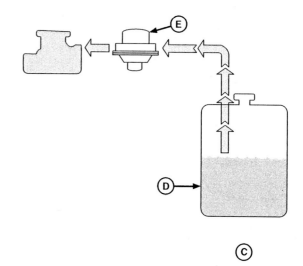

Fuel Supply Systems (Two Types)

A—Gravity-Feed
B—Tank (Above Carburetor)

C—Pressure-Feed
D—Tank (Below Carburetor)

E—Fuel Pump Needed to Pull
Fuel

Fuel is supplied to the gasoline fuel system in two major ways:

- **Gravity-Feed System (A)**
- **Pressure-Feed System (C)**

GRAVITY-FEED SYSTEM

The gravity-feed system has a fuel tank (B) placed above the carburetor: fuel lines, fuel filter, and the gravity-fed carburetor.

A float attached to a valve allows fuel to enter the carburetor at the same rate at which the engine is consuming it. This system maintains a uniform level in the carburetor regardless of the amount of fuel in the tank.

PRESSURE-FEED SYSTEM

The pressure-feed system allows the fuel tank (D) to be located at a level below the carburetor as shown.

However, a fuel pump (E) is required to raise the fuel from the tank to the carburetor. Fuel pump operation is described later in this chapter.

SS40167,00000BB -19-23JUL09-1/1

DXP00359—UN—01JUN09

Fuel Tank

The location of the fuel tank varies on each machine. The tank is usually made of sheet metal and attaches to the frame.

Most systems have the fuel line attached at or near the bottom of the tank and usually have a filter screen at the fuel line connection.

A stand pipe is commonly used on tractor tanks. It provides a fuel outlet that projects above the bottom of the tank to avoid drawing out sediment.

A drain cock at the bottom of the tank allows water and sediment to be drained off periodically.

A shutoff valve may be used to close the fuel outlet before removing the tank.

To ventilate the fuel tank, a vent mechanism may be built into the filter cap or as a separate opening near the top. The vent allows air to replace the fuel as it drawn out and prevents restriction of the fuel flow or a vacuum in the tank.

INSPECTION AND REPAIR OF FUEL TANKS

Cleaning

Flush the tank for 15 minutes with hot water. Run in at the bottom and allow it to overflow at the top.

Steam the tank for 30 minutes. Force in live steam at top of the tank and allow it to escape through the bottom. If live steam is not available, again flush the tank with boiling water continuously for 30 minutes and dry thoroughly with compressed air.

 CAUTION: Cleaning and repairing a fuel tank is very dangerous. Never permit live sparks, smoking, or fire of any nature in the vicinity.

Inspecting for Leaks

For information, see Inspecting a Tank for Leaks in Chapter 7.

Fuel Tank on Modern Machine

Repair of Fuel Tanks

For information, see Repairing a Fuel Tank in Chapter 7. Be sure to observe all the precautions on filling the tank with water and venting it before welding.

Fuel Tank Caps

The fuel tank cap must do three jobs:

- Seal out dust and dirt.
- Keep fuel from splashing out of the tank.
- Allow air to enter the tank to force fuel out (unless the tank has other ventilation).

Be sure that the gasket on the cap seals the tank and that the vents are open so the tank can breathe.

If the fuel filter cap is vented, always replace it with a vented cap. Otherwise, the tank may collapse when a vacuum is created as fuel flows out.

SS40167,00000BC -19-17JUL09-1/1

Fuel Lines

Fuel lines, usually made of steel tubing, transfer fuel from one location to another. Recently, polyethylene lines have become popular where temperatures permit their safe use.

Maintenance involves watching for leaks from loose connections or damaged fittings, and checking for bends or dents that might restrict fuel flow.

SS40167,00000BD -19-17JUL09-1/1

Fuel Gauges

Refer to Chapter 7 for information on fuel gauges.

SS40167,00000BE -19-25JUN09-1/1

Fuel Pumps

The more basic fuel systems depend on gravity or air pressure to get fuel from the tank to the carburetor.

However, for many years, the fuel pump has been widely used on cars, trucks, buses, tractors, stationary, marine, and aircraft engines.

The fuel pump automatically supplies the needed fuel from the tank to the carburetor.

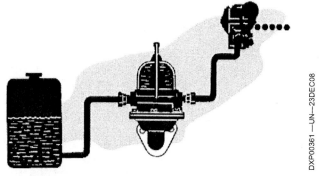

The Fuel Pump Draws Fuel from the Tank to the Carburetor

SS40167,00000BF -19-17JUL09-1/3

MECHANICAL FUEL PUMP

The mechanical fuel pump operates as shown.

Power is applied to the pump rocker arm (E) by an eccentric on the engine camshaft. As the camshaft rotates, the eccentric causes the rocker arm to pivot or "rock" back and forth. The inner end of the rocker arm is linked to a flexible diaphragm (H) located between the upper and lower pump housings. As the rocker arm pivots, it pulls the diaphragm down, then releases it. A spring (G) located under the diaphragm forces it back up. Thus the diaphragm moves up and down as the rocker arm pivots.

When the diaphragm is pulled down, a low vacuum or low pressure area is created above the diaphragm. This causes atmospheric pressure in the fuel tank to force fuel into the pump. The inlet valve (C) opens to admit fuel into the center chamber.

When the diaphragm is released, the spring forces it back up, causing pressure in the area above the diaphragm. This pressure closes the inlet valve and opens the outlet valve (I), forcing fuel from the pump through the outlet to the carburetor.

If the needle valve in the float bowl of the carburetor closes the inlet so that no fuel can enter the carburetor, the fuel pump can no longer deliver fuel.

In this case, the rocker arm continues to pivot but the diaphragm remains at its lower limit of travel so the spring cannot force the diaphragm up. Normal operation of the pump resumes as soon as the needle valve in the float bowl opens the inlet valve, allowing the spring to force the diaphragm up.

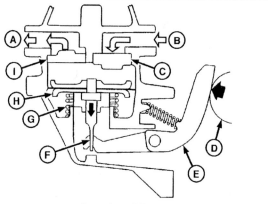

Operation of Fuel Pump

A—Fuel Outlet	F—Pull Rod
B—Fuel Inlet	G—Spring
C—Inlet Valve	H—Diaphragm
D—Engine Cam	I— Outlet Valve
E—Rocker Arm	

COMBINATION PUMP (FUEL AND VACUUM)

These pumps contain not only a fuel pump but also a vacuum pump. Both fuel and vacuum sections of the combination pump are actuated by a single rocker arm. It has a pair of valves and a spring-loaded diaphragm. However, it pumps air instead of fuel, thus creating a vacuum for operating an accessory such as a windshield wiper or a vacuum brake.

Continued on next page   SS40167,00000BF -19-17JUL09-2/3

ELECTRIC FUEL PUMP

This type of fuel pump is used where a mechanical drive is not practical. It contains flexible metal bellows operated by an electromagnet (G). When the electromagnet is connected to the battery (by turning on the ignition switch), it pulls down the armature (E), which extends the bellows (H). This action creates a vacuum in the bellows and fuel from the fuel tank enters the bellows through the inlet valve (I).

When the armature reaches the lower limit of travel, it opens a set of contact points. This disconnects the electromagnet from the battery and thus allows the spring to push the armature upward and collapse the bellows. This in turn forces the fuel from the bellows through the outlet valve (D) to the carburetor.

When the armature reaches the upper limit of its travel, the contact points are closed. This energizes the electromagnet, causing it to again pull the armature down, starting the cycle again.

SERVICING FUEL PUMPS

Testing the Fuel Pump

A fuel pump analyzer can be used to test the fuel pump for delivery and pressure. Usually, however, the pump is visually checked for defects.

For a visual test, disconnect the pump outlet line.

Set the throttle so that the engine will not start, and turn the engine over several times.

If fuel spurts from the line, the pump is operating properly.

If little or no fuel flows, check the following items:

- Primer lever left in upward position (diesel)
- Leaking sediment bowl gasket (if used)
- Plugged screen inside sediment bowl (if used)
- Loose or damaged connections
- Clogged fuel lines
- Loose cover screws on the pump

If the problem is not within these areas, repair or replace the pump.

Repairing the Fuel Pump

1. Disassemble the pump as outlined in the machine technical manual.

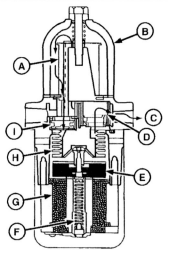

Cutaway View of Electric Fuel Pump

A—Filter Element	F—Return Spring
B—Filter Bowl	G—Electromagnet
C—Fuel Out	H—Bellows
D—Outlet Valve	I— Inlet Valve
E—Armature	

2. Inspect the pump parts as follows:

   a. Look for diaphragm punctures or leaks. Check the slot in the diaphragm pull rod for wear.

   b. Examine the cover and body assembly for cracked or warped gasket surfaces.

   c. Examine the valve and cage assemblies for worn valves or broken springs.

   d. Check the diaphragm and rocker arm spring for sufficient tension.

   e. Inspect the rocker arm link and pin for worn holes and other wear or damage.

   f. Inspect the filter screen for punctures and clogging.

3. Reassemble the pump as instructed in the technical manual. Replace all defective parts with new parts from the repair kit, or replace the fuel pump.

SS40167,00000BF -19-17JUL09-3/3

Fuel Filters

Contamination of fuel is a major cause of excessive engine wear and failures.

Some engines have a separate filter bowl to clean the fuel before it enters the fuel pump. The purpose of the separate filter is to trap water and any foreign objects that

may contaminate the fuel system. These filters should be checked and drained periodically. Most of them have a drain plug that can be loosened; the fuel pump primer lever can be actuated until all deposits are drained out.

If the filter has a sediment bowl and screen, remove and clean them periodically.

SS40167,00000C1 -19-17JUL09-1/1

Carburetors

Engines will not run on liquid gasoline. The gasoline must be vaporized and mixed with air for all types of conditions.

For example:

- Cold or Hot Starting
- Idling
- Part Throttle
- Acceleration
- High-Speed Operation

By mixing fuel with air for each of these conditions, the carburetor regulates the combustion and so the power of the engine.

To get the right air-fuel mixture, the carburetor must "atomize" the fuel and mix the fine particles of fuel with air.

Atomizing is done by adding air to the liquid fuel as it moves through the carburetor passages and then spraying the air-fuel mixture through nozzles or jets into a stream of moving air flowing into the engine's intake manifold.

The fuel in the air-fuel mixture is then vaporized before it enters the combustion chamber of the engine.

The various speed and load conditions demand a different volume of air for the mixture. The air-fuel ratio must be kept within flammable limits to permit combustion.

The average gasoline engine works best when about 15 parts of air are mixed with 1 part fuel.

The primary job of the carburetor is to produce this ratio, or air-fuel mixture, for any operating condition.

THEORY OF PRESSURE DIFFERENTIAL

Since the carburetor operates by pressure differentials, let's look at these terms:

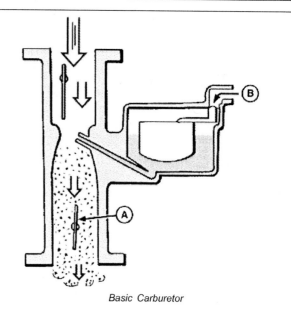

Basic Carburetor

A—Throttle Valve (Wide Open)   B—Fuel

- Vacuum
- Atmospheric Pressure
- Venturi Principle

Vacuum

Absolute vacuum is any closed area completely free of air or atmospheric pressure. We can better understand the carburetor by calling any pressure less than atmospheric a vacuum or low pressure area.

Continued on next page   SS40167,00000C4 -19-23JUL09-1/18

Atmospheric Pressure

Atmospheric pressure is the weight or pressure of the air around us. Air pressure at sea level is 14.7 pounds per square inch (100 kPa). This air tries constantly to occupy all space within our atmosphere.

Piston movement in an engine creates a vacuum or low pressure area (B). Atmospheric pressure forces air to flow into this vacuum.

A—Valve Closed
B—Valve Closed Creating Space (Vacuum)
C—Valve Open
D—Space (Vacuum)

ATMOSPHERIC PRESSURE

AIR PRESSURE AT SEA LEVEL IS 14.7 PSI (100 kPa)
AIR TRIES TO OCCUPY ALL SPACE WITHIN OUR ATMOSPHERE

Atmospheric Pressure

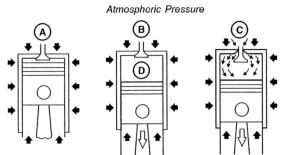

The Engine Creates a Vacuum

SS40167,00000C4 -19-23JUL09-2/18

Venturi Principle

Pressure differentials are basic to how a carburetor works. A venturi is used in the throat or bore of all carburetors to create this pressure differential. The faster the air moves, the lower the air pressure at the venturi. This low pressure is the basic force by which a carburetor works.

BASIC TYPES OF CARBURETORS

There are three basic types of carburetors:

• Natural draft
• Updraft
• Downdraft

In all three, there is a fuel supply in the fuel bowl, a passage or air tube through the carburetor for the stream of air going to the engine, and a nozzle connecting the bowl to the air tube. The venturi is also a feature of all three, creating the pressure drop by which the carburetor works.

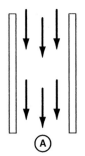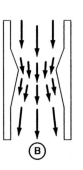

The Venturi Is a Restriction of Air Flow, Which Lowers Air Pressure

A—Air Flow Minus Venturi
B—Air Flow Past Venturi

Continued on next page

SS40167,00000C4 -19-23JUL09-3/18

Uses of the Basic Carburetors

The natural draft carburetor uses a crossdraft to supply the air flow and is used on engines where there is little space on top or where the atomized fuel in the mixture is vaporized by heat from the water in the engine water jacket.

A—Air
B—Low Velocity-Low Vacuum
C—High Velocity-High Vacuum
D—Low Velocity-Low Vacuum
E—Nozzle
F—Fuel Bowl
G—Vent

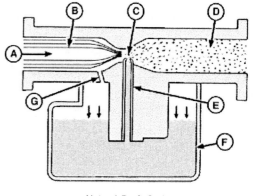

Natural Draft Carburetor

SS40167,00000C4 -19-23JUL09-4/18

The updraft carburetor can be placed low on the side of the engine and supplied fuel by gravity feed. However, the fuel must then be lifted up into the engine. This means that air velocities must be high, which can only be attained by using small passages in the carburetor and manifold.

These carburetors are best adapted for use on most modern farm and industrial machines.

A—Air
B—High Velocity-High Vacuum
C—Low Velocity-Low Vacuum
D—Nozzle
E—Vent
F—Fuel Bowl

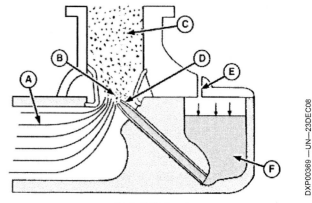

Updraft Carburetor

SS40167,00000C4 -19-23JUL09-5/18

The downdraft carburetor allows for larger volumes, since the fuel will reach the engine even though the air velocity is low. These carburetors can be used when high speeds and high power outputs are required.

A—Air
B—Nozzle
C—Fuel Bowl
D—Low Velocity-Low Vacuum
E—High Velocity-High Vacuum

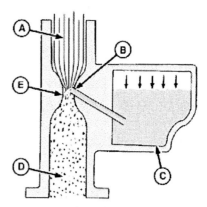

Downdraft Carburetor

Continued on next page

SS40167,00000C4 -19-23JUL09-6/18

Fuel Source

All carburetors require a source of fuel and a means of maintaining the proper level in the fuel bowl.

Fuel enters the bowl under pressure (either gravity or pump) through a valve that is controlled by the float. As fuel rises in the bowl, the float rises with it. When the correct fuel level (E) is reached, the float (A) stops the fuel supply by forcing the valve (B) against its seat.

As fuel is used by the carburetor, the float lowers with the fuel supply, allowing the valve to move from its seat and admit more fuel.

A—Float
B—Valve (Closed)
C—Fuel In
D—Hinge
E—Fuel Level

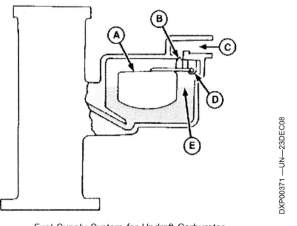

Fuel Supply System for Updraft Carburetor

SS40167,00000C4 -19-23JUL09-7/18

Choke Systems

The choke system provides an extremely rich air-fuel mixture (B) during starting, particularly in cold weather.

To provide this rich mixture, a disk or choke valve (A) is located in the air intake side of the carburetor tube.

When the choke valve (A) is closed, the vacuum created inside the intake manifold extends beyond the nozzle to the choke valve. Because of this vacuum, atmospheric pressure in the fuel bowl can force more fuel into the carburetor, resulting in a richer air-fuel mixture as shown.

A—Choke Valve (Closed)
B—Rich Air-Fuel Mixture
C—Partial Vacuum
D—Air (Shut Off)

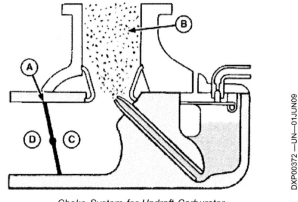

Choke System for Updraft Carburetor

Continued on next page

SS40167,00000C4 -19-23JUL09-8/18

Automatic Chokes

The operation of an automatic choke depends primarily on the unwinding of a thermostatic coil spring (H) as heat is applied. The illustration shows a typical model.

When the engine is cold, the thermostatic spring holds the choke valve closed.

When the engine starts, the intake manifold vacuum acts on the vacuum piston (G) and partially opens the choke valve.

Hot air from the exhaust manifold (J) heats the thermostatic spring. This heat expands the spring, allowing the choke valve to open wide.

A—Fast-Idle Adjusting Screw
B—Throttle Plate
C—Throttle Shaft
D—Countershaft
E—Air Opening (Throat)
F—Choke Plate
G—Piston
H—Thermostatic Spring
I— Vacuum Cylinder
J— Hot Air from Exhaust Manifold

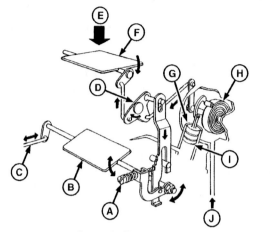

Automatic Choke in Operation

SS40167,00000C4 -19-23JUL09-9/18

Throttle Systems

The throttle system regulates the amount of air-fuel mixture entering the engine cylinders. This is required for two reasons:

- To vary the engine speed.
- To keep a uniform engine speed under varying loads.

On many engines, the throttle valve (A) is connected by a linkage to a governor, which in turn is connected to a speed control lever. When the speed control lever is set to operate the engine at a given speed, the governor will maintain that speed (unless the engine is overloaded).

If the load is increased, the governor will automatically open the throttle valve wider, permitting more air-fuel mixture to enter the engine, thus maintaining a uniform speed.

For more information, see Governing Systems in Chapter 11.

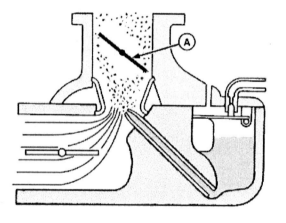

Throttle System for Updraft Carburetor

A—Throttle Valve

Continued on next page

SS40167,00000C4 -19-23JUL09-10/18

Load Systems

The load system delivers the proper air-fuel mixture to the engine in all ranges of speed and load above idling.

The illustration shows a load system for an updraft carburetor. The amount of fuel entering the nozzle is regulated by a load adjusting needle (C).

In many carburetors, a fixed jet or orifice allows the proper amount of fuel for maximum power and economy to enter the nozzle (B).

B—Nozzle   C—Load Adjusting Needle

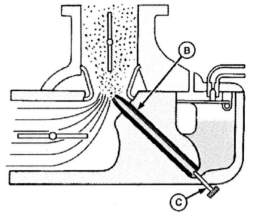

Load System for Updraft Carburetor

SS40167,00000C4 -19-23JUL09-11/18

Accelerating Systems

Whenever the throttle is opened quickly to give extra power for a sudden load, extra fuel is required for a momentarily richer air-fuel mixture.

This extra fuel can be supplied in two ways:

- **Acceleration pump** — Used for rapid acceleration. This can be operated either by the throttle or by a vacuum.
- **Acceleration well (or similar device)** — Used for less rapid acceleration.

ACCELERATION PUMP

The acceleration pump is often a simple piston-type pump that delivers fuel into the airstream at the start of acceleration.

The pump is actuated either by the throttle or by vacuum. The throttle actuated pump is shown.

When the throttle is opened, the plunger (F) descends under the pressure of the plunger spring (G) and fuel is forced out through the accelerating jet (B). When the throttle is fully depressed and at rest, the spring forces the plunger to the bottom of the well, thus continuing the flow of the accelerating charge.

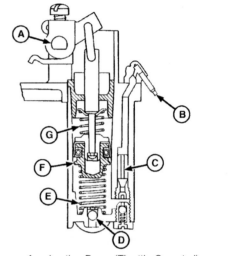

Acceleration Pump (Throttle Operated)

A—Throttle Linkage   E—Pump Spring
B—Jet   F—Plunger
C—Check Needle   G—Plunger Spring
D—Intake Ball

Continued on next page

SS40167,00000C4 -19-23JUL09-12/18

ACCELERATION WELL

When the engine is idling, fuel rises inside the load nozzle and passes through holes in the side of the nozzle into a chamber or well surrounding the nozzle.

When the throttle is suddenly opened, the fuel stored in the accelerating well (A) pushes through the holes in the side of the nozzle (B) without being metered by the adjusting needle and combines with the normal flow in the nozzle.

The two quantities of fuel enter the air stream, making a much richer air-fuel mixture to satisfy the sudden need for more power.

As the fuel supply drops in the accelerating well, the holes that are uncovered become air bleeds.

The accelerating well then remains drained of fuel until the throttle returns to one-fourth load or to fast idle, at which time it refills.

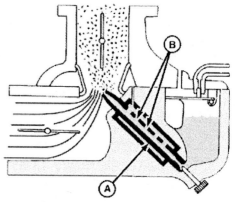

Acceleration Well on Updraft Carburetor

A—Accelerating Well   **B—Side of Nozzle**

SS40167,00000C4 -19-23JUL09-13/18

Idling Systems

When the throttle valve is closed, the idling system supplies just enough air-fuel mixture to keep the engine running at slow idle.

The idling system is different for updraft and downdraft carburetors.

Idling System for Updraft Carburetors

In the updraft carburetor shown, air for the idle system goes around the venturi and enters the fuel system through a drilled passage (see arrows). The idle adjusting needle (D) regulates the amount of air used.

Fuel is forced through drilled passages from the accelerating well to be mixed with air and delivered to the engine through the primary idle orifice (C). The operation of the primary idle orifice and secondary idle orifice (B) is the same as in the natural draft carburetor.

Idling System for Downdraft Carburetors

In the downdraft carburetor (not shown), air for the idle system enters a small passage above the venturi. Atmospheric pressure pushes air and fuel through the passages. Then they mix and flow past the tapered point of an idle adjusting screw. The air-fuel mixture is very rich but leans out somewhat as it mixes with the small amount of air that gets past the closed throttle valve. The idle adjusting screw regulates the amount of air-fuel mixture used.

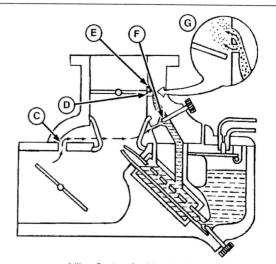

Idling System for Updraft Carburetor

A—Air Pickup for Idle System   **D—Idle Adjusting Needle**
B—Secondary Idle Orifice   **E—At Fast Idle**
C—Primary Idle Orifice

The operation of the primary and secondary orifices is the same as for the natural draft and updraft carburetors.

Continued on next page
SS40167,00000C4 -19-23JUL09-14/18

Economizer Systems

The purpose of the economizer system is to retard the flow of fuel to the engine at part throttle when the full capacity of the nozzle is not required. The basic operation is the same for all three types of carburetors.

The passage providing air for the bowl vent is extended to a point near the throttle valve. When the throttle valve is partially open, the passage is on the engine side of the throttle. Action of the engine draws air through the passage, reducing the air pressure in the bowl.

When the bowl pressure is reduced, the difference between the pressure in the venturi and on the fuel in the bowl is also reduced. This retards the flow of fuel out the nozzle.

In the updraft carburetor, an economizer jet (D) in the passage regulates the air pressure at part throttle.

In some updraft carburetors, the economizer passage (C) is brought around the throttle shaft so that when the throttle valve is closed or partially opened, a slot in the shaft permits air to enter the engine.

As the throttle valve is opened, the shaft rotates until its slot closes the passage completely at full throttle and no air can be drawn out of the fuel bowl.

Summary: Operation of Carburetors

The following chart gives a summary of the job of each major system in the carburetor.

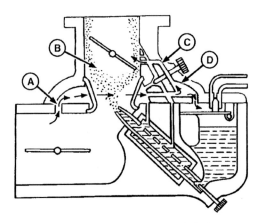

Economizer System for Updraft Carburetor

A—Air Pickup
B—Reduced Air-Fuel Mixture
C—Economizer Passage
D—Economizer Jet

Operation of Carburetor Systems	
System	Function
Float	Controls flow of fuel from tank to carburetor to maintain a constant level of fuel in carburetor.
Choke	Creates a rich air-fuel mixture for starting a cold engine.
Throttle	Allows engine to accelerate smoothly for idle to high speeds.
Load	Allows engine to operate at its maximum power output.
Idle	Allows engine to operate economically at idle speeds when power is not needed.

Continued on next page

SS40167,00000C4 -19-23JUL09-15/18

SERVICING OF CARBURETORS

Often, the carburetor (F) is the first component to be blamed for bad engine operation.

An experienced service technician will first check out other faults:

1. Faulty ignition.
2. Low engine compression.
3. Faulty supply of fuel or air to carburetor.
4. Worn or badly adjusted governor linkage.
5. A faulty carburetor — checked last of all.

Proper carburetion can only be obtained if the pistons, rings, valves, gaskets, manifolds, camshaft, combustion chambers, air cleaner, fuel pump, governor, and ignition system are in good condition and are functioning properly.

The illustration shows some of the things which can affect good carburetion.

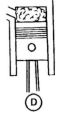

Things that Affect Carburetion

A—Fuel Supply　　　　　　D—Compression
B—Governor Linkage　　　 E—Ignition
C—Air Cleaner　　　　　　 F—Carburetor

Maintenance Tips and Precautions for Carburetors

Below are listed some items which will help you in servicing carburetors:

- Always service the carburetor at these times:
  - After engine valve grinding or major engine overhaul.
  - Every year or at the beginning of each season on seasonal machines. At this time, clean the carburetor, repack the shaft, check the bearings, and replace the seals and gaskets.
- Always adjust the carburetor at the following times:
  - During engine tune-up.
  - After major overhaul of the engine.
  - Whenever the carburetor has been removed for service.
  - Anytime the engine idles badly or requires speed adjustment.
- Repair kits are provided for many carburetors. When repairing, be sure to use ALL of the new parts in the kit.
- Clean all parts thoroughly when repairing the carburetor. Use a carburetor cleaning solution for removing varnishlike deposits from all metal parts and rinse them in solvent.

- Never use small wires or drills to clean out jets or orifices. This may enlarge or burr the precision bores and upset the performance of the carburetor.
- Never use compressed air to clean a completely assembled carburetor. To do so may cause the metal float to collapse.
- To test a metal float for leaks, immerse it in hot water. If air bubbles escape from the float, replace it.
- Always drain the carburetor before any long storage period. Many carburetors have a drain plug in the fuel bowl.
- Most carburetors are vented to avoid vapor lock. Be sure the vent is kept clean and open.
- Never turn adjusting needles too tightly against their seats as you may damage them.
- Use special tool kits, when available, to recondition the carburetor.
- Be sure to tighten the screws that hold the throttle disk in place.

SS40167,00000C4 -19-23JUL09-16/18

- Always check the height (A) of the float when assembling the carburetor. See the machine technical manual for the correct height. If the float is badly bent or warped, replace it.

A—Float Height

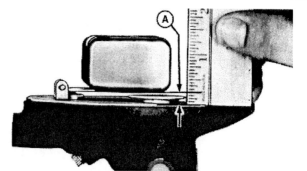

Checking Position of Carburetor Float

Continued on next page　　　SS40167,00000C4 -19-23JUL09-17/18

Adjusting the Carburetor

Keep the carburetor adjusted for smooth, economical operation of the engine.

However, remember that the engine speed adjustments and ignition timing both affect and are affected by carburetor settings. Don't change one without checking the other!

Most carburetors have three basic adjustments:

• Idle Speed Adjustment
• Idle Fuel Adjustment
• Full-Load Fuel Adjustment

Adjust the carburetor as follows:

1. Warm up the engine fully.

2. Set the idle speed with the engine throttle closed. Adjust the screw (B) to get the slowest engine idling without stalling or "roughening" the engine. See the technical manual for the recommended idle rpm.

3. Adjust the idle fuel mixture for smooth engine idling. Turn the screw in until the engine begins to stall; then turn out the screw slightly until the engine idles smoothly.

4. If necessary, readjust the idle speed (step 2) after getting the correct idle fuel adjustment.

5. Adjust the full-load fuel mixture last for good fuel economy under load. For safety, adjust while the

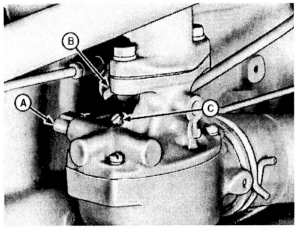

Adjustment and Maintenance Points on a Typical Carburetor

A—Fuel Inlet Strainer   **C—Idle Fuel Adjusting Needle**
B—Idle Speed Stop Screw

machine is under load — but is stationary. Adjust at fast idle by turning in the load screw until the engine starts to lose power, then back it out a little until the engine picks up speed and runs smoothly.

6. Check the operation of the speed control or governor mechanisms. If they are erratic or overspeed, adjust the linkage. For information, see Governing Systems in Chapter 11.

SS40167,00000C4 -19-23JUL09-18/18

Carburetor Troubleshooting Charts

While the basic causes of carburetor trouble will vary, the common difficulties and respective causes are outlined below.

Poor Performance

Possible causes of poor performance generally result from too lean an air-fuel mixture. If the carburetor is correctly adjusted, a lean mixture and poor performance may result from the following:

1. Air leaks at carburetor or intake manifold
2. Clogged engine air cleaner
3. Clogged fuel lines
4. Defective fuel pump
5. Low fuel level
6. Clogged fuel screen
7. Dirt in carburetor jets and passages
8. Worn or inoperative accelerating pump
9. Load needle valve, economizer, or jet not operating
10. Damaged or wrong-size main metering jet
11. Worn idle needle valve and seat
12. Loose jets in carburetor
13. Defective gaskets in carburetor
14. Clogged mufflers, ignition, and poor compression
15. Worn throttle shaft or bushings

Poor Idling

Poor idling is usually caused by a malfunctioning carburetor, a defective ignition system, leaking engine valves, or uneven engine compression. In the carburetor, check the following:

1. Incorrect adjustment of idle needle valve
2. Incorrect float level
3. Sticking float needle valve
4. Defective gaskets between carburetor and manifold
5. Defective gaskets in carburetor
6. Loose carburetor to manifold nuts
7. Loose manifold to cylinder head nuts
8. Idle discharge holes partly clogged
9. Loose jets in carburetor
10. Vacuum leaks that are partly compensated for by a rich idle adjustment
11. Worn main metering jet
12. Restricted or clogged air system

Hard Starting

In addition to the following fuel system troubles, hard starting may be caused by use of engine oil that is too heavy, defective ignition system, low compression, weak starting battery, defective starting motor, or excessive engine friction.

1. Incorrect choke adjustment
2. Defective choke
3. Incorrect float level
4. Incorrect fuel pump pressure
5. Sticking fuel inlet needle
6. Improper starting

Poor Acceleration

Poor acceleration may be caused by defective ignition system, excessive engine friction, lack of compression, and incorrect carburetion.

In the case of the carburetor, check the following:

1. Incorrect fuel level
2. Accelerator pump incorrectly adjusted
3. Accelerator pump not operating
4. Corroded or bad seat on accelerator bypass jet
5. Accelerator pump leather hard or worn
6. Clogged accelerator jets or passages
7. Defective ball checks in accelerator system

Carburetor Floods

The usual causes of carburetor flooding are:

1. Dirt on float valve seat
2. Sticking float valve
3. Leaking float
4. Fuel level too high
5. Defective gaskets in carburetor
6. Excessive fuel pump pressure

Excessive Fuel Consumption

There are many causes of excessive fuel consumption other than defective carburetion. Among the causes are poor engine compression, excessive engine friction, clogged mufflers, and defective ignition. In the carburetor and fuel system, check the following:

1. Poor adjustment of idle mixture
2. Wrong setting of load adjusting needle
3. Fuel leaks in carburetor or lines
4. Clogged air cleaner
5. Defective fuel economizer
6. Defective carburetor gaskets
7. Excessive fuel pressure
8. Sticking fuel inlet needle

SS40167,00000C5 -19-17JUL09-1/1

Test Yourself

Questions

1. What are the three basic components of a gasoline fuel system?
2. What is the purpose of a fuel pump?
3. What is the primary function of a carburetor?
4. (True or false?) The greater the air velocity, the greater the air pressure.
5. Name the three basic types of carburetors.
6. What is the purpose of the choke in a carburetor?
7. What is the purpose of an accelerator pump in a carburetor?
8. Write the correct statement for question No. 4.
9. (Fill in the blank.) On mechanical fuel pumps, power is applied to the pump rocker arm by an eccentric on the _____.

SS40167,00000C6 -19-17JUL09-1/1

LP-Gas Fuel Systems

LP-Gas Fuel Systems — Introduction

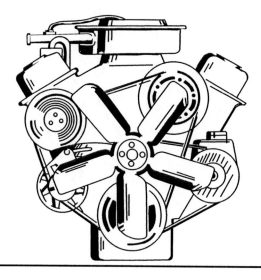

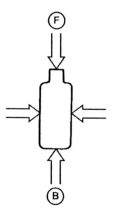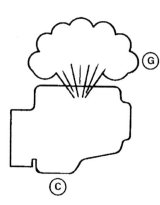

How LP-Gas Becomes Fuel for the Engine

A—Vapor
B—Liquid

C—Vapor Again
D—Butane

E—Propane
F—Pressurized

G—Engine Fuel

Liquefied petroleum (LP) gas vaporizes very easily. In fact it normally remains a liquid only when under pressure. Therefore it must be kept in strong heavy tanks.

LP-gas is vaporized before it reaches the carburetor, while gasoline remains a liquid until this point. This is the basic difference between the two systems.

LP-gas is:

- A byproduct of gasoline manufacture.
- Also obtained from natural gas.
- Made up mainly of propane and butane.
- A vapor unless compressed or cooled.
- Liquefied by compressing many gallons of vapor into one gallon of liquid.
- Easier to handle and store as a liquid.
- Expansive when heated (due to more vaporization).
- Stored in strong tanks with pressure relief valves.

- Converted to vapor again on way to engine.

Fuel Combustion

LP-gas burns slower than gasoline because it ignites at a higher temperature.

For this reason, the spark is often advanced farther on LP-gas engines.

More voltage at the spark plugs may be needed for LP-gas engines than for gasoline.

To solve this, colder plugs or smaller spark plug gaps may be recommended in the engine technical manual.

LP-gas engines do not require as much heat at the intake manifold as gasoline models. This is because LP-gas will vaporize at lower temperatures. The result: Less heat wasted and more heat goes into engine power.

Continued on next page

SS40167,00000C7 -19-17JUL09-1/2

Safety in Handling Lp-Gas

LP-gas is not hazardous if handled safely.

However, there are special methods which must be used for LP-gas.

Be sure you understand LP-gas equipment before handling it.

Many states and local municipalities have regulations for handling LP-gas. Check this with your local and state governments, or LP-gas distributor.

For additional information, see Safety Rules for LP-Gas in Chapter 2.

SS40167,00000C7 -19-17JUL09-2/2

How Fuel Is Withdrawn

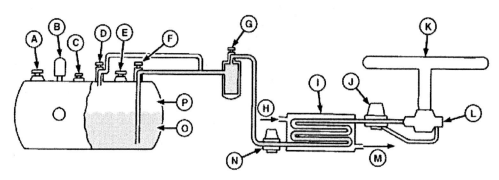

LP-Gas System Using Liquid Withdrawal

A—Filling Valve	E—Vapor Return Valve	I— Vaporizer	M—Return Water
B—Pressure Relief Valve	F—Liquid Line Valve	J— Low-Pressure Regulator	N—High-Pressure Regulator
C—80% Fill Valve	G—Strainer Valve	K—Engine Intake Manifold	O—Liquid Fuel
D—Vapor Line Valve	H—Hot Water from Engine	L—Carburetor	P—Vapor

To withdraw LP-gas fuel from the tank, two methods are used:

- **Liquid withdrawal**
- **Vapor withdrawal**

These two systems are compared in Chapter 1.

Most LP-gas systems using the liquid withdrawal method also use vapor withdrawal for starting a cold engine.

LIQUID WITHDRAWAL

Opening the liquid line valve (F) allows the pressurized liquid fuel (O) to flow out of the tank, through the strainer valve (G), high-pressure regulator (N), then to the vaporizer (I).

The vaporizer is basically a heat exchanger. Hot water from the engine flows through the vaporizer, heating

and vaporizing the liquid fuel as it passes through it. The vaporized fuel then flows through the low-pressure regulator (J) to the carburetor (L), where it is mixed with filtered air and sent to the engine intake manifold (K).

VAPOR WITHDRAWAL

When starting a cold engine, the liquid line valve (F) is closed and vapor line valve (D) is opened to allow vapor to be drawn directly from the tank and flow to the carburetor. This method is used only as starting fuel and is done because in a cold engine there is not enough heat to convert the liquid fuel to vapor. Once the engine is started and warmed up, it can be switched to the liquid withdrawal system again.

SS40167,00000C8 -19-17JUL09-1/1

Parts of LP-Gas Fuel System

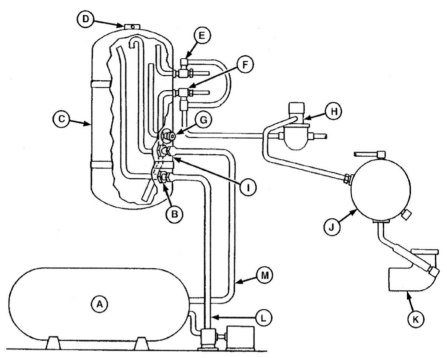

Valves and Gauges in LP-Gas Fuel System

A—Storage Tank
B—Filler Valve
C—Pressurized Fuel Tank
D—Safety Relief Valve

E—Vapor Withdrawal Valve
F—Liquid Withdrawal Valve
G—Liquid Level Gauge
H—Fuel Strainer

I— Vapor Return Valve
J—Converter
K—Carburetor
L—Fill Line

M—Vapor Return to Fuel Tank

The LP-gas fuel system has four basic parts:

• **Pressurized Fuel Tank**
• **Fuel Strainer**
• **Converter**
• **Carburetor**

PRESSURIZED FUEL TANK (C) stores the liquid fuel under pressure. A space for vapor is left at the top of the tank.

FUEL STRAINER (H) cleans the liquid fuel. It normally has a solenoid, which permits flow only when the engine ignition is turned on.

CONVERTER (J) changes the liquid fuel to vapor by warming it and then drops the pressure of the vapor.

CARBURETOR (K) mixes the fuel vapor with air in the proper ratio for the engine.

SS40167,00000C9 -19-17JUL09-1/1

DXP00385 —UN—23DEC08

Fuel Tank, Valves, and Gauges

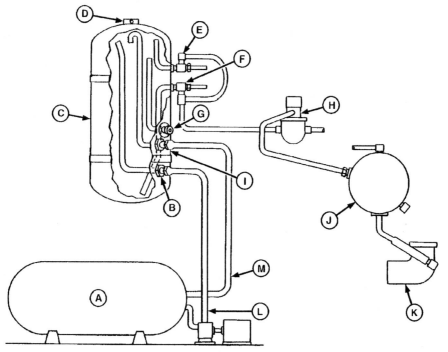

Valves and Gauges in LP-Gas Fuel System

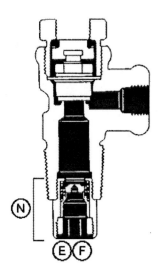
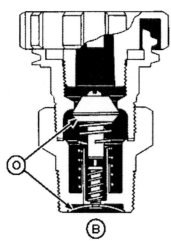
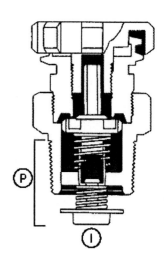

Cross Section of LP-Gas Valves

A—Storage Tank	E—Vapor Withdrawal Valve	I— Vapor Return Valve	M—Vapor Return to Fuel Tank
B—Filler Valve	F—Liquid Withdrawal Valve	J—Converter	N—Excess Flow Mechanism
C—Pressurized Fuel Tank	G—Liquid Level Gauge	K—Carburetor	O—Check Valves
D—Safety Relief Valve	H—Fuel Strainer	L—Fill Line	P—Excess Flow Mechanism

The LP-gas fuel tank must be strong and heavy to withstand the pressure of its fuel.

The fuel tank is never filled completely full of liquid fuel because room must be left for vapors and expansion.

The fuel tank also has these valves and gauges:

- Filler Valve
- Safety Relief Valve
- Vapor Withdrawal Valve

- Liquid Withdrawal Valve
- Liquid Level Gauge
- Vapor Return Valve

FILLER VALVE (B) has a double check valve (O) that prevents fuel from escaping when the filler hose is disconnected or accidentally breaks.

SAFETY RELIEF VALVE (D) will open if pressure in the tank becomes too high.

Continued on next page   SS40167,0000106 -19-23JUL09-1/4

VAPOR WITHDRAWAL VALVE (E) pulls vapor from the top of the tank for starting the engine.

LIQUID WITHDRAWAL VALVE (F) pulls liquid fuel from the bottom of the tank for normal engine operation.

Withdrawal valves (E and F) have safety devices called excess flow mechanisms (N). The valves automatically shut off when too much fuel is flowing out. If a leak develops in the system, the valve instantly closes and limits only a small amount of gas to flow.

LIQUID LEVEL GAUGE (G) tells when the tank is 80% full during filling of the tank. This assures that space is

left for vapors. By opening the gauge, the level can be checked. Before the 80% fill, vapor will issue from the gauge. When the 80% fill is reached, a spray or liquid will issue. Caution must be taken, however, not to leave the gauge open during filling.

VAPOR RETURN VALVE (I) permits vapor to return to the storage tank as the fuel tank is filled. This equalizes the pressures and makes it easier to fill the tank. An excess flow mechanism (P) automatically closes if flow through the vapor return line is too much. This protects against escape of gas if the return line is broken or detached.

SS40167,0000106 -19-23JUL09-2/4

Fuel Gauges

Two types of fuel gauges are used for LP-gas fuel tanks:

- **Magnetic**
- **Electrical**

MAGNETIC fuel gauges use a float (in the tank) with an attached pivoting shaft. The opposite end of the pivoting shaft attaches to a magnet (C). As the float rises and falls with the fuel level, it rotates the magnet, which operates another magnet in the gauge. This moves the pointer (B) to indicate the amount of fuel on the face of the gauge.

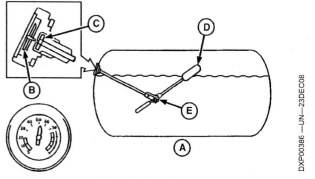

Magnetic Fuel Gauge

A—Fuel Tank
B—Pointer
C—Magnet
D—Float
E—Gears

Continued on next page

SS40167,0000106 -19-23JUL09-3/4

ELECTRICAL fuel gauges (A) are similar to gasoline gauges. For more information, see Chapter 7.

A sender located on the fuel tank signals a receiver on the machine instrument panel.

Auxiliary Fuel Connection

This connection allows a portable fuel tank to be attached for moving the machine without filling the main fuel tank. The connection is usually made at the fuel strainer.

Servicing Fuel Tank, Valves, and Gauges

Be sure that you are familiar with all rules and regulations before servicing LP-gas equipment.

Two rules are important here:

- Never service the fuel tank except to remove and install valves or gauges.
- Never service the valves or gauges except to remove and install them or to replace caps or dust covers.

If a valve or gauge is damaged or worn and must be replaced, first empty all fuel from the tank to remove the pressure. If possible, do this outdoors by running the engine until all fuel is exhausted. If not possible, consult your local LP-gas distributor for the approved method in your locality for emptying the tank.

Permatex is a trademark of Illinois Tool Works

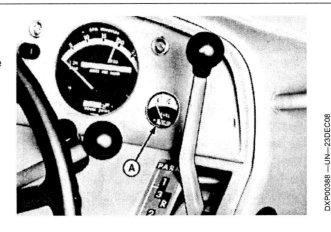

A—Fuel Gauge

When replacing a valve, use a small amount of Permatex ™ or similar sealer on the threads. Install the valve and tighten securely. Test for leaks by using soapy water. **Never use an open flame.**

For additional information, refer to the engine technical manual.

SS40167,0000106 -19-23JUL09-4/4

Fuel Strainer

The fuel strainer is located between the fuel tank and the converter.

The fuel strainer has two functions:

- Strains fuel to clean it.
- Shuts off fuel when system is not operating.

The SHUTOFF is an electrical solenoid coil (C). When the engine ignition is turned on, the solenoid magnetizes the valve plunger (D) and opens it, allowing fuel to flow. When the ignition is turned off or fails, the valve closes automatically. Pressure of fuel holds it tightly on its seat.

Servicing the Fuel Strainer

If the fuel strainer frosts, its filter element (E) is probably clogged and needs cleaning. Before attempting to clean the strainer, make sure both withdrawal valves are closed, the engine is cold, and the lines and strainer are empty of gas.

To clean the strainer, first remove the drain plug (G) and drain out all foreign matter. Then follow instructions on blowing out the strainer. To clean the strainer parts, see the engine technical manual.

If the solenoid fails, it may show up as follows: When the ignition is turned on, gas fails to flow to the converter. In this case, the valve plunger may be stuck or restricted.

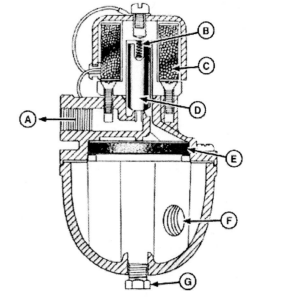

LP-Gas Fuel Strainer

A—Fuel Out
B—Spring
C—Solenoid Coil
D—Valve Plunger
E—Filter Element
F—Fuel In
G—Drain Plug

SS40167,00000CA -19-17JUL09-1/1

Converter

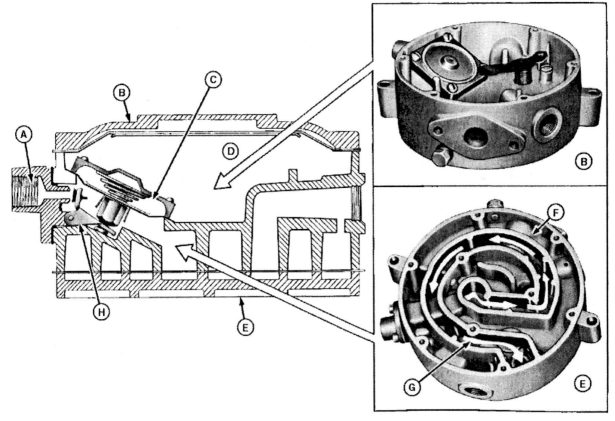

LP-Gas Converter

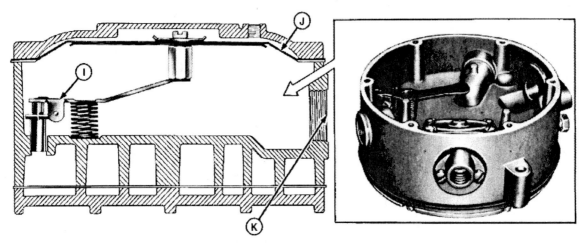

Low-Pressure Chamber of Converter

A—Liquid Fuel Enters Here
B—Low-Pressure Chamber
C—High-Pressure Diaphragm

D—Vapor
E—Heat Exchanger Chamber
F—Coolant Spiral Passage

G—Fuel Spiral Passage
H—High-Pressure Valve and Lever
I— Low-Pressure Valve and Lever

J— Low-Pressure Diaphragm
K—Fuel Outlet to Carburetor

The converter does two things:

• Heats the liquid fuel to vaporize the fuel
• Reduces vapor pressure

Liquid fuel must be converted to vapor (D) before the engine can use the fuel. Fuel pressure must also be reduced in order for the carburetor to handle the fuel.

Continued on next page
SS40167,00000CB -19-17JUL09-1/4

The converter has two chambers separated by a wall.

The heat exchanger chamber (E) is divided into two spiral passages — one for fuel, the other for engine coolant.

The low-pressure chamber contains fuel vapor under low pressure.

Heat Exchanger Chamber

Liquid fuel under pressure enters the converter through the high-pressure valve (H) and passes into the spirals of the heat exchanger where it rapidly changes to a vapor.

As the fuel vaporizes, pressure builds up and pushes on the high-pressure diaphragm (C). This diaphragm is connected to the high-pressure valve as shown. When pressure reaches a preset point, it overcomes the spring pressure and closes the valve. This seals the fuel vapor in the heat exchanger.

Low-Pressure Chamber

The second chamber of the converter is connected to the heat exchanger by a low-pressure valve (I). This valve is held closed until opened by its lever.

A large low-pressure diaphragm (J), as shown, actuates the lever.

OPERATION IS AS FOLLOWS:

1. When the engine piston is on the intake stroke, it draws gas (through the carburetor) from the low-pressure chamber (B), reducing the pressure beneath the large diaphragm to slightly below atmospheric pressure.

2. Since the other side of the diaphragm is vented to atmospheric pressure, the diaphragm is pushed down. As it moves downward, it actuates the low-pressure valve lever, opening the low-pressure valve and permitting gas from the heat exchanger to enter the low-pressure chamber. From here the gas goes to the carburetor and then into the engine cylinder.

3. When the engine intake valve closes and the demand for gas momentarily ceases, pressure on both sides of the large diaphragm becomes equal, and the spring again closes the low-pressure valve.

4. Since some gas has been drawn from the heat exchanger, pressure in the chamber is reduced, permitting the high pressure valve spring to push the high-pressure diaphragm (C) down and open the high-pressure valve (H).

5. Liquid fuel again flows into the heat exchanger, where it vaporizes, and again builds pressure, closing the valve.

SS40167,00000CB -19-17JUL09-2/4

Precaution Against Converter Freeze-Up

If the converter should ever freeze, an expansion device is built into the back cover (A).

As coolant freezes, it expands into a series of spirals that are located along the back cover. As expansion continues, the coolant will push against a flexible gasket (E) located on the cover to protect the converter from damage.

A drain plug at the bottom of the converter allows coolant to be drained to prevent freezing. If the cooling system is drained, also drain the converter. And disconnect the coolant inlet line (D) and coolant outlet line (C).

A—Back Cover
B—Expansion Spirals
C—Coolant Outlet
D—Coolant Inlet
E—Flexible Gasket

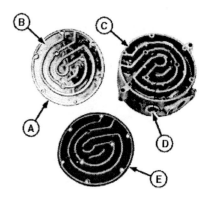

Expansion Device to Protect Converter if It Freezes

Continued on next page

SS40167,00000CB -19-17JUL09-3/4

Problems of Engine Air Pulsations Affecting Converter

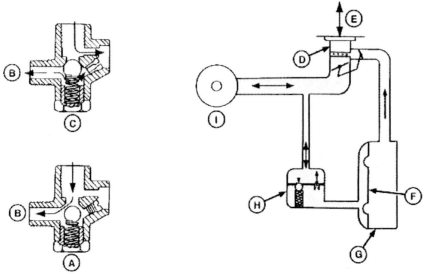

Regulator Valve for Stabilizing Converter Against Engine Air Pulsation

A—Valve During Positive Pulse
B—To Diaphragm
C—Valve During Negative Pulse
D—Carburetor
E—To Engine
F—Low-Pressure Diaphragm
G—Converter
H—Regulator Valve
I— Air Cleaner

In some four-cylinder engines, the air pulsations from the cylinders will affect the low-pressure diaphragm (F) located in the converter (G).

If the diaphragm works in unison with the engine intake cycle, the flow of fuel vapor from the converter will become erratic.

A regulator valve (H) can be used to prevent this. The valve, in effect, helps the diaphragm to breathe normally.

During positive pulses, the valve moves to direct pressure to the atmospheric side of the diaphragm as shown. This overcomes the pulse from inside the diaphragm, keeping the low-pressure valve open for normal operation.

During negative pulses, the diaphragm is allowed to breathe at air inlet pressure through the orifice in the regulator valve as shown.

The regulator valve ball is operated by air pressure changes.

Servicing the Converter

The converter is constructed so that it will seldom cause trouble, providing clean fuel is used in the tank and clean, soft water in the cooling system.

If the converter frosts up when the engine is cold, it is probably due to a leaking high-pressure valve caused by dirt under the valve seat. If frosting occurs when the engine is hot, it is probably due to poor coolant circulation through the heat exchanger or a restriction in the high-pressure valve.

For details on converter service, see the engine technical manual.

SS40167,00000CB -19-17JUL09-4/4

Carburetors

The LP-gas carburetor does much the same job as gasoline models (Chapter 4). The main difference is that when fuel enters the LP-gas model, it is a vapor, while in the gasoline model it is a liquid.

The illustration shows an updraft model, but downdraft models are also used.

The metering valve (E) varies the amount of fuel vapor entering the carburetor.

The spray bar (C) takes the fuel vapor and mixes it with air flow coming in from the bottom.

A—Air In
B—Throttle Disk (Wide Open)
C—Spray Bar
D—To Engine
E—Metering Valve (Wide Open)
F—Fuel Vapor In

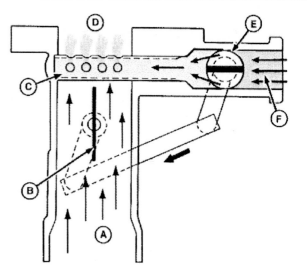

Metering Valve and Throttle Disk Operate Together
(Updraft Model Shown)

SS40167,00000CC -19-23JUL09-1/4

A full-load adjusting screw (C) meters the fuel vapor at the spray bar at full throttle for the best engine power.

The throttle disk controls the incoming air and works with the metering valve to get the best air-fuel mixture for each load and speed.

A—Idle Speed Stop Screw (Throttle)
B—Idle Fuel Adjusting Link
C—Full-Load Fuel Adjusting Screw

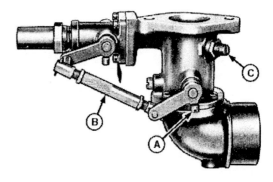

Adjusting Points on Typical LP-Gas Carburetor (Updraft Model Shown)

Continued on next page

SS40167,00000CC -19-23JUL09-2/4

OPERATION

Both the metering valve (E) and throttle disk (B) are operated by the engine throttle.

At slow idle speeds, they close down to limit the air-fuel mixture fed to the engine.

At fast idle speeds, the valve and disk open up to feed more air-fuel to the engine.

Spray Bar Operation

Incoming air flows through the carburetor bore and hits a restriction — the spray bar. This deflects the air and accelerates it, creating a pressure drop (similar to the venturi in gasoline carburetors).

MAINTENANCE

For information, see Maintenance Tips and Precautions for Carburetors in Chapter 4.

A—Air In
B—Throttle Disk (Wide Open)
C—Spray Bar
D—To Engine
E—Metering Valve (Wide Open)
F—Fuel Vapor In

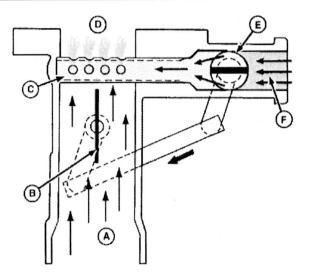

Metering Valve and Throttle Disk Operate Together

SS40167,00000CC -19-23JUL09-3/4

CARBURETOR ADJUSTMENTS

The carburetor must be adjusted for smooth, economical operation of the engine.

However, remember that engine speed adjustments and ignition timing both affect and are affected by carburetor adjustments. Don't change one without checking the other!

Most LP-gas carburetors have these basic adjustments:

• Idle Speed Adjustment
• Idle Fuel Adjustment
• Full-Load Fuel Adjustment

These adjusting points are shown on the typical carburetor.

For additional details, see the engine technical manual.

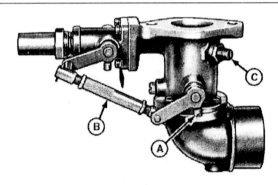

Adjusting Points on Typical LP-Gas Carburetor

A—Idle Speed Stop Screw (Throttle)
B—Idle Fuel Adjusting Link
C—Full-Load Fuel Adjusting Screw

OTHER FACTORS IN USING LP-GAS

Air Cleaners

Too-heavy oil in the air cleaner can be a problem with LP-gas engines. The result is a restriction in air intake, which chokes the carburetor and floods the engine.

Crankcase Oil and Filters

LP-gas is known to burn cleaner than other fuels but this has led to the notion that the engine oil will last much longer.

This is not true, however. The old oil gets thicker with use regardless of how dirty it is, resulting in a lack of lubrication and loss of power from friction. Also, the additives may wear out before the oil.

Always follow the maintenance schedule in the operator's manual for your particular engine on when to change the engine oil and replace the filters.

SS40167,00000CC -19-23JUL09-4/4

LP-Gas Troubleshooting Chart

Refer to the engine technical manual for all repairs suggested below.

HARD STARTING

- Improperly blended fuel — See fuel dealer.
- Over-priming — See operator's manual.
- Improper starting — See operator's manual.
- Fuel strainer solenoid failed — Check electrical circuit.
- Air vent on converter plugged — Open it.
- Defective low-pressure diaphragm in converter — Replace diaphragm.
- Carburetor metering valve or linkage binding — Repair or replace.

LOSS OF POWER

- Throttle not opening fully — Check governor and throttle linkage.
- Air vent on converter restricted — Open it.
- Clogged fuel strainer — Clean the strainer as recommended.

⚠ CAUTION: Do not clean the strainer while the engine is running or when near an open flame.

- Liquid fuel outlet on tank clogged — Clean it.
- Excess-flow valves (vapor and liquid) closed, indicated by frosting at withdrawal valve — Reset by closing valves and opening slowly.
- Fuel strainer frosted, meaning restriction due to dirt, or water, or to stuck valves — Clean or replace valves.
- Lean fuel mixture caused by restricted or altered fuel hose or pipes — Replace with new approved parts.
- Converter high-pressure valve sticking — Remove valve assembly. Clean and reinstall. Be sure valve is free.
- Low-pressure valve in converter restricted — Check for free flow of fuel as recommended. Clean valve as necessary.
- Defective or ruptured diaphragm — Replace it.
- Carburetor in need of adjustment or metering valve binding — Adjust carburetor and free up or replace metering valve.
- Air leak between carburetor throttle body and air horn — Replace gasket or tighten screws.

POOR ECONOMY (Also See Loss of Power)

- Tank not properly filled — Fill as recommended in operator's manual.
- Improper carburetor adjustment — See operator's manual.
- Metering valve worn or damaged — Replace with a new valve.
- Wrong fuel — See fuel dealer.

FREEZE-UP OF CONVERTER

Freeze-up during normal operation is usually due to lack of water circulation through the converter. Possible causes may be:

- Low coolant level in radiator — Fill radiator.
- Defective water pump — Repair pump.
- Thermostat removed — Install thermostat.
- Water circulation reversed — Inlet water pipe connected to wrong point on converter.
- Restriction in coolant pipes or converter — Clean or replace pipes or converter.
- Operating engine on liquid fuel before coolant is warm — Warm up engine on vapor.

FROST ON FUEL STRAINER OR WITHDRAWAL VALVES

Frost on fuel strainer is caused by:

- Dirt, water, or other restrictions — Clean the strainer.
- Excess flow mechanism in the withdrawal valve is closed — Clean or replace valve.

Frost on withdrawal valves is caused by:

- Excess flow mechanism in valve closed — Reset by closing valve and opening slowly.
- Water in fuel tank (if this is cause, ice will form on liquid withdrawal valve and frost will be present between valve and strainer) — Empty strainer hose of gas and flush out with antifreeze solution as recommended.

ROUGH IDLING

- Improper carburetor throttle rod or drag link adjustments — Readjust.
- Defective carburetor-to-manifold gasket — Tighten cap screws or replace gasket.
- Converter-to-converter fuel hose leaking — Replace it.
- Dirty or wrong type of spark plugs — Clean or replace.
- Metering valve in carburetor binding — Remove and clean or replace.

POOR IDLE SPEED ACCELERATION

- Diaphragm punctured — Replace diaphragm.
- Lean fuel mixture due to a restriction in the fuel inlet hose — Replace hose.
- Wrong carburetor idle adjustment — Adjust as recommended.

LACK OF FUEL AT CARBURETOR

NOTE: To check, press primer (if used) at converter with throttle in wide-open position. An audible hiss should be heard if fuel is flowing.

- No fuel supply — Fill tank if necessary.
- Liquid or vapor withdrawal valve closed — Open valve.
- Fuel lines kinked, bent, or restricted — Check fuel lines.
- Broken or loose electrical wiring connections on fuel strainer — Repair.
- Fuel strainer clogged — Clean it. See operator's manual.

Continued on next page   SS40167,00000CD -19-17JUL09-1/2

ENGINE STOPS WHEN SPEED CONTROL LEVER IS MOVED TO SLOW IDLE

- Slow idle speed incorrectly adjusted — See operator's manual.
- Converter-to-carburetor hose leaking — Replace.
- Carburetor gasket leaking — Replace and tighten cap screws securely.
- Converter back cover gasket leaking — Replace gasket.

OVERHEATING

- Too lean fuel mixture — See operator's manual for carburetor adjustment.
- Clogged radiator — Clean.

- Defective Water pump — Repair.
- Overloaded engine — Reduce load.
- Incorrect ignition timing — Check timing.
- Slipping fan belt — Adjust.

FUEL SEEPING INTO WATER PASSAGES IN CONVERTER

The transfer of fuel into the water passages in the converter is caused by failure of the back cover gasket.

To check, remove radiator cap, turn ignition switch on, and watch for bubbles or water blowing out of the radiator filler neck. If so, replace the gasket.

SS40167,00000CD -19-17JUL09-2/2

Test Yourself

Questions

True or False?

1. In both gasoline and LP-gas engines, the fuel is a vapor when it reaches the carburetor. _____

2. LP-gas is liquefied by putting it under high pressure. _____

3. Frosting of LP-gas components is usually caused by a restriction. _____

4. Always fill the LP-gas fuel tank completely full. _____

5. Change the ignition timing when converting an engine from gasoline to LP-gas. _____

6. Keep the doors and windows closed in winter to prevent freezing of LP-gas equipment. _____

7. The converter changes liquid fuel to vapor and raises its pressure. _____

SS40167,00000CE -19-17JUL09-1/1

Natural Gas Fuel Systems

Natural Gas Fuel Systems — Introduction

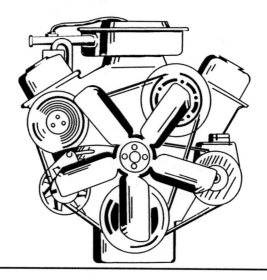

DXP01067 —UN—08JUN09

The natural gas fuel system supplies a precise amount of vaporized gas mixed with air to the engine cylinders for combustion. An electrical spark is used (as in the gasoline engine) to ignite the charge of fuel and air in the cylinders.

The natural gas fuel system offers a clean-burning fuel with low particulate emissions. Natural gas is also a very economical fuel.

Natural gas is a fossil fuel found in the earth's crust. It is a mixture of methane, ethane, propane, butane, and trace amounts of other inert gases. However, methane is by far the largest component — about 90% of the total volume of natural gas.

Methane has very good antiknock characteristics, which means that is does not ignite readily. The octane level of natural gas is approximately 130. Because of this, the engine's compression ratio is typically higher for natural gas than for gasoline.

Natural gas burns slower than gasoline because natural gas ignites at a much higher temperature [1300°F (704°C)] than gasoline [1100°F (593°C)]. For this reason, the ignition spark is often advanced farther for natural gas engines and a higher spark voltage may be required.

Modern natural gas engines typically use a high-energy, coil-on-plug, inductive ignition system.

Natural gas is in a gaseous state at any temperature above −259°F (−126°C). At atmospheric pressure, the temperature of −259°F (−126°C) is the boiling point of natural gas, while water is gaseous (steam) at +212°F (+100°C). Because of this property, natural gas must be stored in a closed fuel tank to prevent vapor from escaping.

SS40167,00000CF -19-17JUL09-1/1

Fuel Storage

Using natural gas to power the engine in a mobile application requires a transportable fuel storage system. The storage system must contain a sufficient supply of gas to provide the vehicle with an operating range comparable to that of a gasoline fuel system. To accomplish this, natural gas is stored in one of the following forms:

- **Compressed Natural Gas (CNG)**
- **Liquefied Natural Gas (LNG)**

Compressed natural gas (CNG) is produced in CNG fueling facilities that compress the pipeline supply of natural gas to pressures of 3000–3600 psi (20 684–24

821 kPa). The compressed gas is stored in high-strength tanks.

Liquefied natural gas (LNG) is produced through a cryogenic (or freezing) process in which the gas is cooled to −259°F (−126°C) to form a liquid. The liquefied gas is then stored in high-strength, insulated tanks.

With the exception of the method of storage, compressed natural gas systems and liquefied natural gas systems have many things in common.

Let's look at each fuel storage system in more detail — first the compressed natural bas (CNG) system.

SS40167,00000D0 -19-17JUL09-1/1

Compressed Natural Gas (CNG) Fuel Storage

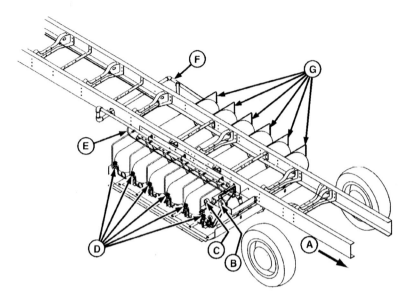

Compressed Natural Gas (CNG) Fuel Storage-Delivery System

A—Vehicle Front
B—1/4-Turn Shutoff Valve
C—Pressure Gauge
D—Manual Shutoff Valves and Pressure Relief Valves
E—Common Fuel Line
F—Common Vent Pipe
G—Fuel Tanks

The major parts of the compressed natural gas (CNG) fuel storage system are:

• Fuel Tanks
• Relief Valves
• Fuel Lock-Off Valve
• Coalescing Filter

Natural gas has a higher heating value per pound than gasoline, but being in a gaseous state, it has a much lower energy storage density than gasoline. Therefore, to provide a sufficient supply of fuel as compared to a gasoline fuel tank, natural gas must be compressed and stored in high-strength cylinders. The result is that storage systems for compressed natural gas are bulkier and heavier than gasoline storage containers due to the increased volume and the need to withstand high pressures.

The fuel tanks (G) may be constructed of high-strength steel or aluminum, reinforced with fiberglass or composite fibers. The gas is stored in the tanks at 3000–3600 psi (20 684–24 821 kPa).

Each tank is equipped with a manual shutoff valve and pressure relief valve (D). The pressure relief valve

is designed to vent at either high pressure or high temperature. Each tank's relief valve is plumbed into a common vent pipe that routes the vented gas out of the vehicle.

A common fuel line (E) connects the tanks to a high-pressure fuel lock-off device and a coalescing filter.

The fuel lock-off device is a valve that shuts off the fuel flow to the engine when the engine is not running, even if the ignition switch is in the ON position. The lock-off device opens when the ignition switch is in the ON position and the engine is either being started or is running.

A coalescing filter removes debris and contaminants from natural gas. The contaminants consist primarily of oil from natural gas compressors. If oil contaminants are allowed to enter the fuel system, the fuel system components can become restricted. The result may be a decrease in engine performance and even damage to engine components.

SS40167,00000D1 -19-17JUL09-1/1

Liquefied Natural Gas (LNG) Fuel Storage

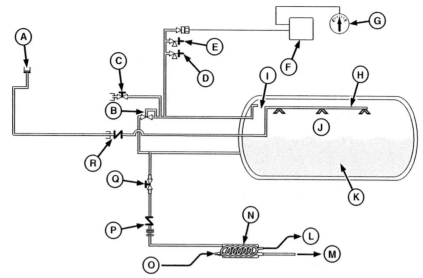

Liquefied Natural Gas (LNG) Fuel Storage-Delivery System

A—Remote Filling Connector
B—Economizer Regulator
C—Vent Valve
D—Primary Relief Valve
E—Secondary Relief Valve

F—Electronic Converter
G—Fuel Gauge
H—Fill Spray Header
I— Vapor
J— Fuel Tank

K—Liquid
L—Engine Coolant Out
M—Fuel Supply to Engine
N—Vaporizer
O—Engine Coolant In

P—Excess Flow Valve
Q—Manual Shutoff Valve
R—Fill Check Valve

The major parts of the liquefied natural gas (LNG) fuel storage system are:

• **Fuel Tank**
• **Fill Check Valve**
• **Relief Valves**
• **Regulator Valve**
• **Fuel Gauge**
• **Vaporizer**

Natural gas is converted to an extremely dense gas, liquefied natural gas (LNG), by a cryogenic (or freezing) process in which the gas is cooled to −259°F (−126°C). The LNG fuel storage system is designed to receive and store the liquefied natural gas, then distribute vaporized gas to the engine.

The fuel tank (J) is a cryogenic container designed to maintain the fuel in a liquid state. The fuel tank (J) is similar to the common Thermos™ bottle. It consists of an inner pressure vessel designed to hold the cold liquid positioned inside a larger outside vessel. The inner vessel is wrapped with special insulating material. The space between the two vessels is evacuated to form a superior insulation system.

This design prevents almost all heat from reaching the liquid (K). This allows the tank to keep the natural gas in a liquefied state for a period of time without excess pressure buildup.

Thermos is a trademark of Thermos LLC.

Each tank is equipped with overfill protection and a one-way fill check valve at the liquid filling connection to provide safety during the fueling operation. During filling of the tank, the liquid is sprayed into the top of the inner vessel through the top fill spray header. The action of the spraying liquid will reduce the pressure within the vessel, eliminating the need to vent or recover vapor.

The tank filling is terminated automatically when the vessel becomes full. As the top spray ports are covered with liquid, the pressure quickly rises in the vessel to that of the delivery pressure. Once these pressures equalize, filling stops.

The fuel level inside a fuel tank is measured with an electronic fuel gauge. A probe inside the tank reads the column height of the liquid and sends a signal to an electronic converter module (F) that is remotely mounted. The electronic module converts the signal and registers it on an electric fuel gauge (G).

LNG fuel tank systems operate at a specific pressure, which is maintained by a series of relief valves. The primary relief valve (D) is designed to safely vent any excess pressure in the tank. The secondary relief valve (E) will open at a slightly higher pressure in the event that the primary relief valve malfunctions or cannot handle the volume of excess pressure.

Continued on next page

SS40167,00000D2 -19-17JUL09-1/2

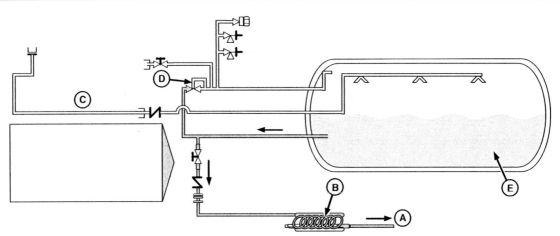

Liquefied Natural Gas (LNG) Fuel Delivery

A—Fuel Supply to Engine
B—Vaporizer
C—Valve Closed
D—Economizer Regulator Valve
E—Liquid

To use LNG in a natural gas engine, two processes must take place:

- The gas pressure in the fuel tank must be regulated in order to supply adequate fuel to the engine.
- The cold liquid must be vaporized.

Typically, LNG fuel storage systems rely on the pressure buildup that occurs as the temperature of the liquefied gas increases to force the fuel from the tank. For this reason, the temperature of the LNG must rise to about −205°F (−96°C) to maintain 100 psi (690 kPa) pressure in the tank.

The economizer regulator valve (D) maintains a specified tank operating pressure.

When the operating pressure is higher than the selected pressure for the tank, the economizer regulator will open. The gas supplied to the engine will be withdrawn from the vapor space in the top of the tank. The gas will travel through the economizer regulator to the vaporizer (B) where it will be warmed before it reaches the pressure regulator.

The action of removing gas from the top of the tank reduces the tank's pressure. When the tank operating pressure is reduced below the selected pressure level, the economizer regulator will close. Liquid (E) will then be forced from the bottom of the tank and into the vaporizer.

NOTE: The illustration shows the LNG system withdrawing liquid.

The vaporizer is nothing more than a heat exchanger located between the fuel tank and the engine. In the fuel vaporization process, warm engine coolant is circulated through the vaporizer. As the LNG flows into the vaporizer, the heat from the coolant turns the liquid into a gas. The gaseous fuel can then be supplied to the engine.

SS40167,00000D2 -19-17JUL09-2/2

Parts of Natural Gas Fuel System

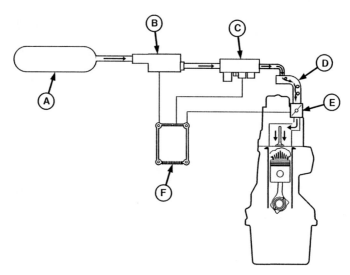

Natural Gas Fuel System

A—Fuel Tank
B—Pressure Regulator
C—Fuel Metering Valve
D—Air-Fuel Mixture
E—Throttle Body
F—Engine Control Unit (ECU)

The major parts of a natural gas fuel system are:

- **Fuel Tanks**
- **Fuel Pressure Regulator**
- **Fuel Metering Valve**
- **Air-Fuel Mixer**
- **Throttle Assembly**

Fuel tanks store natural gas in either a compressed or liquefied state.

Fuel pressure regulator reduces the fuel pressure to the pressure required by the specific engine. Natural gas fuel systems vary in the types and number of regulators used. There are three basic systems:

- Two-Regulator System
  - Uses a high-pressure (first stage) regulator and a low-pressure (second stage) regulator.
  - Controls the gas flow and pressure to the air-fuel mixer (carburetor).
  - Is typically used with an air valve type air-fuel mixer.
- Single Regulator System
  - Uses one regulator with three stages.
  - Controls the gas flow and pressure to the air-fuel mixer.
  - Is typically used with a venturi type air-fuel mixer.
- Fuel Injection System
  - Uses a single regulator that drops the fuel pressure to 80–150 psi (552–1034 kPa).
  - Fuel flows to a metering valve body that electronically controls the amount of fuel delivered to the engine.

A fuel metering valve is used with natural gas fuel injection systems. For more information, see Electronic Fuel Injection System later in this chapter.

An air-fuel mixer is located upstream of the throttle body. The air-fuel mixer combines the natural gas with air to form the air-fuel mixture for combustion.

The two basic types of air-fuel mixers are:

- Air Valve Mixer
- Venturi Mixer

Air valve mixers use a spring-loaded diaphragm (air valve) to meter the correct amount of fuel to the engine. The diaphragm is positioned over the top of the mixer throat so that air passing around it creates a pressure differential between the top and bottom of the diaphragm. The pressure differential causes the air valve to move off its seat, opening the gas metering valve to allow the gas to mix with the air.

The system meters fuel in direct and constant proportion to the amount of air flow. Thus, the greater the air flow, the more fuel is delivered. For more information, see Carburetor Air-Fuel Mixer System in this chapter.

Venturi mixers use the venturi principle. The venturi shape of the mixer causes the air flow to speed up, creating a low-pressure area in the mixer.

Small holes or slots below the venturi allow fuel to pass into the air stream and mix with the air. These mixers combine with the regulator to deliver fuel to the engine on demand.

Throttle assembly controls the volume of air flow through the air-fuel mixer. The amount of natural gas introduced into the air flow will be proportional to the amount of air flowing through the mixer. Thus, varying the air flow to the engine controls the power output of a natural gas engine.

SS40167,00000D3 -19-22JUL09-1/1

Types of Natural Gas Fuel Delivery Systems

Two basic types of natural gas fuel delivery systems are in common use today. They are:

- **Air-Fuel Mixer System Carburetor**
- **Electronic Fuel Injection System**

The carburetor air-fuel mixer system uses either an air valve mixer or a venturi mixer combined with the pressure regulator to deliver fuel to the engine on demand.

The carburetor air-fuel mixer systems meter fuel in direct and constant proportion to the amount of air flowing through the mixer.

We'll examine the carburetor air-fuel mixer fuel system in more detail later in this chapter.

The electronic fuel injection system uses an electronic engine control unit (ECU) to optimize the engine air flow, natural gas fuel delivery, and ignition spark timing.

The ECU monitors engine performance and, based on monitored operating conditions, controls electrically pulsed fuel injectors to precisely meter natural gas to the engine.

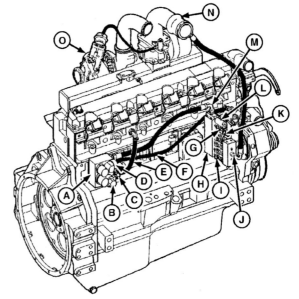

Natural Gas Electronic Fuel Injection System

A—Fuel Pressure Regulator
B—Fuel Filter
C—Regulator Fuel Inlet
D—Pressure Relief Valve
E—Coolant Outlet
F—Coolant Inlet
G—Gas Temperature Sensor
H—Fuel Metering Block
I—Fuel Injectors
J—Common Outlet Port
K—Gas Pressure Sensor
L—Fuel Lock-Off Solenoid
M—Metering Block Fuel Inlet
N—Air-Fuel Mixer
O—Turbocharger

SS40167,00000D4 -19-22JUL09-1/1

Electronic Fuel Injection System

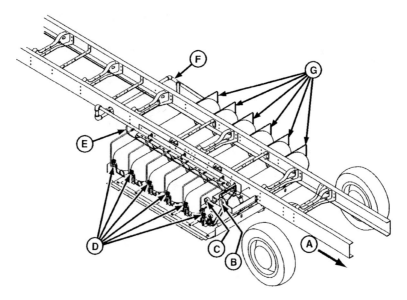

Compressed Natural Gas (CNG) Fuel Storage-Delivery System

A—Vehicle Front
B—1/4-Turn Shutoff Valve
C—Pressure Gauge
D—Manual Shutoff Valves and Pressure Relief Valves
E—Common Fuel Line
F—Common Vent Pipe
G—Fuel Tanks

To better understand how a natural gas fuel injection system operates, let's trace the fuel flow through a typical system.

In operation, compressed natural gas flows from the individual fuel tanks into a common high-pressure fuel line. A pressure gauge and 1/4-turn shutoff valve (B) is plumbed into the common fuel line (E). The pressure gauge (C) measures the common fuel line pressure. The 1/4-turn shutoff valve shuts off the fuel flow to the common fuel line.

Fuel flows through the common fuel line into a high-pressure lockoff and a coalescing filter.

The high-pressure fuel lock-off device is a valve that shuts off the fuel flow to the engine when the engine is not running, even if the ignition switch is in the ON position. The lock-off device opens when the ignition switch is in the ON position and the engine is either being started or is running.

Continued on next page   SS40167,0000107 -19-20JUL09-1/2

DXP00396 —UN—23DEC08

The fuel flows through the coalescing filter to the fuel pressure regulator (A). The pressure regulator reduces the high-pressure gas to the proper working pressure, approximately 130 psi (896 kPa), for the fuel system.

As the pressure drops, the gas expands and increases in volume. The rapid expansion of the gas through the regulator causes the gas to act as a refrigerant, absorbing heat and possibly causing icing of the regulator. To prevent icing, engine coolant is circulated through the pressure regulator. The heat from the engine coolant is transferred to the regulator.

A regulator pressure relief valve protects against overpressure problems. The pressure relief valve (D) is connected to the vehicle vent line.

From the pressure regulator, fuel flows to the fuel metering valve. The fuel metering valve contains a low-pressure lock-off solenoid that is activated by the ignition switch and the electrically pulsed injectors that deliver a precise amount of fuel to the engine.

The lock-off solenoid controls a normally closed valve that opens when the solenoid is energized. The lock-off solenoid will be closed anytime the ignition is OFF or is ON but hasn't been in the START position for a specific period of time.

An electronic engine control unit (ECU) controls the on time (pulse width) of each pair of injectors to deliver the correct amount of fuel to the engine.

Natural gas pressure and temperature are measured at the metering valve. The ECU uses these measurements to adjust the injector pulse width to compensate for variations in the natural gas density.

A common outlet port from the fuel metering valve delivers fuel to the air-fuel mixer (N). The air-fuel mixer combines the fuel with air to form the air-fuel mixture for combustion.

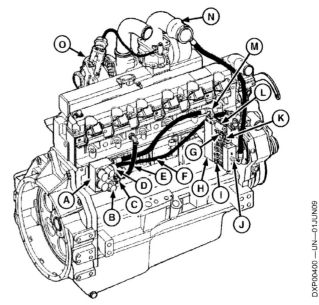

Natural Gas Electronic Fuel Injection System

DXP00400 —UN—01JUN09

A—Fuel Pressure Regulator	I— Fuel Injectors
B—Fuel Filter	J— Common Outlet Port
C—Regulator Fuel Inlet	K—Gas Pressure Sensor
D—Pressure Relief Valve	L—Fuel Lock-Off Solenoid
E—Coolant Outlet	M—Metering Block Fuel Inlet
F—Coolant Inlet	N—Air-Fuel Mixer
G—Gas Temperature Sensor	O—Turbocharger
H—Fuel Metering Block	

The air-fuel mixer is located upstream of the throttle body. The throttle valve controls the volume of air and fuel flowing through the air-fuel mixer.

SS40167,0000107 -19-20JUL09-2/2

Combustion Theory

A stoichiometric combustion engine burns an air-fuel mixture that contains just enough air to theoretically burn all the fuel. Most gasoline engines are stoichiometric combustion engines. The air-fuel ratio is approximately 15 parts air to one part gasoline. For natural gas engines, the air-fuel ratio is approximately 17:1.

A lean-burn combustion engine burns an air-fuel mixture that contains more air than is theoretically needed to completely burn the fuel. Many natural gas engines operate on the lean-burn principle. A lean-burn natural gas engine may run approximately 35% lean of stoichiometry. Typically, the air-fuel ratio may be as lean as 25:1.

The excess air in lean-burn combustion results in a more complete combustion than what is achievable in a stoichiometric engine. This complete combustion yields lower emissions of hydrocarbon (HC) and carbon monoxide (CO). In addition, lean air-fuel mixtures burn at a lower temperature, resulting in less formation of oxides of nitrogen (NOx) emissions.

The amount of energy in a lean mixture is less than the amount of energy in the same volume of a stoichiometric mixture. A turbocharger may be used on the lean-burn engine to increase the density of the air-fuel mixture, allowing the engine to develop a high power output.

Lean air-fuel mixtures don't ignite as readily as stoichiometric mixtures. Ignition is accomplished using a powerful inductive ignition system to provide a high-energy, long-duration spark.

Accurate control of the air-fuel ratio in a lean-burn engine is extremely important. If the mixture is too rich, the advantages of lean-burn will be lost. If the mixture is too lean, the mixture will not ignite and a misfire will result.

Accurate control of the air-fuel mixture is accomplished using an electronic engine control unit (ECU). The ECU commands the electronic fuel injection system to deliver a precise amount of natural gas to the engine.

SS40167,00000D5 -19-20JUL09-1/1

Electronic Engine Controls

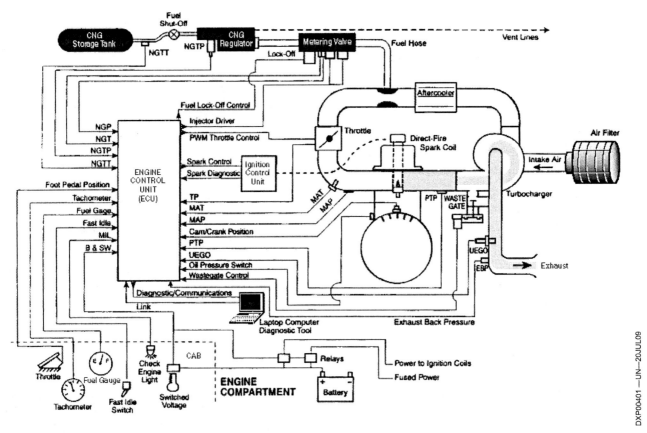

Electronic Engine Control System

The following describes the basic operation of a typical electronically controlled natural gas fuel system.

The microprocessor-based system controls engine air flow, engine fuel delivery, and ignition spark timing to optimize the engine exhaust emissions, driveability, and fuel economy. The system shown is for an on-highway bus.

The electronic engine control unit (ECU) uses sensors to monitor many performance characteristics of the engine. The ECU uses the inputs from the sensors in conjunction with engine-specific calibration data stored in a programmable read-only memory (PROM) to determine the air flow rate to the engine.

Once the air flow rate is determined, the desired fuel flow rate is calculated by the ECU. The ECU sends a command signal to the electrically pulsed natural gas fuel injectors to meter the required fuel flow to the engine.

The ECU may also have self-diagnostic capabilities.

SS40167,00000D6 -19-22JUL09-1/1

Types of Sensors

Sensors are simple devices. Most use a variable resistor to supply a signal to the engine control unit (ECU). The ECU makes decisions on the optimum air flow, fuel flow, and ignition timing based on the information from the sensors.

Sensor inputs may include:

- Manifold Absolute Pressure
- Exhaust Back Pressure
- Natural Gas Pressure
- Natural Gas Tank Pressure
- Engine Coolant Temperature
- Manifold Air Temperature
- Natural Gas Temperature
- Natural Gas Tank Temperature
- Throttle Position
- Foot Pedal Position
- Exhaust Gas Oxygen
- Camshaft Position

Pressure Sensors

Pressure sensors are pressure-sensitive variable resistors. Changes in pressure can be determined by monitoring the resistance of the sensor. As the pressure changes, the sensor resistance changes.

The engine control unit (ECU) sends a 5-volt reference voltage to the sensor, then monitors the voltage returning on the sensor signal wire to determine the voltage drop across the sensor. The ECU then translates this voltage to a pressure value.

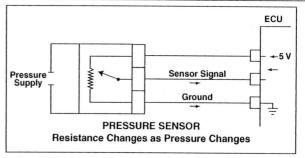

PRESSURE SENSOR
Resistance Changes as Pressure Changes

Typical Pressure Sensor

The system uses pressure sensors to measure:

- **Manifold Absolute Pressure and Pre-Turbine Pressure** — These pressures are used in conjunction with several other sensors to determine the amount of air flow into the engine.
- **Exhaust Back Pressure** — This pressure is used to aid the ECU in determining air density and the closed-loop fuel correction, and to enhance turbocharger wastegate control.
- **Gas Pressure** — This pressure is used in conjunction with the gas temperature to allow the ECU to determine the density of the natural gas at the injectors. The ECU uses the gas density information when calculating the injector pulse width (fuel flow).
- **Gas Tank Pressure** — This pressure is used along with the natural gas tank temperature to determine the amount of fuel in the storage tanks. The ECU transfers this information to the vehicle's fuel gauge.

SS40167,00000D7 -19-22JUL09-1/3

Temperature Sensors

Temperature sensors are temperature-sensitive variable resistors. Changes in temperature can be determined by monitoring the resistance of the sensor. As the sensor temperature increases, the resistance decreases.

The engine control unit (ECU) sends a 5-volt supply signal to the sensor, then monitors the voltage drop across the sensor. The ECU compares the sensor voltage to preprogrammed values to determine the temperature.

The system uses temperature sensors to measure:

- **Engine Coolant Temperature and Manifold Air Temperature** — These temperatures are used by the ECU when making engine air flow calculations.
- **Gas Temperature** — This temperature is used in conjunction with gas pressure to allow the ECU to determine the density of the natural gas at the injectors.

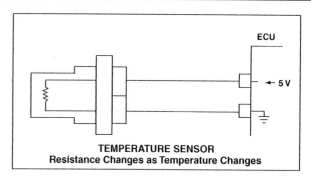

TEMPERATURE SENSOR
Resistance Changes as Temperature Changes

Typical Temperature Sensor

The ECU uses this information when calculating the injector pulse width (fuel flow).

Continued on next page   SS40167,00000D7 -19-22JUL09-2/3

Position Sensors

Position sensors are variable resistors (potentiometers). The resistance of these sensors varies depending on the sensor position. A movable contact slides along a resistor as the sensor position is changed.

The engine control unit (ECU) sends a 5-volt reference signal to the sensor, then monitors the voltage value returning on the sensor signal wire. The sensor signal wire voltage varies depending on the sensor position. The ECU translates this sensor voltage reading to sensor position.

The system uses position sensors to measure:

- **Throttle Position** — The position of the throttle is monitored so that the ECU can determine if the throttle is opening as commanded.
- **Foot Pedal Position** — The position of the foot pedal is used by the ECU to determine the operator's desired engine load.

Universal Exhaust Gas Oxygen Sensor

The universal exhaust gas oxygen Sensor (called UEGO) measures the amount of oxygen in the exhaust gas. The amount of free oxygen in the exhaust is directly related to the air-fuel ratio. A rich mixture results in little free oxygen; a lean mixture results in considerable free oxygen.

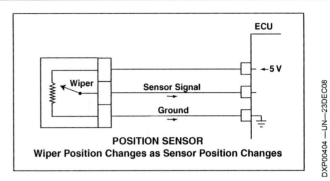

Typical Position Sensor

The UEGO sensor is composed of a sense cell, a pump cell, a heater and a standard calibration resistor. The sense cell voltage varies according to the sensed oxygen concentrations. A low oxygen concentration causes a relatively high voltage. Higher concentrations of oxygen cause a relatively low voltage.

The engine control unit (ECU) monitors the sense voltage and the pump voltage to determine the air-fuel ratio, then adjusts the command signal to the fuel injectors to deliver the correct amount of fuel to the engine.

SS40167,00000D7 -19-22JUL09-3/3

Ignition Timing

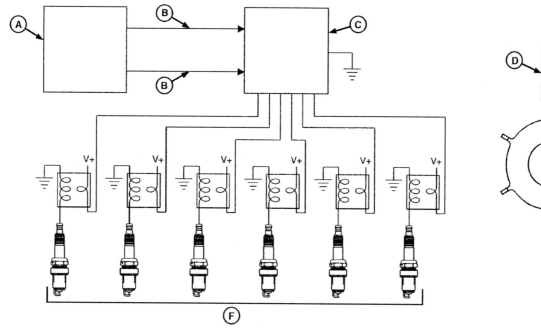

Electronic Ignition

A—Engine Control Unit (ECU)
B—Trigger
C—Ignition Control Unit (ICU)
D—Camshaft Position Sensor
E—Camshaft Trigger Wheel
F—Spark Plugs

The ignition system is a microprocessor-based, camshaft-referenced, non-waste spark, inductive ignition system. Being a non-waste spark system, each cylinder has its own ignition coil, and the coil fires on the combustion stroke only.

The overall operation of the ignition system is controlled by the engine control unit (ECU) (A). The spark timing is controlled by the ECU based on engine speed, manifold absolute pressure, and engine coolant temperature.

The manifold absolute pressure sensor sends a signal to the ECU to indicate the manifold (boost) pressure. The ECU interprets this signal as an indication of the load on the engine.

The ECU monitors signals from the camshaft position sensor (D) to determine engine speed and piston position.

The ECU determines the desired timing advance, then outputs a trigger signal to the ignition control unit (ICU) (C) when a coil should be fired. A synchronization pulse is also sent by the ECU to the ICU so that the ICU fires the correct coil.

The ignition coil consists of a primary coil and a secondary coil. The primary coil receives the primary charge from the ICU and induces a high-voltage, low-current charge in the secondary coil. The secondary charge discharges through the ignition coil output terminal, spark plug wire, and into the spark plug (F) to produce a spark.

For more information on electronic ignition systems, refer to the FOS manual on Electronic and Electrical Systems

SS40167,00000D8 -19-20JUL09-1/1

Engine Control Unit (ECU)

IMPORTANT: DO NOT connect or disconnect the ECU to or from the wiring harness without first removing the negative (–) battery cable from the battery.

The engine control unit (ECU) is the brains of the electronic engine control system. The ECU performs the following functions:

- Converts the electrical signals from the sensors into digital signals that the ECU central processing unit can use.
- Central processing unit determines the best air flow, fuel flow, and ignition timing based on information from the sensors.
- Performs self-diagnosis on the engine control system.
- Stores trouble codes in its memory.

The ECU consists of three subsystems:

- **Analog/Digital Converter** — Converts analog electrical signals from the various sensors into digital data that the ECU central processing unit can use.
- **Central Processing Unit (CPU)** — Performs mathematical computations and logical functions necessary for controlling air flow, fuel delivery, and ignition timing.
- **Memory** — The ECU contains three types of memory:
  - Random Access Memory (RAM)
  - Read Only Memory (ROM)
  - Programmable Read Only Memory (PROM)

Engine operational data is temporally stored in RAM. Information in RAM is erased when battery voltage to the ECU is removed.

ROM contains information that can only be read, not changed. ROM information is retained when battery voltage is removed.

PROM contains information programmed in at the factory specific to the engine. Information in PROM is retained when battery voltage is removed.

FUEL CONTROL MODES

The Engine Control Unit (ECU) operates in two different fuel control modes:

- **Open-loop mode**
- **Closed-loop mode**

OPEN-LOOP MODE is utilized during engine start-up and for the first 30 to 60 seconds of run time. During open-loop fuel control, fuel flow rate is determined by the ECU based on calibration data stored in the Programmable Read-Only Memory (PROM) of the microprocessor and information from several sensors. This is speed-density fuel control.

CLOSED-LOOP MODE provides improved fuel delivery accuracy. After the engine has been running for 30 to 60 seconds, the ECU changes from open-loop fuel control to closed-loop fuel control. In closed-loop mode, the ECU has the ability to correct for gas composition variation and for changing operating conditions.

When operating in closed-loop fuel control mode, an Exhaust Gas Oxygen (EGO) sensor measures the amount of oxygen in the exhaust gas. The amount of free oxygen in the exhaust is directly related to the air-fuel ratio. A rich mixture results in little free oxygen; a lean mixture results in a high level of free oxygen.

The ECU uses the oxygen measurement to determine the air-fuel ratio, then adjusts the amount of fuel delivered until the optimum air-fuel ratio is achieved.

The ECU has the capability for adaptive learn control. The adaptive learn function acts as a fuel correction memory to maintain optimum performance and emissions calibration during changing operating conditions.

The adaptive learn is enabled when the engine reaches operating temperature and is in closed-loop fuel control. The adaptive learn function stores fuel flow data for different combinations of engine speed and engine load.

Each time a closed-loop fuel correction is "successful" in achieving the optimum air-fuel ratio for a certain engine speed and load, the adaptive learn stores that fuel correction in a "cell". The next time the engine operates at that same engine speed and load, the fuel correction stored in that cell is used.

Continued on next page

ECU Controls the Turbocharger Boost Pressure

A—Pre-Throttle Intake Air
    Pressure
B—Vent
C—Wastegate Control Solenoid
D—To Throttle

E—ECU
F—Boost Control
G—Spring
H—Vent

I— Diaphragm
J—Wastegate
K—Open
L—Closed

M—Engine Exhaust
N—Turbocharger
O—Orifice

AIR CONTROL

Varying the air flow to the engine controls the power output of a spark-ignited natural gas engine. More air flow results in more power. The Engine Control Unit (ECU) (E) determines the engine air flow based on various engine sensor readings, then adjusts the fuel delivery to provide the correct air-fuel ratio.

The ECU controls the engine air flow by controlling both the throttle opening and the turbocharger boost pressure.

The throttle is a drive-by-wire electronically-controlled throttle/actuator. The ECU monitors the signals from the foot pedal position sensor. The ECU translates these signals to determine the pedal position, and thus the operator's desired engine load. The ECU then provides a pulse width modulated signal to command the throttle opening based on the foot pedal position.

Turbocharger boost control is achieved by using a conventional diaphragm wastegate actuator that is

controlled by a pulse width modulated wastegate control solenoid (C). The ECU controls the amount of boost that the turbocharger (N) is allowed to generate by sending a signal to the wastegate solenoid. The wastegate solenoid routes pre-throttle intake air pressure (A) to either the wastegate diaphragm (J) or to a vent (H).

When the solenoid is energized, air pressure is routed to the vent. The diaphragm spring (G) force holds the wastegate closed and the turbocharger generates maximum boost.

When the solenoid is not energized, air pressure is routed to the wastegate diaphragm. The air pressure overcomes the spring force, opening the wastegate. As the wastegate opens, the engine exhaust gases are allowed to bypass the turbocharger turbine, resulting in a reduction of turbocharger boost pressure.

SS40167,00000D9 -19-21JUL09-2/2

Electronic Engine Control Diagnostics

The electronic engine control system has self-diagnostic capabilities. Most problems that affect emissions or operation of the engine will set a Diagnostic Trouble Code (DTC). The Engine Control Unit (ECU) will store that code in memory and cause the CHECK ENGINE light to come on.

The CHECK ENGINE light will come on when the ignition is ON and the engine is off. This verifies that the light is working. Once the engine is in the START or RUN mode, the light should go off. If the light comes on while the engine is in the START or RUN mode, the ECU has detected a problem and will store the diagnostic trouble code in memory.

For more information about electronic engine control systems, refer to the FOS manual on Electronic and Electrical Systems.

Diagnostic Trouble Codes (DTC)

Some malfunctions that set Diagnostic Trouble Codes (DTCs) are "intermittent" that may seem to "go away," such as bad wiring connections. Some intermittent malfunctions will cause the CHECK ENGINE light to come on, then go off after a short period when the problem goes away.

Other intermittent malfunctions that could seriously affect the engine operation will cause the CHECK ENGINE light to come on and remain on even though the problem may have gone away. In either case, the diagnostic trouble code will be stored in memory.

Other malfunctions are "hard" malfunctions that do not go away. This type of malfunction will set a diagnostic trouble code. The code will be stored in memory, and the CHECK ENGINE light will come on and remain on until the problem is repaired.

The diagnostic trouble codes are numbered in order of importance, the lower numbers being more important than higher numbers. See the engine technical manual for a listing of the diagnostic trouble code numbers and their definitions.

Diagnostic trouble codes can be read by using either the CHECK ENGINE light or a diagnostic scan tool, such as a portable laptop computer. See the engine technical manual for detailed instructions.

SS40167,00000DA -19-20JUL09-1/1

Servicing Natural Gas Fuel Systems

Fuel Pressure Leak-Off Procedure

⚠ **CAUTION: Natural gas is highly flammable. DO NOT smoke while working on or near natural gas equipment. Natural gas fumes may cause sickness or death. Work in a well-ventilated area. Shut off the natural gas supply and relieve internal regulator pressure before servicing the equipment.**

Compressed natural gas (CNG) fuel systems operate at high pressures. The safety rules outlined in Chapter 2 for CNG must be observed when servicing the natural gas fuel system.

NOTE: The illustration shows a typical application and does not show all the shutoff valves. Refer to the vehicle's operator's manual and technical manual to help locate all the fuel tank shutoff valves.

CNG fuel system pressure must be relieved before disconnecting any fuel system component. To relieve fuel system pressure, close all fuel tank manual shutoff valves (A). A shutoff valve is closed when turned fully clockwise.

IMPORTANT: Always relieve natural gas pressure by closing each fuel tank manual shutoff valve (A). DO NOT relieve pressure by closing the 1/4-turn shutoff valve (B).

If the engine **will start**, let it run until it stops by running out of fuel. Make certain the fuel pressure is completely relieved by attempting to start the engine several times.

If the engine **will not start**, make sure the vehicle is outside, away from heat, flames, or sparks. **Slowly** loosen the hose from the fuel metering block inlet fitting just until escaping gas is heard. Let the fuel escape until fuel pressure is completely relieved.

Fuel Leak Check Procedure

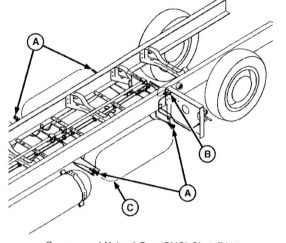

Compressed Natural Gas (CNG) Shutoff Valves

A—Manual Shutoff Valves
B—1/4-Turn Shutoff Valve
C—Fuel Tank

If any components of the natural gas fuel system have been replaced or if a leak is suspected in the natural gas fuel system, the fuel system must be checked for leaks as follows:

1. Open the fuel tank manual shutoff valves (A).

2. Turn the ignition switch to the ON position for 3 to 4 seconds, then back to the OFF position.

3. Use a non-ammonia soap solution or a commercial leak detector solution and wet all lines and fittings. If bubbles form, the fuel line or fitting is leaking.

4. If a leak is detected, relieve the natural gas pressure in the fuel system. Repair or replace the faulty component. Repeat the leak check procedure until no leaks are detected.

SS40167,00000DD -19-20JUL09-1/1

Fuel Tanks, Valves, and Gauges

Compressed natural gas (CNG) fuel tanks (C) must be high strength to withstand the pressure of the gas. CNG is stored in the tanks up to 3600 psi (24 821 kPa). Each fuel tank is connected to a common high pressure fuel line.

Each fuel tank contains:

- **Manual Shutoff Valve (A)**
- **High-Pressure Relief Valve**

The MANUAL SHUTOFF VALVE is closed when turned fully clockwise, and open when turned counterclockwise. Any tank with its shutoff valve open feeds fuel into the common high pressure fuel line.

With its shutoff valve closed, the tank doesn't feed the common line, but the valve's tee is exposed to common line high pressure.

IMPORTANT: **When relieving the natural gas pressure in the fuel system, always close the manual shutoff valves.**

During refilling of the fuel tanks, fuel is routed through a filter, then backflows into the common high-pressure fuel line and into any tank with an open shutoff valve.

A 1/4-turn shutoff valve (B) near the filter assembly shuts off the fuel flow to the common fuel line.

The HIGH-PRESSURE RELIEF VALVE is designed to vent at either high pressure or high temperature. Each tank's relief valve is plumbed into a common vent line. If the relief valve opens, the vent line routes the vapors out of the vehicle.

Servicing Fuel Tanks and Valves

Be sure that you are familiar with all rules and regulations before servicing natural gas equipment.

Two rules are important here:

- Never service a fuel tank except to remove and install a valve.
- Never service a valve unless it is replaced.

If a valve is damaged or faulty and must be replaced, first empty all fuel from the tank to release the pressure. If

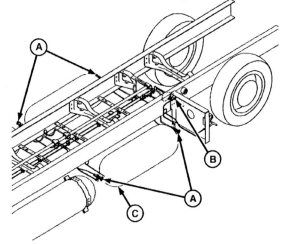

Compressed Natural Gas (CNG) Shutoff Valves

A—Manual Shutoff Valves   C—Fuel Tank
B—1/4-Turn Shutoff Valve

possible, do this outdoors by running the engine until all fuel is spent. If it is not possible to run the engine, contact your compressed natural gas distributor for the approved method to empty your tank.

Gauges

A pressure gauge is plumbed into the common high-pressure fuel line. The pressure gauge measures the common fuel line pressure.

The fuel gauge for the compressed natural gas (CNG) fuel system operates on a completely different principle than the float operated fuel gauge used for gasoline and LP-Gas fuel systems. The compressed natural gas fuel gauge is operated by the engine control unit (ECU).

The ECU measures fuel pressure at the fuel tanks and at the pressure regulator. It also monitors fuel temperature at each tank and at the fuel metering valve. Using this information, the ECU calculates the CNG fuel volume and operates the fuel gauge on the dash of the vehicle accordingly.

Continued on next page   SS40167,00000DE -19-23JUL09-1/3

Fuel Pressure Regulator

Compressed natural gas (CNG) is supplied to the fuel pressure regulator at pressures up to 3600 psi (24 821 kPa). The diaphragm-type pressure regulator reduces the fuel pressure to approximately 130 psi (896 kPa).

Engine coolant circulates through the pressure regulator to prevent freeze-up as a result of a rapid drop in pressure.

The pressure regulator also includes a pressure relief valve (B) and a replaceable fuel filter. The relief valve is connected to the vehicle vent system.

The pressure regulator is constructed for reliability. Other than the fuel filter, serviceable parts may not be available for the pressure regulator. For information on pressure regulator service, see the engine technical manual.

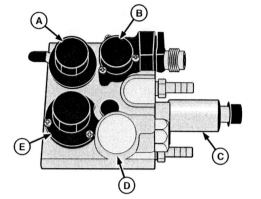

Fuel Pressure Regulator

A—Second-Stage Diaphragm
    Regulator
B—Pressure Relief Valve
C—Natural Gas Pressure
    Sensor
D—Natural Gas Filter
E—First-Stage Diaphragm
    Regulator

SS40167,00000DE -19-23JUL09-2/3

Fuel Metering Valve

The typical fuel metering valve contains electrically pulsed injectors, a natural gas pressure sensor (C), a natural gas temperature sensor (A), and a low pressure fuel lock-off solenoid (B).

Fuel is metered to the engine by the electrically pulsed injectors located on the fuel metering block. The on time (pulse width) determines the amount of fuel delivered by the injectors. The engine control unit (ECU) controls the injector pulse width of each pair of injectors to deliver the correct amount of fuel to the engine.

The number of fuel injectors used is determined by the particular engine application and can range from two to eight injectors.

The density of natural gas will vary with changes in gas pressure and temperature. Sensors on the metering block measure the natural gas pressure and temperature at the metering valve. The ECU then adjusts the on time (pulse width) of the PWM fuel injectors (D) to compensate for variations in the natural gas density to deliver the precise amount of fuel to the engine.

The low pressure fuel lock-off solenoid stops the flow of fuel upstream from the injectors. The solenoid controls a normally closed valve that opens when the ECU activates it.

The lock-off solenoid will be closed anytime the ignition switch is in the OFF position. The lock-off solenoid will also be closed anytime the ignition switch is in the ON position but hasn't been in the START position for the past 4 to 5 seconds.

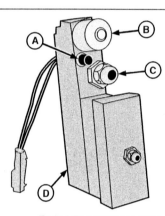

Typical Fuel Metering Valve

A—Natural Gas Temperature
    Sensor
B—Fuel Lock-Off Solenoid
C—Natural Gas Pressure
    Sensor
D—Pulse-Width-Modulated
    (PWM) Fuel Injectors

Servicing the Fuel Metering Valve

The fuel metering valve is constructed so that it will seldom cause trouble. Serviceable parts may not be available for the fuel metering valve. In this case, replacement of the fuel metering valve assembly is required for repair. For information on metering valve service, see the engine technical manual.

SS40167,00000DE -19-23JUL09-3/3

Carburetor Air-Fuel Mixer System

The natural gas carburetor air-fuel mixer system is commonly used on stationary engines.

Operation

In operation, natural gas flows through the natural gas filter and the electric shutoff valve (E) into the pressure regulator valve.

The pressure regulator reduces the natural gas pressure to maintain the proper pressure differential between the regulated gas pressure and the air pressure at the carburetor inlet. The differential pressure must be maintained at a constant value above the air intake pressure (M) under all operating conditions to enable the carburetor (B) to maintain a consistent air-fuel mixture.

Gas flows from the pressure regulator to the carburetor. The natural gas carburetor acts as a mixing valve and is located above the throttle valve. In the carburetor, gas is mixed with air to form the air-fuel mixture for combustion. The air-fuel mixture flows from the carburetor, through the throttle valve (A), and into the engine where it is burned.

The throttle valve controls the volume of air and fuel flowing through the carburetor.

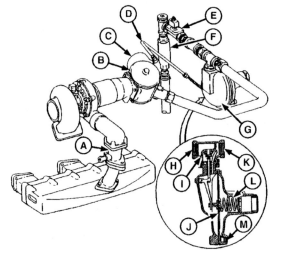

Carburetor Air-Fuel Mixer System

A—Throttle Valve
B—Carburetor
C—Air Inlet
D—Balance Line
E—Shutoff Valve
F—Gas Filter
G—Gas Regulator Valve
H—Supply Pressure
I— Ring Valve
J— Diaphragm
K—Regulated Pressure
L—Adjusting Spring
M—Air Intake Pressure

SS40167,00000DF -19-20JUL09-1/1

Pressure Regulator Valve

The function of the gas regulator valve (G) can be likened to that of the float in the gasoline carburetor (Chapter 4). They both must provide a constant fuel level (one gaseous, the other liquid) to enable the carburetor (B) to maintain a consistent air-fuel mixture.

The carburetor float opens or closes the fuel inlet valve to control the flow of fuel into the carburetor and maintain the correct liquid fuel level in the carburetor fuel bowl.

The pressure regulator uses a diaphragm (J) linked to a ring valve (I) to control the flow of gas into the regulator.

Controlling the gas flow enables the pressure regulator to maintain a constant pressure differential between the gas pressure and the air inlet pressure at the carburetor.

One side of the pressure regulator diaphragm is connected to the carburetor air inlet by a balance line (D). A pilot tube connects the opposite side of the diaphragm to the regulated gas pressure line.

The air intake pressure (from the balance line) (M) and the regulator control spring apply a force on one side of the diaphragm. The regulated gas pressure on the opposite side of the diaphragm opposes this force.

The regulated gas pressure must be maintained at a higher value (differential pressure) than the air intake pressure. This differential pressure is determined by the setting of the regulator control spring.

A change in pressure on either side of the diaphragm causes the diaphragm to move, resulting in a corresponding opening and closing movement of the ring valve.

During engine operation, the ring valve is continuously adjusted, allowing natural gas to flow into the regulator

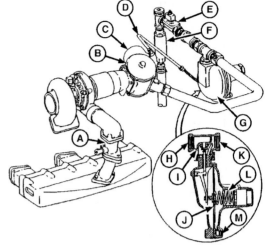

Carburetor Air-Fuel Mixer System

A—Throttle Valve
B—Carburetor
C—Air Inlet
D—Balance Line
E—Shutoff Valve
F—Gas Filter
G—Gas Regulator Valve

H—Supply Pressure
I— Ring Valve
J—Diaphragm
K—Regulated Pressure
L—Adjusting Spring
M—Air Intake Pressure

until the pressure on each side of the diaphragm reaches a point of equilibrium.

Servicing the Pressure Regulator Valve

Pressure regulator parts are subject to normal wear and must be inspected and replaced as necessary. Check on the availability of parts before attempting to service the regulator. For details, see the engine technical manual.

Continued on next page   SS40167,00000E0 -19-20JUL09-1/2

Adjusting Regulator Gas Differential Pressure

The regulator has been preset at the factory to provide the proper differential pressure between the air inlet pressure and the regulated gas pressure. However, the regulator pressure setting can be adjusted as necessary if the pressure is not within specifications.

To check and adjust pressure:

1. Connect a water manometer to the balance line fitting (A) and the regulated gas supply line fitting (B) between the regulator and the carburetor.

2. Start the engine and read the differential pressure on the manometer. See the engine technical manual for the specified differential pressure.

3. If pressure is incorrect, stop engine and remove the cap (C) from the regulator cover. Turn the control spring adjusting screw (D) clockwise to increase pressure, counterclockwise to decrease pressure.

4. Repeat steps 2 and 3 until the correct differential pressure is obtained.

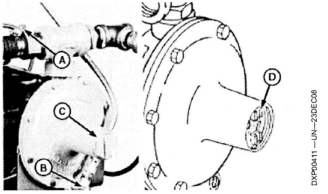

Adjusting Point on a Typical Pressure Regulator

A—Balance Line Fitting
B—Regulated Gas Supply Line Fitting
C—Regulator Cover Cap
D—Control Spring Adjusting Screw

5. Stop engine. Install cap on regulator cover, and remove the manometer.

SS40167,00000E0 -19-20JUL09-2/2

Carburetor

The natural gas carburetor does much the same job as gasoline models (Chapter 4). The main difference is that when the fuel enters the natural gas carburetor, it is a vapor, while in the gasoline carburetor it is a liquid. For this reason, the construction and operation of the natural gas carburetor and gasoline carburetor differ.

The illustration shows a typical natural gas carburetor.

The diaphragm (E) actuates a fuel metering valve (D).

The fuel metering valve varies the amount of fuel vapor entering the carburetor.

The idle mixture screw (A) controls the amount of air that flows through the metering valve.

The power valve (G) controls the fuel flow under full-load conditions.

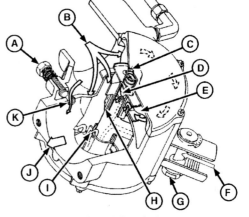

Natural Gas Carburetor

A—Idle Mixture Screw
B—Intake Air
C—Spring
D—Metering Valve
E—Diaphragm
F—Regulated Gas Supply
G—Power Valve
H—Bleed Hole
I— Gas Flow
J— Air-Fuel Mixture
K—Idle Air Passage

Continued on next page   SS40167,00000E1 -19-22JUL09-1/5

Operation

The air flow through the carburetor (B) is controlled by the position of the throttle valve (A), which is located downstream from the carburetor (B).

As the throttle valve opening is increased, the air pressure at the outlet side of the carburetor is reduced.

A—Throttle Valve
B—Carburetor
C—Air Inlet
D—Balance Line
E—Shutoff Valve
F—Gas Filter
G—Gas Regulator Valve
H—Supply Pressure
I— Ring Valve
J— Diaphragm
K—Regulated Pressure
L—Adjusting Spring
M—Air Intake Pressure

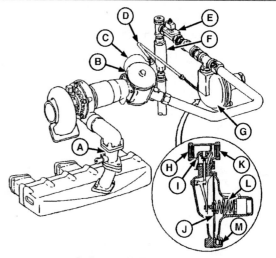

Carburetor Air-Fuel Mixer System

Continued on next page

SS40167,00000E1 -19-22JUL09-2/5

Bleed holes (H) in the metering valve (D) allow the air pressure at the carburetor outlet to equalize on the back side of the diaphragm (E). The outlet air pressure plus the diaphragm spring (C) apply a force on the back side of the diaphragm (E). The inlet air pressure on the opposite side of the diaphragm opposes this force.

The air pressure at the carburetor outlet is reduced as the throttle valve opening is increased. This reduction in pressure on the back side of the diaphragm creates a pressure differential across the diaphragm. The inlet air pressure on the opposite side of the diaphragm is able to overcome the spring pressure and move the diaphragm. As the diaphragm spring is compressed, the metering valve is lifted off its seat, allowing the fuel and air to mix.

A change in the air pressure on either side of the diaphragm causes the diaphragm to move, resulting in a corresponding change in the metering valve opening. The metering valve is continually adjusted to balance the forces on each side of the diaphragm.

At slow idle, the throttle valve is nearly closed, allowing only a small amount of air to be pulled through the carburetor. As a result, the pressure differential between the carburetor air inlet and outlet is reduced. This enables the diaphragm spring to move the diaphragm, decreasing the metering valve opening and reducing flow of fuel into the carburetor.

Since fuel flow at idle is limited by the position of the of the metering valve, the air-fuel mixture (J) at idle is controlled by limiting the amount of air that flows through the metering valve. The idle mixture screw (A) functions as an air bypass valve.

Turning the idle mixture screw out allows more air to bypass the metering valve through the idle air passage (K). As a result, the metering valve does not open as far and less fuel is metered, causing the air-fuel mixture to be leaner.

Turning the idle mixture screw in allows less air to bypass the metering valve. As a result, the metering valve opens

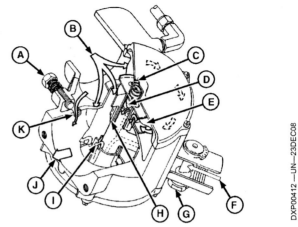

Natural Gas Carburetor Components

A—Idle Mixture Screw
B—Intake Air
C—Spring
D—Metering Valve
E—Diaphragm
F—Regulated Gas Supply
G—Power Valve
H—Bleed Hole
I— Gas Flow
J— Air-Fuel Mixture
K—Idle Air Passage

further, allowing more fuel to be metered. The idle mixture becomes richer.

Under full load, the throttle valve is nearly wide open, allowing a large amount of air to be pulled through the carburetor. As a result, the pressure differential between the carburetor air inlet and outlet will be at its maximum. Since the metering valve will be wide open to allow the large volume of air to flow, restricting the fuel flow at the power valve (G) controls the air-fuel ratio. As the power valve is closed, fuel flow is restricted, resulting in a leaner mixture. As the valve is opened, more fuel flows into the carburetor, resulting in a richer mixture.

Continued on next page   SS40167,00000E1 -19-22JUL09-3/5

Servicing the Carburetor

The natural gas carburetor is constructed so that it will seldom cause trouble.

Many of the problems with the gasoline carburetor, such as water or other contaminant in the fuel, or restricted fuel passages due to the accumulation of dirt or varnish, do not occur in the natural gas carburetor. Because of this, servicing the carburetor is very basic, consisting of replacing a faulty gasket or diaphragm (A). For details, see the engine technical manual.

A—Diaphragm   C—Spring
B—Cover

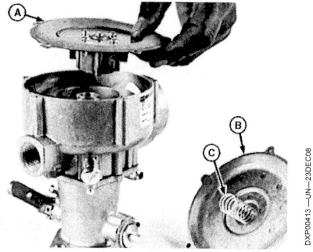

Typical Natural Gas Carburetor

SS40167,00000E1 -19-22JUL09-4/5

Carburetor Adjustments

The carburetor must be adjusted for smooth, economical operation of the engine.

Most natural gas carburetors have these basic adjustments:

- **Slow Idle Fuel Adjustment**
- **Full-Load Fuel Adjustment**

The idle mixture screw (A) should be adjusted until the engine idles smoothly.

The power valve (B) controls the full-load mixture. The power valve should be adjusted with the engine operating at full-load condition. Adjust the power valve to achieve the leanest mixture possible while still maintaining full power output.

NOTE: The engine will operate at higher efficiency and lower exhaust temperatures with a leaner fuel mixture. If too lean, the engine won't produce full power, or could begin to misfire.

For the specific details of the carburetor adjustments, see the engine technical manual.

A—Idle Mixture Screw   C—Throttle
B—Power Valve

Slow Idle Fuel Mixture Adjustment

Full-Load Fuel Mixture Adjustment

SS40167,00000E1 -19-22JUL09-5/5

Test Yourself

Questions

1. (True or False?) At atmospheric pressure, natural gas is in a gaseous state at any temperature above −259°F (−126°C).

2. (True or False?) Compressed natural gas (CNG) fuel is stored in high strength tanks up to 3600 psi (24 821 kPa).

3. (True or False?) Natural gas is liquefied by putting it under high pressure.

4. (True or False?) Power output of a spark ignited natural gas engine is controlled by varying the air flow to the engine.

5. Name the five major parts of the natural gas fuel system.

6. What function does the vaporizer perform in the liquefied natural gas (LNG) fuel system?

7. Match each item on the left with the correct one on the right.

a. A stoichiometric combustion engine...

1. burns an air-fuel mixture that contains more air than is theoretically needed to completely burn the fuel.

b. A lean-burn combustion engine...

2. burns an air-fuel mixture containing just enough air to theoretically burn all of the fuel.

8. (True or False?) A lean-burn natural gas engine may run approximately 35% lean of stoichiometry.

9. Some natural gas engines are equipped with an engine control unit (ECU) that controls the intake air flow to the engine. Name two other functions that are controlled by the ECU.

10. The engine control unit (ECU) operates in two different fuel control modes. Name the two modes.

11. What safety procedure must be performed before disconnecting any natural gas fuel system component?

SS40167,00000E2 -19-20JUL09-1/1

Diesel Fuel Systems — Introduction

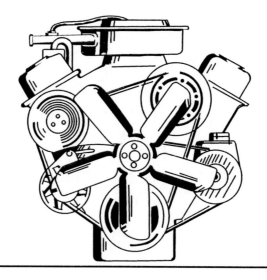

7

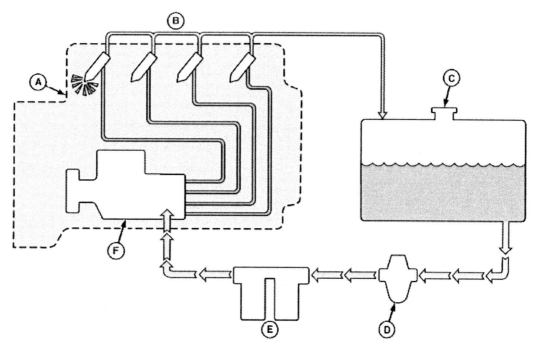

Diesel Fuel Systems (Distributor Type Shown)

A—Combustion Chamber
B—Injection Nozzles
C—Fuel Tank
D—Fuel Transfer Pump
E—Fuel Filters
F—Injection Pump

The prime job of the diesel fuel system is to inject a precise amount of atomized and pressurized fuel into each engine cylinder at the proper time.

Combustion in a diesel engine occurs when a charge of fuel is mixed with hot, compressed air. No electrical spark is used (as in the gasoline engine).

The major parts of a typical diesel fuel system are:

- **Fuel Tank (C)** — Stores fuel.
- **Fuel Transfer Pump (D)** — Pushes fuel through filters to the injection pump (F).
- **Fuel Filters (E)** — Clean the fuel.
- **Injection Pump (F)** — Times, measures, and delivers fuel under pressure to cylinders.
- **Injection Nozzles (B)** — Atomize and spray fuel into combustion chambers (A).

SS40167,00000E3 -19-30JUN09-1/1

Operation

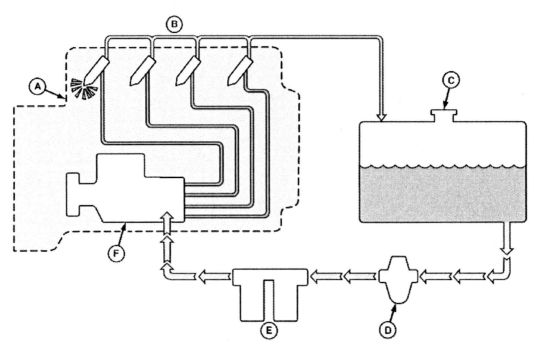

Diesel Fuel Systems

A—Combustion Chamber C—Fuel Tank   E—Fuel Filters
B—Injection Nozzles   D—Fuel Transfer Pump F—Injection Pump

Fuel flows from the fuel tank (C) to the transfer pump (D).

The transfer pump pushes the fuel through the filters (E), where it is cleaned.

The fuel is then pushed on to the injection pump (F), where it is put under high pressure and delivered to each injection nozzle (B) in turn.

The injection nozzles (B) atomize the fuel and spray it into the combustion chamber (A) of each cylinder.

Later we will discuss in detail how each part of a diesel fuel system works.

SS40167,00000E4 -19-21JUL09-1/1

What Fuel Injection Must Do

The diesel fuel injection system must:

- Supply the correct quantity of fuel.
- Time the delivery of the fuel.
- Control the fuel delivery rate.
- Break up or atomize the fuel.
- Distribute fuel evenly through the cylinder.

Let's take a closer look at these requirements.

Fuel Quantity — The fuel system must supply the correct amount of fuel to each cylinder, for every power stroke.

Fuel Delivery Timing — Fuel delivered too early or too late during the power stroke causes a loss of power. Fuel

must be injected into the cylinder at the instant maximum power can be realized.

Fuel Delivery Rate — Smooth operation from each cylinder depends on the length of time it takes to inject the fuel. The higher the engine speed, the faster the fuel must be delivered.

Fuel Atomization — The fuel must be thoroughly mixed with the air for complete combustion. For this reason the fuel must be broken up into fine particles.

Fuel Distribution — The fuel must be spread evenly throughout the cylinder and mixed with all available oxygen. This makes the engine run smoothly and helps achieve maximum power.

SS40167,00000E5 -19-21JUL09-1/1

Fuel Tanks

Fuel tanks (C) come in many shapes and sizes, in plastic or steel. Listed below are the main design elements found in most fuel tanks:

Capacity — A fuel tank must be capable of storing enough fuel to operate the engine for a reasonable period of time.

Size — The size, of course, depends on the required capacity. The size of a tank also depends on the available amount of space on the machine.

Shape — A fuel tank can be tall or short, round or square. This allows for a larger capacity tank in a restricted space. It is common to find fuel tanks made of plastic when there are space constraints and weight limitations.

Venting — The tank must be sealed to prevent dirt from contaminating the fuel, yet it must also be allowed to vent to allow air to enter, replacing the fuel used.

Openings — Three or four tank openings are also required: one to fill, one to discharge, and one to drain. Sometimes a fourth opening is required for the leak-off fuel returning from the injection system.

FUEL TANK SERVICE

 CAUTION: Cleaning or repairing a fuel tank can be dangerous. Sparks, smoking, or an open flame near the tank or near the area where the tank is being cleaned or repaired is extremely hazardous.

Leaks, condensation, and dirt are the typical problems found with fuel tanks.

Removing a Tank

A fuel tank should be removed from the machine when checking it for leaks or making repairs.

Cleaning a Tank

Once the tank is drained, it must be thoroughly cleaned.

1. Flush the tank with hot water or steam until all fuel and vapors are removed.
2. Dry the tank with clean, compressed air.

Inspecting a Tank for Leaks

Use one of the following methods to locate a leak in a fuel tank:

1. Wet method
2. Air pressure method

WET METHOD

Plug all fuel outlets tightly. Dry the entire outer surface of the tank thoroughly with clean compressed air and a clean, dry rag. Position the tank so that all surfaces are easily visible, such as putting the tank on top of blocks. Fill the tank with water. Then insert the end of the air hose in the filler neck and apply approximately 3 psi (21 kPa) of air pressure against the water. Inspect the entire surface

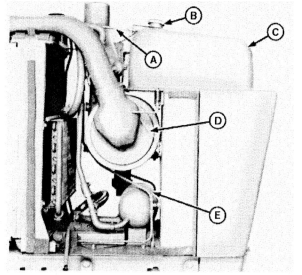

Typical Fuel Tank on a Tractor

A—Fuel Return Line
B—Filler Cap
C—Fuel Tank
D—Fuel Gauge Sender Wire
E—Fuel Shutoff

of the tank for moist areas where water may have been forced through a hole.

AIR PRESSURE METHOD

Plug the filler neck and attach an air hose to the fuel outlet. Submerge the fuel tank in clean water and apply approximately 3 psi (21 kPa) of air pressure. Draw a ring around each spot on the fuel tank where bubbles appear. Bubbles indicate a leak in the tank, which requires repair.

Repairing a Fuel Tank

 CAUTION: Fuel tanks made of plastic cannot be repaired. If a problem develops with a plastic fuel tank, the tank must be replaced. Only fuel tanks made of steel can be repaired and placed back in service.

SOLDERING

To help ensure a repair job holds up, repair a steel fuel tank as soon as possible after it has been cleaned. When using a soldering iron, the iron should not be **red** hot. A red-hot soldering iron can ignite any explosive mixture remaining in the fuel tank.

WELDING

 CAUTION: Before welding a fuel tank, evacuate all fumes from inside the tank by filling it with water.

A fuel tank can be safely welded, if the following precautions are taken:

1. To eliminate pockets of fuel vapor, plug the outlet of the tank and then fill the tank with water.

Continued on next page   SS40167,00000E6 -19-23JUL09-1/2

2. Remove the filler cap from the tank to allow steam to escape when welding.

   If the tank must be rolled over to reach the area to be welded, you can make an extended steam vent by drilling a hole in an old filler cap and welding an appropriate length of pipe to it. The open end of the pipe must then extend above the waterline to prevent water from draining out of the pipe.

3. After welding the tank, test it for leaks.

In summary: Always fill a fuel tank with water and ventilate it before welding.

SS40167,00000E6 -19-23JUL09-2/2

Fuel Gauges

There are two types of **electric** fuel gauges:

- **Balancing Coil Gauge**
- **Thermostatic Gauge**

Both types of electric fuel gauges have a sending unit located at the fuel tank and a display unit located in the instrument panel.

Balancing Coil Gauge

To understand how a balancing coil gauge works, keep in mind that current takes the path of least resistance.

The fuel level float (A) in the fuel tank is connected to a sliding contact (C) located in the sending unit, which is connected to ground. As the level of fuel moves up and down, the sliding contact moves across a variable resistor accordingly. When the tank is full of fuel (float up), the sending unit is at its maximum resistance (minimum current flow). As the fuel level drops, resistance in the sending unit drops proportionally. And should the fuel tank become empty (float down), the sending unit will have little resistance (maximum current flow).

The display unit (E) uses two coils: an E coil and an F coil. When the ignition switch (J) is moved to the ON (or RUN) position and the fuel tank is full, very little current from the vehicle's battery (K) will flow to ground through the sending unit because the sending unit is at maximum resistance. Instead, most of the gauge circuit's current flows through both the E coil and the F coil, then to a ground. This results in a magnetic field surrounding each coil.

Because the F coil is designed to produce a slightly larger (stronger) magnetic field than the E coil when maximum current is flowing through both coils (full tank), the larger magnetic field surrounding the F coil pulls pointer (F) toward the "F" mark on the display.

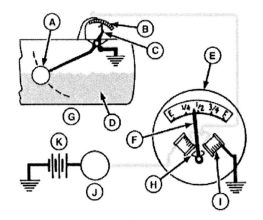

Balancing Coil Gauge

A—Float
B—Resistance
C—Sliding Contact
D—Fuel
E—Dash Unit
F—Pointer
G—Fuel Tank
H—Armature
I— Coil
J—Ignition Switch
K—Battery

As the fuel level drops, resistance in the sending unit drops proportionally, resulting in an increase flow of current through the sending unit to ground. The resulting split in current (through sending unit and F coil) allows the amount of current through the E coil to remain the same, but reduces the current flow through the F coil, thus reducing the magnetic field of the F coil. The reduced magnetic field of the F coil allows the magnetic field of the E coil to pull the pointer towards the "E" mark on the display relative to the movement of the fuel level float.

Continued on next page

SS40167,00000E7 -19-21JUL09-1/2

Thermostatic Fuel Gauge

A thermostatic fuel gauge uses a pair of thermostatic bimetal blades (C) that are individually wrapped in a heating coil (B). A thermostatic blade consists of two strips of different metals that are temperature sensitive and designed to bend with the change in temperature.

One thermostatic blade is associated with the fuel tank, the other with the fuel gauge. The heating coils for both blades are connected in series with the ignition switch (H) and the vehicle's battery (G).

The sending unit, located in the fuel tank, contains a float (I) that actuates a cam (J). The float rotates the cam (J) against a "grounding" lever, which in turn applies pressure against a contact point on the sending unit's thermostatic blade. This ground lever completes the electrical circuit for both heating coils, as long as the lever is making contact with the blade.

The more fuel (K) in the tank, the more pressure the cam puts against the "grounding" lever and the thermostatic blade. Conversely, as the fuel level drops, so drops the pressure against the lever and the blade.

When the ignition switch is moved to the ON (or RUN) position, current flows through both heating coils. Eventually, the heating coil associated with the sending unit heats to the point where its thermostatic blade bends and moves away from the ground lever, which in turn opens the circuit to stop the flow of current through the coils.

Once the thermostatic blade cools to the point where it again makes contact with the lever, current flows and the cycle is repeated. This heating and cooling cycle continues for as long as the key switch is in the ON position.

Because both heating coils are connected in series, the thermostatic blades will simultaneously heat and bend in a like amount. And because the pointer (D) and the thermostatic blade in the display unit (F) are linked, any movement in this blade will directly affect the position of the pointer.

When the fuel tank is full, the float is fully up and maximum tension is applied to the "grounding" lever. This

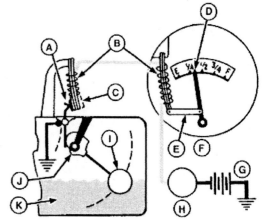

Thermostatic Type Gauge

A—Contacts
B—Heating Coils
C—Thermostat Blades
D—Pointer
E—Linkage
F—Dash Unit
G—Battery
H—Ignition Switch
I— Float
J— Cam
K—Fuel

in turn applies maximum force against the tank unit's thermostatic blades, thus requiring a higher temperature to open the contacts (A).

Because the contacts are closed longer during this time, a higher temperature is produced by the heating coils, which causes the dash unit's thermostatic blades to bend to maximum position. This then moves the pointer to indicate "full" on the gauge.

When the tank is nearly empty, the float is down and less tension is applied to the sending unit's thermostatic blades. As a result, only a small amount of heating is required to bend the blades enough to open the contacts. Thus the dash unit thermostatic blades bend only a little and the pointer indicates "empty" on the gauge.

NOTE: For diagnosing and repairing fuel gauges, refer to the FOS manual on "Electrical Systems" .

SS40167,00000E7 -19-21JUL09-2/2

Fuel Lines

Types of Diesel Fuel Lines

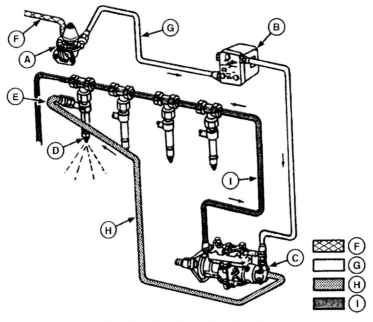

Fuel Lines in a Diesel Fuel System

A—Fuel Supply Pump
B—Fuel Filter
C—Fuel Injection Pump
D—Fuel Injection Nozzle
E—Fuel Pressure Line
F—Gravity Pressure
G—Fuel Supply Pump Pressure
H—Fuel Injection Pump Pressure
I— Return Fuel Pressure

Refer to the illustration for the following types of fuel lines.

Heavy-Weight Lines — For high-pressure fuel lines between the injection pump (C) and the injection nozzles (D).

Medium-Weight Lines — For low- to medium-pressure fuel lines between the fuel tank and the injection pump (C).

Light-Weight Lines — For little- to no-pressure fuel lines such as the leak-off fuel line between the injection nozzles and the fuel tank or between the fuel tank and the pump.

NOTE: Heavy fuel injection lines should be of the same length and inside diameter for proper injection timing in most systems.

SS40167,00000E8 -19-21JUL09-1/2

SERVICING FUEL LINES

Periodically inspect all fuel lines for loose connections, breaks, damage, or leaks.

Keep all line connectors tight enough to prevent leaks and weeping, but not so tight as to strip the threads. Tighten all connections until they are snug. To avoid bending the fuel lines or overtightening the connectors, use the one-hand technique to hold both wrenches, as shown.

When replacing an injection line, the new line must be identical in size, length, and type as the line being replaced. It is critical that the replacement line also have the identical **inside** diameter as the line being replaced.

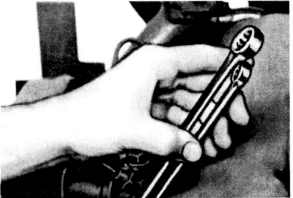

One-Hand Technique to Avoid Bending Fuel Lines

SS40167,00000E8 -19-21JUL09-2/2

Fuel Transfer Pumps

Simple fuel systems use gravity or air pressure to transfer fuel from the tank to the injection pump.

On modern high-speed diesel engines, a fuel transfer pump is normally used. This pump is driven by the engine and supplies fuel to the diesel system. The pump often has a hand primer lever used for bleeding air from the system.

Operation, Testing, and Servicing

For related information on mechanical and electrical fuel pumps, see Fuel Pumps in Chapter 4. Also consult the technical manual of the machine associated with a particular fuel pump for the actual testing and servicing procedures for that pump.

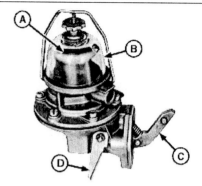

Fuel Transfer Pump (Diaphragm Type)

A—Fuel Strainer   C—Rocker Arm
B—Fuel Bowl   D—Hand Primer Lever

SS40167,00000E9 -19-21JUL09-1/1

Fuel Filters

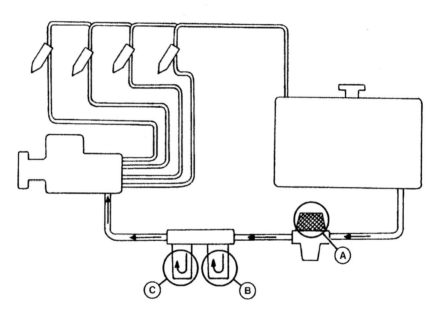

Diesel Fuel Filtration System

A—Filter Screen   B—Primary Filter   C—Secondary Filter

Fuel filtration is critical in diesel operation because:

- Diesel fuel tends to contain particles and moisture
- Injection parts are precision-fitted

As a result, diesel fuel must be filtered not once, but several times in most systems.

A typical filtration system may have three stages of progressive fuel filters:

1. **Filter screen (A) (at tank or transfer pump)** — Removes large particles.

2. **Primary filter (B)** — Removes most small particles.
3. **Secondary filter (C)** — Removes tiny particles.

Some systems also utilize a final filter as a watchdog for the system.

Most diesel filters have a trap to separate water and sediment from the fuel system. Traps must be drained periodically.

Continued on next page SS40167,00000EA -19-21JUL09-1/3

Types of Fuel Filtration

Fuel filtration removes suspended matter from the fuel. Some filters will also remove soluble impurities such as water.

Fuel filtration can be done three ways:

- **Straining**
- **Absorption**
- **Separation**

STRAINING (A) is a mechanical way of filtering fuel. It uses a screen, which blocks and traps particles larger than the openings in the screen. The screen may be of wire mesh for coarse filtering, or of paper or cloth for fine filtering.

ABSORPTION (B) is a way of trapping solid particles and water by passing the fuel through various materials such as cotton waste, cellulose, woven yarn, or felt.

SEPARATION is a method of removing water from fuel. By treating a paper filter with chemicals, water droplets are formed and separated into a water trap. (This type of filter also removes solid particles by one of the other methods mentioned previously.)

A—Straining
B—Absorption
C—Outlet
D—Air Vent
E—Inlet
F—Drain
G—Spring Clamp

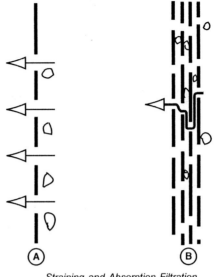

Straining and Absorption Filtration

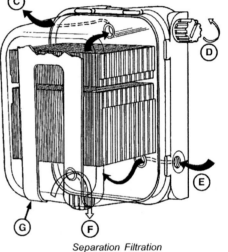

Separation Filtration

Continued on next page   SS40167,00000EA -19-21JUL09-2/3

Dual Filters

Dual filters function in two ways:

- **Series** — All fuel passes through one filter, then through another filter.
- **Parallel** — A portion of the fuel passes through multiple filters.

Advantages of Series and Parallel Filters:

- SERIES FILTERS — By using progressively finer filters, a series filter system can economically remove small particles from the fuel. This is because the coarser, primary filter removes the larger particles, which prevents the finer, secondary filter from becoming clogged.
- PARALLEL FILTERS — A parallel filter system can clean a larger flow of fuel.

Servicing Filters

Although fuel filters help clean fuel, they are not meant to clean dirty fuel from a bad fuel supply. Check and drain all water traps frequently. If there is an excessive accumulation of water or sediment in a trap, verify the fuel supply is clean.

Under normal operation, fuel filters should be changed at regular intervals as recommended in the operator's manual.

Fuel filters should be changed more often when operating in extremely dusty or dirty environments. Filters should also be replaced anytime there is a noticeable loss of engine power.

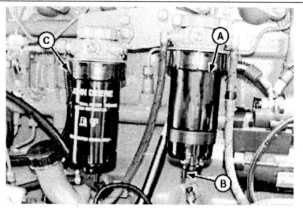

Series Filters

A—Primary Filter
B—Water/Sediment Drain
C—Secondary Filter

On series filters, normally the primary filter is changed more frequently than the secondary filter.

NOTE: When replacing a screw-on, canister-type fuel filter, fill the canister with clean diesel fuel before installing the filter. This will displace much of the air from the filter and quicken the fuel system bleeding process.

Air must be bled from a fuel system each time a new fuel filter is installed.

SS40167,00000EA -19-21JUL09-3/3

Fuel Injection Systems

The fuel injection system is the heart of a diesel engine. Let's repeat the tasks a fuel injection system must perform:

- Supply correct quantity of fuel
- Time the delivery of fuel
- Control the fuel delivery rate
- Break up or atomize the fuel
- Deliver fuel evenly through the cylinder

All of the above functions are performed by a fuel injection system. But as detailed in the following paragraphs, these functions are handled a bit differently by each of the various types of fuel injection systems.

TYPES OF INJECTION SYSTEMS

Some early diesel engines used compressed air to blow fuel into the cylinders. But the high-speed engines of today use a form of forced-liquid fuel injection.

Below are the major types of injection systems:

- **Early Common Rail**
- **Accumulator**
- **Injection Pump**
- **Electronic Unit Injector (EUI)**
- **Electronic Injector/High-Pressure Common Rail**

The EARLY COMMON RAIL system maintains constant pressure in a fuel supply rail that serves the injection nozzles. The pump and nozzles are mechanically actuated by rotating cams. The quantity of fuel injected into each nozzle depends on how long the nozzle valve stays open. This makes it difficult to accurately control fuel delivery at any engine speed or load.

In ACCUMULATOR systems, the quantity of fuel injected can be varied regardless of pump speed by using an adjustable spring or a hydraulic accumulator to control the fuel pressures.

With an INJECTION PUMP system, the fuel injection pump monitors, measures, and forces fuel at high pressure through the injection nozzles. There are two types of injection pumps: in-line and distribution. An in-line pump system offers reliable control of fuel dispersion at various speeds and loads.

The ELECTRONIC UNIT INJECTOR (EUI) fuel system uses a combination pump and nozzle for each cylinder. An EUI is either mechanically actuated by revolving cams or hydraulically actuated by high-pressure engine oil. Injection timing control and fuel quantity metering are electronically controlled by the engine control unit (ECU).

An ELECTRONIC INJECTOR/HIGH-PRESSURE COMMON RAIL system utilizes a high-pressure pump and a common fuel rail to deliver fuel at the required pressure for fuel injection. It also uses an electronic injector for each cylinder — with the operation of each injector controlled by a two-way electromagnetic valve (TWV). Because each TWV is controlled by the engine control unit (ECU) that is constantly monitoring the various engine parameters, each valve is controlled to deliver a precise amount of fuel at the exact moment necessary for smooth and efficient engine operation.

Let's take look at each system in more detail.

SS40167,00000EB -19-21JUL09-1/1

Early Common Rail Fuel System

The common rail system has a high-pressure, constant-stroke, constant-delivery pump that discharges into a common rail to which each injection nozzle (E) is connected by tubing.

A spring-loaded bypass valve on the header maintains constant pressure in the system, returning all excess fuel to the fuel tank.

The injection nozzles are operated mechanically and the amount of fuel injected into the cylinder at each power stroke is controlled by the lift of the needle valve (D). The operation of the injection system is detailed in the illustration.

The fuel cam (H) lifts the push rod (A); this motion is transferred to the needle valve through the rocker arm (B) and an intermediate lever (C). The fuel area above the needle valve seat is connected to the fuel header.

When the needle valve is lifted from its seat, fuel is forced into the combustion chamber through small holes drilled in the injector tip. Passing through these tiny holes, the fuel is divided into small streams, which are then atomized.

The amount of fuel injected is controlled by a wedge (F), which modifies the length of the fuel valve operating linkage to control fuel valve lash.

When the wedge is pushed in (made tighter), the length of the operating linkage is increased, resulting in the motion of the cam follower (G) transferring sooner to the push rod, which in turn moves the rocker arm sooner. The needle valve will open earlier and close later to increase its upward motion slightly — resulting in more fuel per cycle.

When the wedge is pulled out slightly (made looser), the length of the fuel valve operating linkage is decreased, causing the needle valve to lift later and close sooner — resulting in less fuel per cycle.

Fuel injection pressure can be adjusted to suit the operating conditions of the machine by changing the spring pressure in the bypass valve. Fuel injection pressures from 3200 psi (22 000 kPa) to 5000 psi

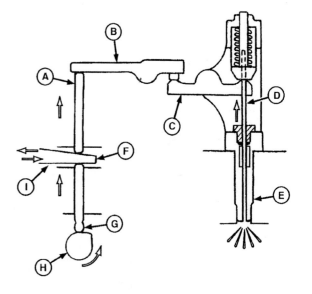

Common Rail Injection System

A—Push Rod
B—Rocker Arm
C—Lever
D—Needle Valve
E—Injection Nozzle

F—Control Wedge
G—Cam Follower
H—Fuel Cam
I— To Governor

(34 000 kPa) are used at rated load and speed, depending on the type of engine.

To reduce fluctuations with fuel injection pressure as a result of intermittent fuel discharge from the pumps or withdrawals by the fuel valves, the volume of fuel in the system may be increased by attaching a large-capacity fuel container (called an accumulator) to the header.

The early common rail system is not suitable for high-speed, small-bore engines because it is difficult to accurately control small quantities of fuel injected into the cylinders during each power stroke.

SS40167,00000EC -19-23JUL09-1/1

Accumulator System

The quantity of fuel injected per stroke can be varied, regardless of pump speed in the accumulator system.

Regulating is by spring pressure or by hydraulic pressure (accumulator).

A cam-driven pump sends fuel through the nozzle check valve (B) into the accumulator (C) as well as through the spill duct (H) into the spring chamber (G).

When the pump is too full and starts to bypass fuel, the nozzle check valve closes (as inlet pressure drops).

Fuel in the spring chamber is then vented through the spill duct back to the pump, while fuel now trapped in the accumulator passes through the discharge duct (D) to the nozzle (E).

Since accumulator pressure is higher than the nozzle opening pressure, the nozzle valve lifts and injects fuel.

Injection stops when the accumulator pressure drops to the nozzle closing pressure. The maximum injection pressure depends upon the accumulator volume and the quantity of fuel metered to it by the pump. Therefore, it is independent of the pump speed and nozzle orifices.

In summary, the quantity of fuel injected per stroke can be varied in the accumulator system.

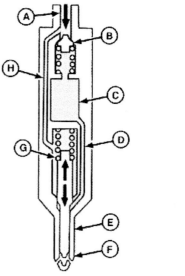

Nozzle in Accumulator System

A—Fuel Inlet
B—Check Valve
C—Accumulator
D—Discharge Duct
E—Nozzle
F—Spray Tip
G—Spring Chamber
H—Spill Duct

SS40167,00000ED -19-30JUN09-1/1

Fuel Injection Pump Systems

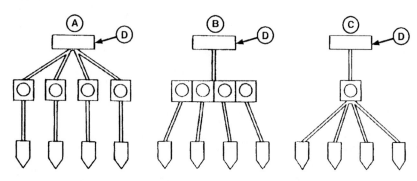

Electromechanically-Controlled Injection Pumps

A—Individual Pump and Nozzle for Each Cylinder

B—Pumps In Common Housing, Nozzles for Each Cylinder (In-Line)

C—Single Pump Serving Injection Nozzles for Several Cylinders (Distributor)

D—Fuel

Fuel injection pumps have evolved over the years. Early injection pumps were mechanical devices that injected varying amounts of fuel into each cylinder, regardless of the load conditions of the engine. Today's fuel injection pumps are electronically controlled to dispense precise amounts of fuel, as required by the operating conditions of the engine.

The most common fuel injection pumps are:

- **Individual pump and nozzle for each cylinder (A)**
- **Pumps in common housing, nozzles for each cylinder (in-line) (B)**
- **Single pump serving injection nozzles for several cylinders (distributor) (C)**

The individual pump and nozzle injection system is used primarily on smaller engines. The in-line and distributor type injection systems are widely used, especially on farm and industrial machines.

We will use the distributor pump and the in-line pump in our examples of how an injection pump works — first, the distributor type.

SS40167,00000EF -19-21JUL09-1/1

Distributor Injection Pump

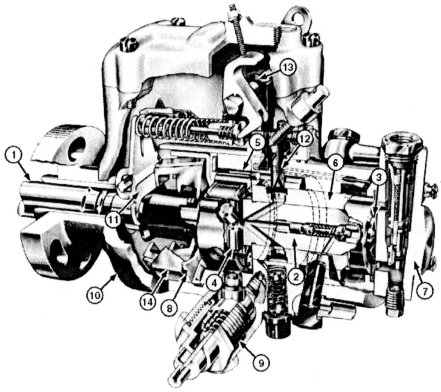

Distributor-Type Injection Pump
DXP00429 —UN—01JUL09

MAIN PARTS OF DISTRIBUTOR INJECTION PUMP

A distributor injection system normally uses one pump to distribute fuel to all cylinders. A typical distributor-type injection pump and its main parts are shown in the illustration.

The main rotating parts of the pump are the drive shaft (1), distributor rotor (2), and transfer pump (3). All rotate on a common axis.

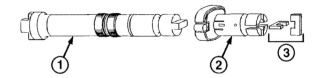

Main Rotating Parts of Distributor Injection Pump

1— Drive Shaft
2— Distributor Rotor
3— Transfer Pump
4— Pumping Plungers
5— Internal Cam Ring
6— Hydraulic Head
7— End Plate
8— Governor
9— Automatic Advance
10— Housing
11— Governor Arm
12— Metering Valve
13— Shutoff Lever
14— Governor Weight Retainer

Continued on next page

SS40167,00000F7 -19-22JUL09-1/15

DXP00430 —UN—23DEC08

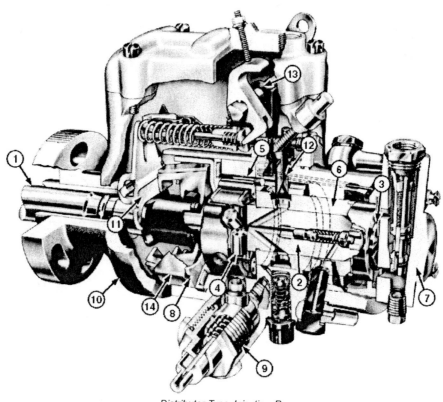

Distributor-Type Injection Pump

1— Drive Shaft
2— Distributor Rotor
3— Transfer Pump
4— Pumping Plungers
5— Internal Cam Ring
6— Hydraulic Head
7— End Plate
8— Governor
9— Automatic Advance
10— Housing
11— Governor Arm
12— Metering Valve
13— Shutoff Lever
14— Governor Weight Retainer

HOW A DISTRIBUTOR INJECTION PUMP WORKS

In the illustration, the drive shaft (1) engages the distributor rotor (2) located in the hydraulic head (6). The drive end of the rotor has two cylinder bores, each containing two plungers (4).

The plungers are forced toward each other by the cams along the inside of the cam ring (5). This action pumps the fuel. The cam ring has as many lobes as there are cylinders. (However, a three-cylinder engine uses a pump with six cam rings.)

The transfer or supply pump (3) in the end of the rotor opposite the pumping cylinders is a positive-displacement, vane-type pump. The pump is covered by the end plate (7).

The distributor rotor (2) has two angled inlet passages for charging, along with an axial bore to serve all outlets.

The hydraulic head (6) contains the bore in which the rotor revolves the metering valve bore (12), the charging

ports, and the head outlets. These outlets are connected through appropriate fuel line fittings to the injection lines, which lead to the nozzles.

The end plate (7) covers the transfer pump on the outer end of the hydraulic head. This assembly houses the fuel inlet connection, fuel strainer, and transfer pump pressure regulating valve.

The pump contains its own mechanical governor (8) capable of close-speed regulation.

The action of the weights in their retainer (14) is transmitted through a sleeve to the governor arm (11) and through a positive linkage to the metering valve (12).

External linkage connected to the shutoff lever (13) can be used to stop the flow of fuel to the metering valve.

Continued on next page

SS40167,00000F7 -19-22JUL09-2/15

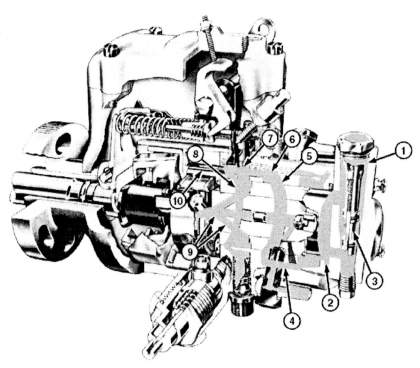

Fuel Flow in Distributor Injection Pump

1— Inlet Strainer
2— Vane-Type Fuel Transfer
 Pump
3— Regulating Valve

4— Drilled Passage
5— Annulus
6— Connecting Passage

7— Metering Valve
8— Charging Ring
9— Inlet Passages

10— Rollers

FUEL FLOW

To understand the operation, let's trace the fuel flow through the pump during a complete pump cycle.

Fuel is drawn from the fuel tank into the pump through the inlet strainer (1) by the vane-type fuel transfer pump (2).

Since the transfer pump displacement greatly exceeds the injection requirements, a large percentage of fuel is bypassed through the regulating valve (3) back to the inlet side of pump. The flow of this positive-displacement pump increases with speed, and the regulating valve is designed so transfer pump pressure also increases with speed.

Fuel, under transfer pump pressure, is forced through the drilled passage (4) in the hydraulic head into the annulus (5). It then flows around the annulus to the top of the sleeve and through a connecting passage (6) to the metering valve (7).

The radial position of the metering valve, controlled by the governor, regulates the flow of fuel into the charging ring (8), which includes the charging ports.

As the rotor revolves, two inlet passages (9) align with two charging ports in the hydraulic head, allowing fuel to flow into the pumping cylinders.

With further rotation, the inlet passages move out of alignment with the charging ports and into alignment with one of the discharge ports. Rollers (10) on the pump plungers meet the lobes on the inside surface of the cam ring, and subsequent rotation forces the plungers together. Fuel trapped between the plungers is forced from the pumping cylinder, out of the discharge port corresponding to an injector, through the injector line, and into the injector.

The injection pump is self-lubricating. As fuel at the transfer pump reaches the charging ports, slots on the rotor shank allow fuel and any trapped air to flow into the pump-housing cavity. In addition, an air bleed in the hydraulic head connects the outlet side of the transfer pump with the pump housing cavity. This allows air and some fuel to bleed back to the fuel tank by way of the return line. Any fuel bypassed helps lubricate the internal parts.

Continued on next page   SS40167,00000F7 -19-22JUL09-3/15

DXP00431 —UN—23DEC08

100709

PN=236

EXHAUST SMOKE REDUCTION FOR TURBOCHARGED SYSTEMS

To help reduce exhaust smoke on turbocharged engines, the engine can be equipped with an **aneroid**. An aneroid is an air pressure sensing device that is normally mounted on the housing of the fuel pump governor. The aneroid controls the flow of fuel into the in-line- or distributor-type injection pump and is a fuel-flow inhibitor regulated by mechanical or hydraulic pressure. For more information on aneroids, see Governors in this chapter.

If air pressure in the manifold is insufficient during a rapid acceleration of a turbocharged engine, the aneroid limits the fuel supply. This prevents a rich fuel mixture from entering the cylinders (due to an insufficient air supply from the intake turbocharger) — thereby reducing exhaust smoke. Aneroids also reduce exhaust smoke during engine startup.

A—Intake Manifold Pressure Inlet   B—Adjusting Shaft

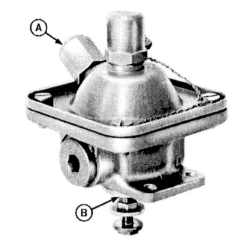

Aneroid

SS40167,00000F7 -19-22JUL09-4/15

CHARGING AND DISCHARGING

The illustrations show the fuel flow through the pump during the charging and discharging cycles.

Charging Cycle

As the distributor rotor (C) revolves, the angled inlet passage (E) aligns itself with the charging ports along the inside edge of the charging ring (A). The metering valve (B) directs fuel from the transfer pump into the pumping cylinders, forcing all plungers (F) apart.

The plungers move outward at a distance proportional to the amount of fuel required to inject fuel into the cylinder at the next stroke. If only a small quantity of fuel is injected into the pumping cylinders, when idling for example, the plunger will move very little. Maximum plunger travel with maximum fuel delivery is limited by adjusting leaf springs, which contact the edge of roller shoes. Only when the engine is operating at full load will the plungers move to the outward-most position.

For the charging cycle, the angled inlet passages in the rotor are aligned with the ports in the charging ring, but the rotor discharge port is not aligned with the head outlet.

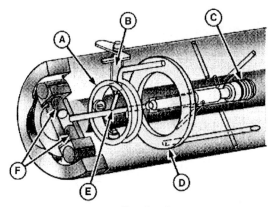

Charging Cycle

A—Charging Ring
B—Metering Valve
C—Distributor Rotor

D—Annulus in Hydraulic Head Barrel
E—Inlet Passage
F—Plungers

Also note that in both charging and discharging cycles, the rollers are between the cam lobes.

Continued on next page   SS40167,00000F7 -19-22JUL09-5/15

Discharge Cycle

As the rotor continues to revolve, the inlet passages move out of alignment with the charging ports. For a brief interval, the fuel is trapped until the rotor discharge passage aligns with one of the head outlets.

As this alignment takes place, both sets of rollers contact the cam lobes and are forced together. During this stroke, the fuel trapped between the plungers is forced through the axial passage of the rotor and flows through the rotor discharge passage (A) to the head outlet (B) and ultimately to the injection line.

Delivery to the injection line continues until the rollers pass the highest point on the cam lobe and are allowed to move outward.

Pressure in the axial passage, the outlet, and the injection line is then relieved, allowing the injection nozzle to close.

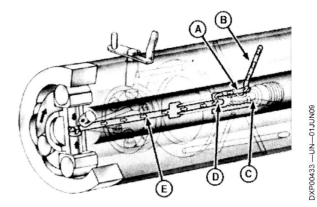

Discharge Cycle

A—Discharge Passage
B—Head Outlet
C—Delivery Valve Stop
D—Delivery Valve
E—Passage

SS40167,00000F7 -19-22JUL09-6/15

Delivery Valve

The major function of a delivery valve (F) is to rapidly decrease injection line pressure once the injection of fuel is complete. Line pressure that is less than nozzle closing pressure is referred to as a controlled line retraction.

The reduction in pressure causes the nozzle valve to return rapidly to its seat. This sharp fuel delivery cut-off prevents fuel from flowing into the combustion chamber. The delivery valve is located in the center of the distributor rotor (B). It requires no seat — only a shoulder to limit travel.

Because the same delivery valve performs the function of retraction for each line, valve retraction does not vary between cylinders. This results in a smooth-running engine at all loads and speeds.

When the injection of fuel begins, fuel pressure moves the delivery valve. Then fuel from the delivery valve's displacement (A) flows into an enlarged cavity of the rotor, occupied by the delivery valve spring (E). This displaces a similar volume of fuel located in the spring cavity, before fuel is delivered through the valve port.

At the end of the injection cycle, the pressure on the plunger side of the delivery valve is quickly reduced by allowing the cam rollers to drop into a retraction step on the cam lobes. The volume of cam retraction is slightly greater than the delivery valve retraction volume.

As the valve returns to its closed position, its displacement is removed from the spring cavity. And because the rotor

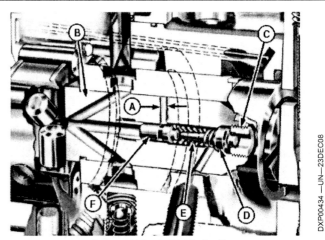

Delivery Valve in Distributor Pump

A—Delivery Valve's Displacement
B—Distributor Rotor
C—Retainer Screw
D—Delivery Valve Stop
E—Delivery Valve Spring
F—Delivery Valve

discharge port remains partially aligned, fuel rushes out of the injection line to fill the void of the retreating delivery valve.

The rotor ports then close completely, and the remaining fuel in the injection line is trapped.

Continued on next page

SS40167,00000F7 -19-22JUL09-7/15

RETURN OIL CIRCUIT

Fuel under transfer pump pressure is discharged into an annular cavity located in the hydraulic head.

The upper half of this cavity connects with a vent passage. Its volume is restricted by a wire to prevent pressure loss.

The vent passage is located behind the metering valve bore and connects with a short vertical passage entering the governor linkage compartment.

Should air enter the transfer pump because of suction-side leaks, it immediately passes to the air vent cavity (C) and then to the vent passage (B) as shown.

Air and a small quantity of fuel then flow from the housing to the fuel tank by way of the return line.

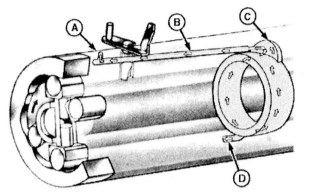

Return Oil Circuit

A—To Housing Interior
B—Air Vent Passage
C—Air Vent Cavity
D—To Metering Valve

END PLATE OPERATION

The end plate has three basic functions:

- Provides fuel inlet passages and houses the pressure regulating valve.
- Covers the transfer pump.
- Absorbs end thrust of drive and governor.

The illustration shows the operation of the pressure-regulating valve while the pump is running.

Pressurized fuel from the transfer pump forces the piston up the sleeve and against the regulating spring (9). As pressure increases, the regulating spring is compressed slightly until the lower edge of the regulating piston starts to uncover port B (12). Since the pressure on the piston is opposed by the regulating spring, the delivery pressure of the transfer pump is controlled by the spring rate and size and number of regulating ports.

High-pressure relief port C (11), just above the regulating port, prevents excessively high transfer pump pressures if the engine or pump operates at an excessive speed.

During hand priming, fuel flows into the inlet side of the transfer pump through port A (1). Priming the remainder of the system is accomplished when the pump is rotated by the engine starting motor or on a test bench.

1— Port A
2— Inlet Screen
3— Adjusting Plug
4— Pressure Regulating Sleeve
5— Regulating Piston
6— Piston Seal
7— Transfer Pump Liner
8— Transfer Pump Blades
9— Regulating Spring
10— Orifice
11— High-Pressure Relief Port C
12— Port B

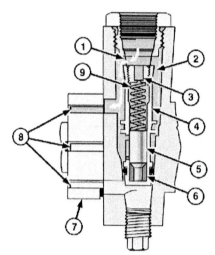

End Plate on Distributor Pump

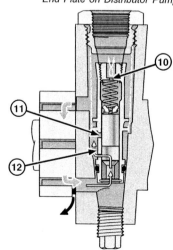

Pressure Regulating Valve for Viscosity Compensation

Continued on next page   SS40167,00000F7 -19-22JUL09-9/15

VISCOSITY COMPENSATION

The distributor pump must work equally well with different fuels and varying temperatures that affect fuel viscosity. A feature of the pressure regulating valve offsets pressure changes caused by viscosity differences. Located in the bottom of the spring adjusting plug is a thin plate containing a sharp-edged orifice (10). This orifice allows fuel leakage past the piston to return to the inlet side of the pump.

Flow through such a short orifice is unaffected by viscosity changes. For this reason, fuel pressure on the spring side of the piston will vary with viscosity changes. The pressure exerted on top of the piston is determined by the flow past the designed clearance between the piston and the sleeve.

With cold or heavy fuels, very little leakage occurs past the piston, while flow through the adjusting plug is a function of the orifice size. The resulting downward pressure on the piston is minimal.

Leakage past a piston is higher when hot or light fuels are used. Spring cavity pressure is also higher because the flow through the short orifice remains the same as it does with cold fuel. Therefore, downward pressure, assisting the regulating spring, positions the piston so less regulating port area is uncovered below the piston.

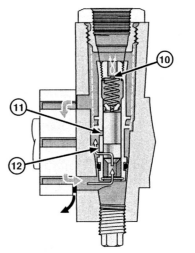

Pressure Regulating Valve for Viscosity Compensation

10— Orifice
11— High-Pressure Relief Port C
12— Port B

Pressure is controlled and may actually over-compensate to offset other leakage in the pump, caused by the use of thinner fuels.

Continued on next page   SS40167,00000F7 -19-22JUL09-10/15

CENTRIFUGAL GOVERNOR

In the centrifugal governor, the movement of the flyweights (G) against the governor thrust sleeve (H) rotates the metering valve (E).

This rotation varies the alignment of the metering valve opening with the passage from the transfer pump, thus controlling fuel flow to the engine.

This type of governor derives its momentum from flyweights pivoting on their outer edge in the retainer. Centrifugal force tips them outward, moving the governor thrust sleeve against the governor arm (J), which pivots on the knife edge of the pivot shaft (I) and, through a simple positive linkage, rotates the metering valve.

The force on the governor arm, caused by flyweights, is balanced by the compression-type governor spring. To regulate engine speed, the governor spring (A) is manually controlled by the throttle shaft linkage.

A slow idle spring (C) is provided for more sensitive regulation at the low speed range.

The limits of throttle travel are set with slow idle and fast idle adjusting screws.

A light tension spring takes up any slack in the linkage joints and allows the stopping mechanism to close the metering valve without having to overcome the governor spring force. Only a very light force is required to rotate the metering valve to the closed position.

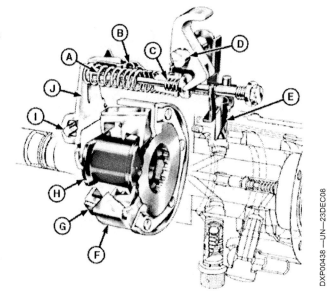

Centrifugal Governor

A—Governor Spring
B—Linkage Hook
C—Slow Idle Spring
D—Throttle Shaft
E—Metering Valve
F—Weight Retainer
G—Flyweight
H—Thrust Sleeve
I— Pivot Shaft
J— Governor Arm

Continued on next page

SS40167,00000F7 -19-22JUL09-11/15

AUTOMATIC LOAD ADVANCE

Pumps equipped with automatic load advance permit the use of a simple hydraulic servo-mechanism, powered by fuel pressure from the transfer pump, to advance injection timing.

Transfer pump pressure operates the advance piston against spring pressure, along a predetermined timing curve.

The purpose of the speed advance device, which responds to speed changes, is to advance the timing for the best engine power and combustion throughout the speed range.

Movement of the cam in the pump housing is limited by the action of the automatic advance bearing piston against the cam advance pin.

During cranking, the cam is in the retard position, since the force exerted by the advance spring is greater than that of transfer pump pressure. As the engine speed and transfer pump pressure increase, the pressurized fuel entering the advance housing behind the power piston moves the cam.

The amount of advance is limited by the length of the advance pistons. A ball check is provided to offset the normal tendency of the cam to return to the retard position during injection.

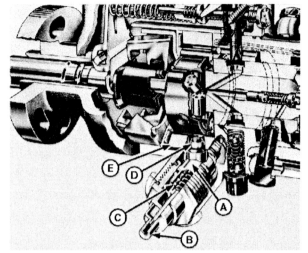

Automatic Advance Mechanism

A—Piston
B—Advance Trimmer Screw
C—Spring
D—Advance Pin
E—Pump Cam

SS40167,00000F7 -19-22JUL09-12/15

Automatic advance fuel circuit

Transfer pump pressure forces fuel through the drilled passage (A) in the hydraulic head into the annular ring (B).

Fuel then flows around and to the top of the annular ring, where it aligns with the bore leading to the metering valve. The metering valve is designed to allow a quantity of fuel to flow into a second annular ring (C), which aligns at the bottom with the bore of the advance clamp screw assembly.

As transfer pump pressure increases, the ball check is lifted off its seat, allowing fuel to pass through the clamp screw into the passage located behind the power piston.

A—Drilled Passage
B—Annular Ring
C—Second Annular Ring
D—Metering Valve
E—Cam Ring
F—Orifice
G—Pin
H—Piston
I— Ball Check Assembly

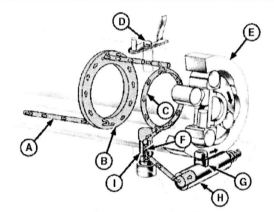

Automatic Advance Fuel Circuit

Continued on next page

SS40167,00000F7 -19-22JUL09-13/15

TORQUE CONTROL

Torque is defined as the rotational strength of an engine.

Since torque increases with load, a predetermined point at which maximum torque is desired may be selected for the engine. As engine speed decreases, torque increases toward this pre-selected point. This desirable feature is called "torque reserve."

Three basic factors that affect torque reserve are:

- Metering valve opening area
- Time allowed for charging
- Transfer pump pressure curve

The only control between engines for purposes of establishing a desired torque curve is the transfer pump pressure curve, since the other factors are common to all engines.

Torque control works as follows:

When the engine is operating at a fast idle speed without a load, the quantity of fuel delivered is controlled only by the metering valve as positioned by the governor.

NOTE: At this point, the torque screw (A) and maximum fuel adjustment have no effect.

As load is applied, the quantity of fuel delivered remains dependent on governor action and metering valve position until full load governed speed is reached. At this point, further opening of the metering valve is prevented by its contact with the previously adjusted torque screw (A). Thus, the amount of fuel delivered at full load governed speed is controlled by the torque screw (A).

Normally, the maximum fuel flow at full speed with a full load is limited by the maximum roller-to-roller setting of the injection pump.

Torque Control Screw

A—Torque Screw   **B—Automatic Speed Advance Trimmer Screw**

Where a torque screw is used, fuel flow is restricted at full load and full speed by limiting the metering valve opening. When an additional load is applied to the engine, the engine slows down. When the engine slows down, the injection pump also slows down, allowing more time between pumping strokes for fuel to flow past the metering valve.

At this slower operating speed, the pumping plungers are forced out to the maximum roller-to-roller setting and maximum fuel is delivered.

Torque adjustment can be done properly only during a dynamometer or bench test. Never attempt it on a unit without means of determining actual fuel delivery.

An external screw is provided to adjust for proper engine torque reserve.

SS40167,00000F7 -19-22JUL09-14/15

ELECTRICAL SHUTOFF

The electrical shutoff device is located under the governor control cover. It is an energized-to-run type of device.

De-energizing the coil allows the shut down coil spring to release the armature. The lower end of the armature moves the governor linkage hook, rotates the metering valve to the closed position, and cuts off the fuel supply.

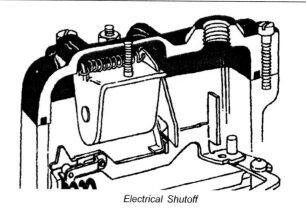

Electrical Shutoff

SS40167,00000F7 -19-22JUL09-15/15

In-Line Injection Pump

In-Line Injection Pump

A—Aneroid	D—Leak-Off Line	G—Sediment Bowl	J— Control Rack
B—Individual Pumping Element	E—Pump Housing	H—Fuel Transfer Pump	K—Governor
C—Injection Line	F—Hand Primer	I— Camshaft	

The in-line injection pump uses an individual pump for each cylinder. The individual pump elements (B) are usually mounted together in "packs" or "in-line" as shown. Because they can supply larger volumes of fuel, in-line pumps are typically found on large diesel engines.

Main Parts of the Pump

Main parts of the pump as shown in the illustration are: individual pumping element, pump housing (E), fuel transfer pump (H), camshaft (I), control rack (J), and governor (K).

The pump shown is a single-acting, plunger-type pump driven by the engine.

Fuel Flow

Fuel flows from the fuel tank to the fuel transfer pump. The transfer pump sends low-pressure fuel to external fuel filters and then back to the injection pump.

Each pumping element then meters the fuel at high pressure to its engine cylinder.

A hand primer (F) is located on the fuel transfer pump. This primer is operated by hand to pump fuel when bleeding the system after the fuel lines have been disconnected.

Continued on next page SS40167,00000F5 -19-23JUL09-1/11

Fuel Transfer Pump .

The fuel transfer pump (M) supplies fuel to the injection pump at all times. Along the way, fuel is pumped through filters to ensure only clean fuel reaches the injection pump.

The transfer pump is a single-acting, piston-type pump mounted on the side of the injection pump. It is driven by a cam (L) on the injection pump camshaft (J). All fuel flows through a preliminary filter located in the transfer pump sediment bowl.

A hand primer (O) is attached to the transfer pump housing. The pump is operated by unscrewing a knurled knob and working the knob up and down. When the knob is pulled out, the pump cylinder fills with fuel. When the knob is pushed in, fuel is forced through the filters and into the injection pump.

Injection Pump Fuel Flow

The single-action, plunger-type injection pump has an engine-driven camshaft that rotates half as fast as the engine. Roller cam followers, riding on the camshaft lobes, operate the plungers to supply high-pressure fuel through delivery valves to the injection nozzles.

A governor-operated control rack is connected to the plungers to regulate the amount of fuel to the engine.

Engine lubricating oil is piped into the injection pump to provide splash lubrication to the working parts.

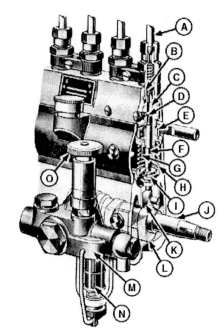

DXP00444 —UN—01JUN09

Elements of an In-Line Injection Pump

A—Delivery Line
B—Delivery Valve
C—Barrel
D—Plunger
E—Control Rack
F—Control Sleeve
G—Plunger Vane
H—Spring
I— Spring Plate
J—Camshaft
K—Cam Follower
L—Cam
M—Fuel Transfer Pump
N—Sediment Bowl
O—Hand Primer

Continued on next page   SS40167,00000F5 -19-23JUL09-2/11

Pumping Elements

A pumping element, consisting of a barrel (C) and a matching plunger (D), supplies fuel to an associated engine cylinder. The plunger is precision-fitted to the barrel by way of lapping to ensure a tight fit between the two components [0.0001 in. (0.002 mm)]. And because of this tight fit, special seals are not required. But if either part ever needs servicing, it will be necessary to replace the entire element.

Each plunger is operated at a constant stroke; that is, they each move when a cam (L) actuates each plunger.

To vary the amount of fuel delivered per stroke, for satisfying various load demands, the upper part of the plunger includes a vertical channel that extends from the top face to an annular groove, the top edge of which is milled in the form of a helix (called the control edge). On top of the plunger is a machined notch called the retarding notch, which retards the injection timing for starting the engine.

The barrel has either a single or double control port, depending on the design. An annular groove inside the barrel routes any fuel leakage, between the plunger and barrel, back to the fuel gallery. The top of the barrel is closed by a spring-loaded valve called the delivery valve (B).

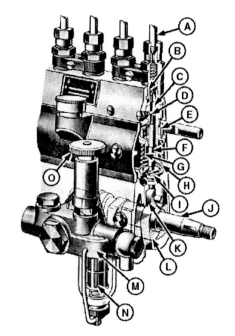

Elements of an in-Line Injection Pump

A—Delivery Line	I— Spring Plate
B—Delivery Valve	J—Camshaft
C—Barrel	K—Cam Follower
D—Plunger	L—Cam
E—Control Rack	M—Fuel Transfer Pump
F—Control Sleeve	N—Sediment Bowl
G—Plunger Vane	O—Hand Primer
H—Spring	

SS40167,00000F5 -19-23JUL09-3/11

OPERATION OF PUMPING ELEMENTS

When a plunger is set at the bottom of its stroke, fuel fills the vertical slot, the cut-away area below the helix, and the space above the plunger. Fuel flows into these areas by way of the fuel gallery, through ports located in the barrel.

As the plunger moves upward, it closes off the barrel ports and discharges the trapped fuel that is located in the pressure area. This trapped fuel flows to the injection nozzle, through a delivery valve and fuel line. Delivery of fuel stops when the control edge of the helix exposes the control port. Fuel flows out through the vertical slot and annular groove, back to the fuel gallery — which effectively bypasses the pumping portion of the plunger.

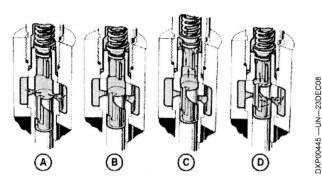

Plunger Operation (Maximum Fuel Delivery Rate)

A—Before Delivery (BDC)	C—Delivery
B—Beginning of Delivery	D—End of Delivery

Continued on next page   SS40167,00000F5 -19-23JUL09-4/11

Plunger Positions

NO FUEL DELIVERY

Fuel delivery stops when the plunger is rotated to a position where the vertical slot in the plunger aligns with the control port. Since the vertical slot prevents the control port from being covered, pressure cannot build up and fuel cannot flow to the injection nozzles.

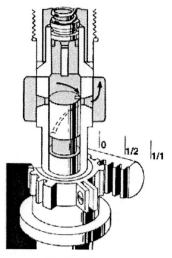

No Fuel Delivery

SS40167,00000F5 -19-23JUL09-5/11

PARTIAL FUEL DELIVERY

Partial fuel delivery occurs at any rotational position of the plunger, between no delivery and maximum delivery, depending on the position of the helix in relation to the control port.

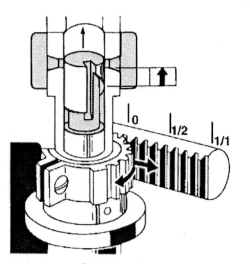

Partial Fuel Delivery

Continued on next page

SS40167,00000F5 -19-23JUL09-6/11

MAXIMUM FUEL DELIVERY

Maximum fuel delivery occurs when the plunger is rotated to a position where the control port is covered by the plunger as it moves upward, for the greatest possible distance (effective stroke) permitted by the control rack.

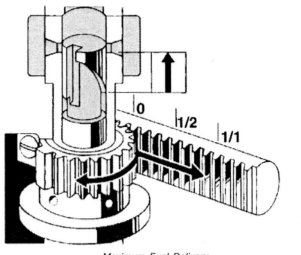

Maximum Fuel Delivery

SS40167,00000F5 -19-23JUL09-7/11

EXCESS FUEL DELIVERY

For excess fuel delivery, the plunger (D) is rotated to a point where the extreme edge of the helix (C) or annular groove provides the maximum effective pumping stroke permitted by design. This excess fuel position should not be confused with the maximum fuel position, since excess fuel is only available when starting the engine, while maximum fuel is used when the engine is running.

To obtain the excess fuel position, move the speed control lever to the slow idle position while the engine is stopped. Doing so allows the rack to move to its extreme forward position at cranking speed. This moves the plunger to a position to deliver excess fuel for starting.

Excess fuel aids in engine starting, especially when the engine is cold. As soon as the engine has started, the control rack moves the plungers from the excess fuel position to a fuel delivery position corresponding to the settling of the speed control lever.

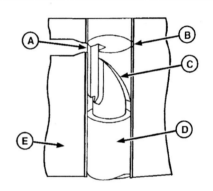

Excess Fuel Delivery and Retarding Notch (Exaggerated)

A—Retarding Notch
B—Control Port
C—Helix
D—Plunger
E—Barrel

Continued on next page

SS40167,00000F5 -19-23JUL09-8/11

Retarding Notch

A retarding notch (A) on the upper face of the plunger (D) also aids in starting a cold engine. This notch is aligned with the control port (B) when starting the engine, which means the plunger must move farther upward before the control port will be completely closed. This delay in port closing results in the desired timing retardation.

When the retarding notch is aligned with the control port, the plunger helix (C) is designed to be in the excess fuel position. This way, both excess fuel and retardation of injection timing are working together when starting the engine.

A—Retard Notch
B—Control Port
C—Helix
D—Plunger
E—Barrel

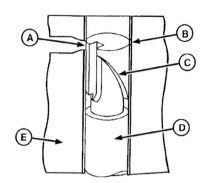

Excess Fuel Delivery and Retarding Notch (Exaggerated)

DXP00449 —UN—01JUN09

SS40167,00000F5 -19-23JUL09-9/11

Control Rack and Sleeve

A control rack (C) is connected to the governor by way of linkage that controls the rack used to regulate the speed of the engine.

A notched sleeve (D) is fitted over a barrel to mesh with the teeth on the plunger (B). As the sleeve is rotated by the control rack, the plunger is rotated as well.

Some fuel transfer pumps have a clamped-on tooth segment that meshes with the teeth on the control rack. Other transfer pumps have a pin at the upper end of the control sleeve, which engages a slot in the control rack.

Also affecting the travel of the control rack is the starting fuel control shaft. It permits the rack to move to the excess fuel position described earlier.

A—Delivery Valve
B—Plunger
C—Control Rack
D—Control Sleeve

Control Rack, Sleeve, and Delivery Valve

DXP00450 —UN—23DEC08

Continued on next page

SS40167,00000F5 -19-23JUL09-10/11

Delivery Valve

The delivery valve is guided by its stem, which is located in the valve housing.

During the fuel delivery stroke, the stem pushes the valve off its seat to send fuel past the face of the valve and into the fuel delivery line.

When the helix on the pump plunger uncovers the control port, pressure in the pump barrel drops suddenly. The pressure in the delivery line and valve spring causes the valve to close. When this occurs, movement of the valve relieves the pressure in the delivery line, which in turn prevents the injection nozzle from dribbling fuel.

To do this, the delivery valve includes a relief plunger that fits into the valve holder (C). When the delivery stroke ends, the valve starts to resume its seated position and the relief plunger moves into the bore of the valve holder to seal the delivery line from the pressure chamber. Once the relief plunger enters the bore of the valve housing, the valve will seat firmly.

At this point, the space for fuel in the delivery line is increased by an amount equal to the volume of the relief plunger. The effect of this increase in volume is a sudden drop of pressure in the delivery line, resulting in the injection nozzle closing instantly.

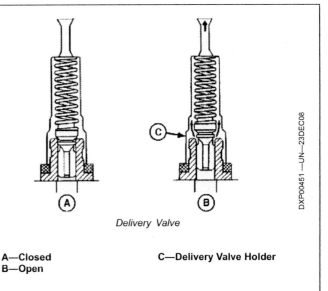

Delivery Valve

A—Closed
B—Open

C—Delivery Valve Holder

NOTE: *For service instructions on a particular in-line injection pump, refer to the technical manual for that pump or contract an authorized diesel repair station.*

SS40167,00000F5 -19-23JUL09-11/11

Governors

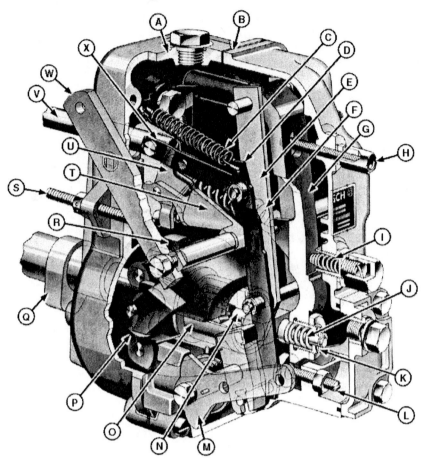

Mechanical Governor in Injection Pump

A—Governor Housing
B—Governor Cover
C—Starting Spring
D—Fulcrum Lever
E—Guide Lever
F—Governor Main Spring
G—Tensioning Lever
H—Shutoff or Idling Stop
I— Supplementary Idling Spring
J— Torque Capsule
K—Shims
L—Full Load Stop (Delivery Rate)
M—Shutoff Lever
N—Link
O—Thrust Sleeve
P—Flyweight
Q—Injection Pump Camshaft
R—Carrier
S—Maximum Speed Stop
T—Swiveling Lever
U—Rocker
V—Control Rod
W—Control Lever
X—Link Member

INTRODUCTION

A governor is a vital component of a diesel fuel system. It maintains a nearly constant engine speed, at any point between idling and maximum speed settings.

The desired speed is set by the operator, using the speed control lever or foot pedal. The governor, by adjusting the position of the injection pump control rack, maintains a nearly constant speed by varying the amount of fuel supplied to the engine to satisfy varying load demands.

There are two basic types of governor systems:

• **Mechanical Governor System**
• **Electronically Controlled System**

The **mechanical governor** shown in the illustration is a centrifugal, variable-speed type. It uses weights and springs to move the fuel injection pump rack. The mechanical governor is capable of maintaining a steady speed in the range between idle and maximum speed settings.

The **electronically controlled governor** determines the precise amount of fuel to deliver to the engine, based on information it receives from various sensors. It then controls an actuator solenoid to move the fuel injection pump rack.

An electronically controlled governor provides a more precise control of the injection pump than a mechanical governor. For more information on the electronically controlled governor, see Electronically Controlled Fuel Injection Pump in this chapter.

Now let's take a look at the mechanical governor in more detail.

Continued on next page
SS40167,00000F4 -19-21JUL09-1/12

DXP00452 —UN—23DEC08

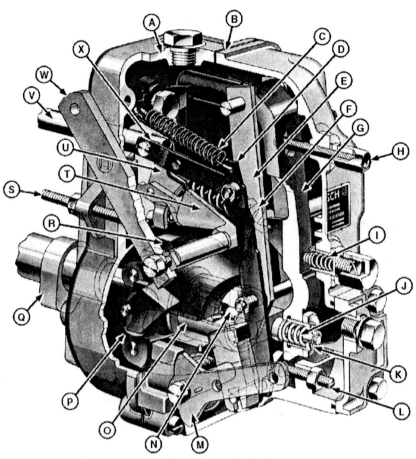

Mechanical Governor in Injection Pump

A—Governor Housing
B—Governor Cover
C—Starting Spring
D—Fulcrum Lever
E—Guide Lever
F—Governor Main Spring

G—Tensioning Lever
H—Shutoff or Idling Stop
I— Supplementary Idling Spring
J—Torque Capsule
K—Shims
L—Full Load Stop (Delivery Rate)

M—Shutoff Lever
N—Link
O—Thrust Sleeve
P—Flyweight
Q—Injection Pump Camshaft
R—Carrier

S—Max. Speed Stop
T—Swiveling Lever
U—Rocker
V—Control Rod
W—Control Lever
X—Link Member

MECHANICAL GOVERNOR DESCRIPTION AND OPERATION

DESCRIPTION

The governor is mounted near the back of the injection pump housing. A governor flyweight assembly (P) is mounted on the injection pump camshaft (Q). Centrifugal force created by the rotation of the camshaft causes the flyweights to move outward. The main spring (F) of the governor counters the force of the flyweights. Movement of the flyweight assembly, acting through the governor linkage, moves the injection pump control rack to provide the desired speed regulation.

The governor is completely enclosed to permit splash lubrication of the working parts, using oil in common with the injection pump.

An operating lever overrides the governor and shuts off fuel delivery to the engine by moving the control rack to the STOP position.

OPERATION

As engine speed increases, the governor flyweights (P) move outward until the force created by the whirling weights equals the counterforce of the governor main spring (F).

As engine speed decreases, centrifugal force diminishes, resulting in the spring forcing the weights to swing inward.

Movement of the flyweights is transmitted to the injection pump control racks through the governor linkage to obtain fuel delivery corresponding to the desired speed setting.

Movement of the governor linkage under varying engine requirements is described on the following pages.

Continued on next page   SS40167,00000F4 -19-21JUL09-2/12

Starting the Engine

With the engine stopped, and the speed control lever (A) in the slow idle position and then advanced slightly, the starting spring (B) will pull the control rack (F) to the excess fuel position. At the same time, the tensioning lever is moved up against the full-load stop, which also moves the guide lever, knuckle, and thrust sleeve (D) forward. The flyweights (E) then come to rest against the thrust sleeve (inner-most position).

While the starter is cranking the engine, the injection pump begins supplying excess fuel to the engine.

Once the engine starts, the centrifugal force produced by the whirling flyweights overcomes the starting spring tension (even before idling speed is reached).

The engine speed increases until the centrifugal force of the flyweights and the governor main spring are balanced.

A—Speed Control Lever Advanced for Starting
B—Starting Spring
C—Full Load Stop
D—Thrust Sleeve, Knuckle, and Guide Lever
E—Flyweights
F—Control Rack

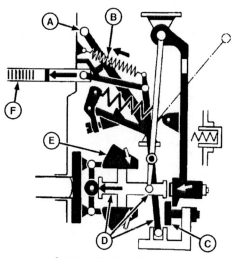

Governor Position When Starting Engine

SS40167,00000F4 -19-21JUL09-3/12

Engine Idling

When the engine is idling, the governor functions automatically. At idle, the main spring (A) of the governor is nearly free of tension, resulting in the spring having very little effect on the linkage. This means the flyweights (H) can swing outward (even at low speed) with little resistance.

As the control lever and guide lever are moved, the fuel control rack (I) is also moved, increasing the tension on the governor main spring.

Since the centrifugal force and spring tension are relatively low when the engine is at idle, the torque capsule in the tensioning lever (D) is only slightly compressed. The gap between the knuckle (G) and tensioning lever is therefore greater at low speed than at high speed, causing the tensioning lever to contact the supplementary idling spring (C), resulting in the desired speed regulation.

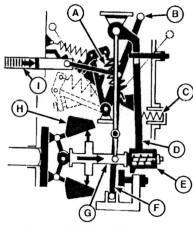

Governor Position with Engine at Idle

A—Main Governor Spring
B—Speed Control Lever
C—Supplementary Idling Springs
D—Tensioning Lever
E—Torque Capsule
F—Guide Lever
G—Knuckle
H—Flyweights
I—Control Rack

Continued on next page

SS40167,00000F4 -19-21JUL09-4/12

Engine at Medium Speed

Any movement of the speed-control lever (A) above idle will cause the control rack (E) to move to the maximum fuel delivery position, with the tensioning lever (B) moving to the full-load stop (C).

The injection pump delivers more fuel to the engine, resulting in increased speed. As soon as the centrifugal force exceeds the force of the governor main spring (D) (determined by the position of the speed-control lever), the governor linkage moves the control rack to a position where the centrifugal force is equal to the spring force. This results in a lower fuel delivery rate with the governor maintaining steady engine speed.

A—Speed-Control Lever
B—Tensioning Lever
C—Full-Load Stop
D—Main Governor Spring
E—Control Rack

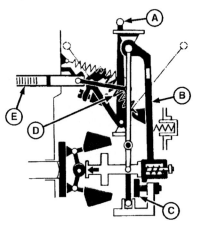

Governor Position (Engine at Medium Speed)

SS40167,00000F4 -19-21JUL09-5/12

Engine at Maximum Speed

The governor operation at maximum rated speed is about the same as at medium speed, except that the tension lever (B) stretches the governor main spring (E) to the fullest.

With the governor main spring fully tensioned, the tension lever is moved against the full-load stop (D) with greater force. This means that the control rack (A) is moved into the maximum fuel delivery position.

The torque capsule is always compressed, once the tensioning lever is moved against the full-load stop. It will remain compressed until the engine speed is reduced enough to reduce the centrifugal force of the governor flyweights. This forces the fuel control rack into a position to provide adequate torque reserve.

Once governed full-load (maximum) speed is reached, governor response will regulate fuel delivery between full load and fast idle to handle varying loads so long as there is no overload.

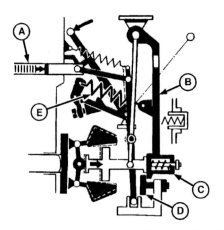

Governor Position (Engine at Maximum Speed)

A—Control Rack
B—Tensioning Lever
C—Torque Capsule
D—Full-Load Stop
E—Main Governor Spring

Continued on next page

SS40167,00000F4 -19-21JUL09-6/12

Stopping the Engine

Actuating the engine shutoff lever (C) moves the stop device (D), which in turn moves the control rack to stop the fuel supply to the engine. This movement takes place independently of the flyweight and speed-control lever positions.

A—Flyweights   C—Shutoff Lever
B—Speed-Control Lever   D—Stop Device

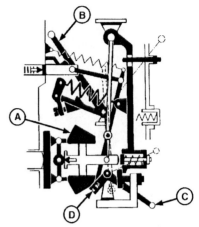

Governor Position (with Stop Device)

SS40167,00000F4 -19-21JUL09-7/12

The supporting lever (H) of the stop device is coupled to the shaft (I) and shutoff lever (E) by three pressure springs (J). The supporting lever continues to pivot until the control rack (A) is in the "stop" or "no fuel" delivery position.

At this position, the supporting lever stops moving and the three pressure springs become tensioned, as the shutoff lever moves to the limit of its travel.

As the engine speed decreases, the tension on the three springs lessens, pushing the supporting lever (H) (with the lower end of the fulcrum lever) farther forward, as the flyweights close. The upper end of the fulcrum lever (D) remains nearly stationary to the fuel control rack in the "stop" position.

Speed Droop

Speed droop is the variation of engine speed between a full-load and no-load speed range, and it's usually expressed as a percentage of rated speed.

For example, if the engine has been operating at full load, and the load is suddenly removed, the engine speed will increase to fast idle speed.

The amount of permissible speed increase is determined by governor design, but is usually no more than 10%. For example, if the maximum full-load engine speed is 2200 rpm, the no-load speed may be 2420 rpm.

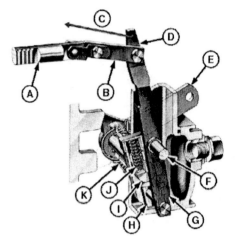

Governor Stop Engine Device

A—Control Rack   G—Stop
B—Link Member   H—Supporting Lever
C—Starting Spring   I—Shaft
D—Fulcrum Lever   J—Pressure Spring
E—Shutoff Lever   K—Return Spring
F—Pivot Point

Continued on next page   SS40167,00000F4 -19-21JUL09-8/12

Aneroid

The aneroid is a device mounted on top of the pump governor housing. It limits the fuel supply to the engine under low manifold pressure conditions to reduce excessive exhaust smoke.

A—Intake Manifold Pressure Inlet

B—Adjusting Shaft

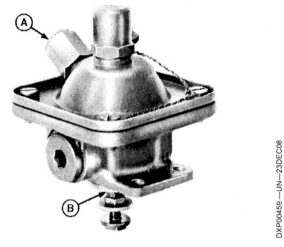

Aneroid

SS40167,00000F4 -19-21JUL09-9/12

The device is operated by the action of intake manifold pressure on a diaphragm (B). The aneroid fuel control shaft (F) engages the control rack either mechanically or hydraulically.

Adjustment is provided to obtain satisfactory acceleration with a minimum amount of exhaust smoke.

A—Adjusting Screw
B—Diaphragm (Operated by Manifold Pressure)
C—Spring
D—Adjusting Shaft
E—Spring
F—Fuel Control Shaft
G—Fuel Control Link
H—Arm

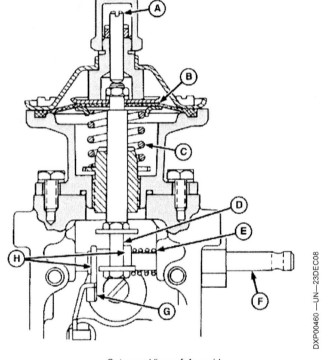

Cutaway View of Aneroid

Continued on next page
SS40167,00000F4 -19-21JUL09-10/12

OPERATION OF ANEROID (MECHANICAL ACTIVATOR)

Without aneroid control, when the speed-control lever is moved to increase the engine speed, even if only partway, the injection pump rack immediately moves to maximum fuel delivery position. Since there is no immediate increase in exhaust pressure to increase speed of the turbocharger (to supply more air), there is not enough air in the engine cylinders to burn all of the injected fuel. The result is a dense cloud of black exhaust smoke.

As we said, the job of the aneroid is to prevent this smoke.

The aneroid works as follows:

When the engine stop knob is pulled, the pump fuel control rack (G) is moved to a position where fuel is no longer injected into the engine. At the same time, the aneroid fuel control link is moved to where it no longer contacts the arm on the fuel control rack.

When the operator advances the engine speed-control lever to start the engine, the starter spring (F) moves the fuel control rack to the excess fuel position.

While the starter is cranking the engine, the injection pump is supplying excess fuel to the engine cylinders.

Once the engine starts, the centrifugal force of the governor flyweights overcomes the starter spring tension, even before slow idle speed is reached.

As the fuel control rack is moved from the excess fuel position, the spring on the aneroid fuel control lever shaft moves the control link back to its original position.

The arm (H) on the fuel control rack contacts the aneroid fuel control link before the rack has moved much over one-half of its full travel. This prevents more fuel from being injected into the engine than there is air to burn it.

The lever holds the rack in this position until the engine turbocharger creates ample intake manifold pressure. It does this through the diaphragm to move the lever to

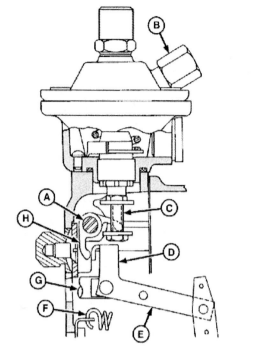

Cutaway View of Aneroid Operating Mechanism

A—Starting Fuel Control Shaft
B—Manifold Pressure Inlet
C—Adjusting Shaft
D—Link Contacts Arm (Limiting Fuel)
E—Governor Linkage
F—Starter Spring
G—Fuel Control Rack (From Pump)
H—Arm

release it. This action occurs every time the engine is accelerated.

As a result, this action limits the amount of fuel injected into the engine during the acceleration period, thereby limiting the amount of black smoke emitted from the exhaust.

Continued on next page   SS40167,00000F4 -19-21JUL09-11/12

OPERATION OF ANEROID (HYDRAULIC ACTIVATOR)

This mechanism hydraulically disengages the aneroid from the control rack during engine start-up. It also engages the aneroid (C) with the control rack (D) during acceleration.

With the engine off, oil pressure to the hydraulic activator is zero. Spring pressure moves the piston (I) and control shaft (J) to the left, disengaging the aneroid arms (E) from the injection pump control rack. This prevents aneroid operation.

During initial start-up, the capillary momentarily restricts pressure oil movement into the aneroid and delays aneroid engagement. The length of delay depends on ambient temperature and engine oil viscosity.

When engine oil pressure reaches 9 psi (62 kPa), oil moves through the capillary valve (H) into the aneroid. The pressure causes the piston to move to the right. This movement forces the fuel control shaft and aneroid arms to contact the control rack, preventing more fuel from being injected into the engine than there is air to burn it.

Again, the arms hold the rack in this position until the turbocharger creates adequate pressure to move the arms and release the rack. This occurs each time the engine accelerates.

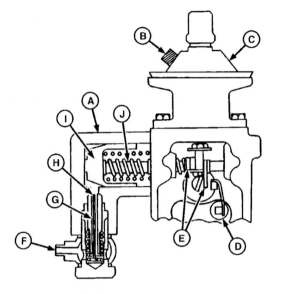

Hydraulic Activator Operation (Engine Off)

A—Activator Housing
B—From Intake Manifold
C—Aneroid
D—Control Rack
E—Aneroid Arms
F—Engine Oil Less Than 9 psi (62 kPa)
G—Restrictor Wire
H—Capillary Valve
I— Piston
J— Control Shaft

SS40167,00000F4 -19-21JUL09-12/12

Electronically Controlled Fuel Injection Pump

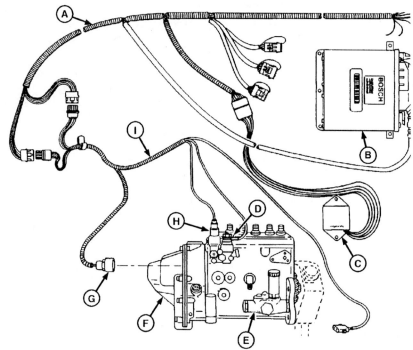

Electronically Controlled Fuel Injection System with Wiring Harness

A—Vehicle Wiring Harness
B—Engine Control Unit (ECU)
C—TVP Module
D—Fuel Shutoff Solenoid
E—Fuel Injection Pump
F—Electronic Governor Actuator
G—Actuator Connector
H—Fuel Temperature Sensor
I— Engine Wiring Harness

An electronically controlled fuel injection pump (E) serves as an engine governor by controlling fuel delivery to the individual fuel injectors. This system uses an electronic actuator (F) to move the fuel injection pump rack. The electronic actuator replaces the mechanical governor on in-line injection pumps.

OPERATION

The following paragraphs describe the basic operation of a typical electronically controlled fuel injection system.

When the engine key switch is turned to the ON position, power is supplied to the engine control unit (ECU) (B) and the fuel shutoff solenoid (D). When energized, the shutoff solenoid opens the fuel shutoff valve. The fuel shutoff valve closes when current flow to the shutoff solenoid is stopped.

When the engine key switch is turned to the START position, the ECU commands the actuator solenoid to move the injection pump rack to the starting fuel position. Starting fuel quantity is not affected by throttle position.

In the starting mode, the ECU monitors fuel temperature and engine speed. Based on this information, the ECU

will regulate the position of the pump rack to deliver the correct amount of fuel for starting. This permits the use of excess fuel and retarding the timing for cold temperature, but less fuel and no retard for hot starts. Thus, cold weather starting is improved and black smoke can be greatly reduced on hot starts.

Once the engine has started, fuel delivery is controlled by the ECU based on various inputs (primarily throttle position and engine speed). The ECU controls the pump rack position by adjusting the current level to the actuator solenoid.

The electronic actuator (F) responds to the signals received from the ECU to control the positioning of the fuel pump rack. The actuator also provides feedback to the ECU on engine speed and rack position. The ECU adjusts the current level to the actuator solenoid until the rack position signal from the actuator matches the commanded signal.

When no fuel is desired, the ECU stops current to the actuator solenoid. If a problem occurs where the ECU cannot control the rack position, it will stop current to the fuel shutoff solenoid in addition to the actuator solenoid.

Continued on next page

SS40167,00000F2 -19-23JUL09-1/14

DXP00464—UN—23DEC08

SUMMARY OF MAJOR COMPONENTS

Most electronic fuel injection systems consist of the following major components:

- Engine Control Unit (ECU)
- Injection Pump/Actuator Assembly
- Actuator Solenoid
- Speed Sensor
- Fuel Temperature Sensor

Engine Control Unit (ECU)

The ECU is a self-contained module containing electronic circuitry and computer software. It is used to perform governor and diagnostic functions. The ECU is mounted in a protected environment away from the engine. All connections between the ECU and the engine are made through a wiring harness.

The ECU controls the fuel delivery as a function of engine speed and throttle command. It also controls the fuel limiting for torque curves and the governing for speed control.

Engine Control Unit (with Wiring Harness and Connector)

Additionally, the ECU performs self-diagnosis on the control system. In most cases, a trouble code will be stored in memory if a problem occurs. The diagnostic capabilities of the system are covered in more detail later in this chapter.

SS40167,00000F2 -19-23JUL09-2/14

Injection Pump/Actuator Assembly

The electronically controlled in-line injection system uses the same basic hydraulic pumping mechanism used in mechanically governed in-line pumps. The mechanical governor mechanism is replaced with the electronically controlled actuator assembly (A). The actuator assembly includes the actuator solenoid to move the pump control rack, rack position sensor, primary speed sensor, and a toothed speed wheel.

The throttle lever mechanism used on mechanical pumps is removed and its function is implemented by a throttle position sensor. The engine control unit (ECU) converts the input signals from the sensor into a percentage of full throttle command and controls the engine speed and fuel quantity accordingly.

Fuel is transferred from the vehicle or engine fuel tanks with a fuel supply pump. The injection pump fuel inlet connection is located at the rear of the pump on the fuel inlet assembly. The fuel inlet assembly also includes the fuel shutoff solenoid and the fuel temperature sensor.

Typical Injection Pump/Actuator Assembly

A—Electronic Governor Actuator

Continued on next page

SS40167,00000F2 -19-23JUL09-3/14

Actuator Solenoid

The actuator solenoid (A) is contained within the actuator housing mounted on the rear of the injection pump. The actuator solenoid responds to signals received from the engine control unit (ECU) to control the position of the fuel pump rack.

The injection pump control rack is spring-loaded to the fuel shutoff position or "zero rack." When the engine key switch is turned to the START position, the ECU powers the actuator solenoid, causing the rack (C) to move to the starting fuel position.

Once the engine has started, the ECU determines the optimum fuel delivery based on various inputs (primarily throttle position and engine speed). The ECU adjusts the current level to the actuator solenoid, causing a corresponding movement of the rack to provide the correct fuel delivery.

The ECU has the ability to control the current to the solenoid in order to position the control rack anywhere between zero rack (fuel shutoff) and full rack (maximum fuel delivery).

The actuator solenoid can only be serviced by an authorized shop. The injection pump will also require re-calibration after the actuator housing is removed.

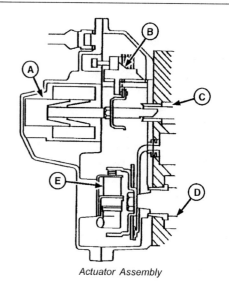

Actuator Assembly

A—Actuator Solenoid
B—Rack Position Sensor
C—Rack
D—Pump Camshaft
E—Speed Sensor

SS40167,00000F2 -19-23JUL09-4/14

Rack Position Sensor

The rack position sensor (B) is located within the actuator housing. The sensor supplies rack position information to the engine control unit (ECU) so that a specific rack position can be controlled.

The sensor includes an electronic module mounted in the actuator housing that provides an input signal to the ECU indicating the position of the rack. The ECU constantly monitors the input signal in order to determine the rack position. The ECU then controls the rack position by adjusting the current level to the actuating solenoid (A) until the rack position signal matches the commanded signal.

The rack position signal is used to control the rack position for all operating conditions. If this critical sensor were to fail, the controller would be forced to shut down the engine due to loss of control of the fuel delivery.

The rack position sensor produces an output voltage which is available for diagnostic purposes and is not used by the governor system. This voltage signal is proportional to the pump rack position, and therefore can be used to determine if the governor control system is actually moving the rack (C) to the proper range for a given operating condition. As the rack moves from low fuel delivery to maximum fuel delivery, the voltage signal should vary from a low voltage to a higher voltage.

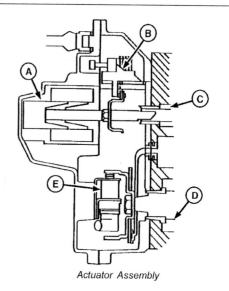

Actuator Assembly

A—Actuator Solenoid
B—Rack Position Sensor
C—Rack
D—Pump Camshaft
E—Speed Sensor

The relationship between the rack position voltage and the rack position may vary depending on the system application. Consult the engine technical manual for complete instructions and specifications.

Continued on next page
SS40167,00000F2 -19-23JUL09-5/14

Primary Speed Sensor

The primary speed sensor (E) is located in the actuator assembly. The sensor is a magnetic pickup that generates electrical pulses as the teeth on the speed wheel move past a sensor. The engine control unit (ECU) monitors the electrical pulses to determine the speed of the engine. The ECU continually adjusts current to the actuator solenoid (A) to maintain the desired engine speed.

If this sensor were to fail completely, the ECU would use the signal from the auxiliary speed sensor to govern the engine.

Because the primary speed sensor is located inside the actuator housing, it can only be serviced by an authorized repair shop.

A—Actuator Solenoid
B—Rack Position Sensor
C—Rack
D—Pump Camshaft
E—Speed Sensor

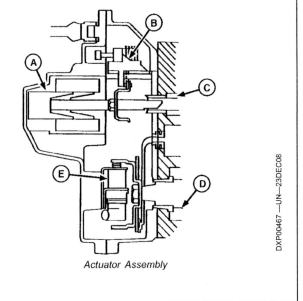

Actuator Assembly

SS40167,00000F2 -19-23JUL09-6/14

Auxiliary Speed Sensor

The auxiliary speed sensor (A) is typically located on the front of the engine. It is also a magnetic pickup that generates voltage pulses to the engine control unit (ECU) as the teeth on the timing gear pass by the tip of the sensor. The frequency of the voltage pulses is proportional to the engine speed.

The ECU receives the input signal from the auxiliary speed sensor, and then transmits the auxiliary speed output signal for use by other electronic modules such as the tachometer. The auxiliary speed sensor serves as a back-up speed sensor in the event of a complete failure of the primary speed sensor.

The auxiliary speed sensor is a serviceable part. No re-calibration is required when this sensor is replaced.

Auxiliary Speed Sensor

A—Auxiliary Speed Sensor

Continued on next page

SS40167,00000F2 -19-23JUL09-7/14

Fuel Temperature Sensor

The fuel temperature sensor (A) is located at the fuel inlet to the pump. This sensor monitors the temperature of the fuel entering the injection pump.

The sensor is a temperature-sensitive variable resistor. As temperature goes up, resistance goes down. Even small changes in temperature can be determined by monitoring the resistance of the sensor. The engine control unit (ECU) continuously sends a voltage signal to the sensor, monitors the voltage drop across the resistor, then compares the sensor voltage to preprogrammed values to determine the fuel temperature.

The ECU uses the fuel temperature to determine the optimum fuel delivery for starting and, depending on the application, to maintain constant power over a predetermined temperature range.

The ECU uses engine speed and initial fuel temperature to control the rack position during starting. This permits the use of excess fuel and retarding the timing for cold temperatures but less fuel and no retard for hot starts. Thus, cold starting is improved and black smoke can be greatly reduced on hot starts.

The ECU provides nearly constant fuel delivery to the engine by compensating for changes in the fuel density

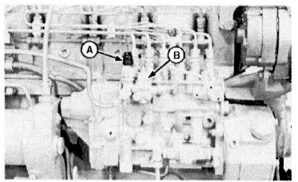

Fuel Shutoff Solenoid and Fuel Temperature Sensor

A—Fuel Temperature Sensor   B—Fuel Shutoff Solenoid

over any desired temperature range. The fuel temperature compensation characteristic is dependent on engine application and is programmed at the factory.

If the fuel temperature sensor were to fail, a low temperature would be assumed by the engine controller. In warm weather this might result in a slight drop in maximum torque and increased smoke on hot starts.

SS40167,00000F2 -19-23JUL09-8/14

Fuel Shutoff Solenoid

The fuel shutoff solenoid (B) is energized (opening the valve) when the engine key switch is turned to the ON position. The shutoff solenoid will stop the fuel supply to the injection pump when the key switch is turned to the OFF position.

If a problem occurs where the engine control unit (ECU) cannot control the rack position, the fuel shutoff solenoid is de-energized in addition to the actuator solenoid.

TRANSIENT VOLTAGE PROTECTION MODULE

The main function of the Transient Voltage Protection (TVP) module is to limit high-energy voltage spikes (from the charging system) to 40 volts. This protects the electronic circuitry in the engine control unit (ECU).

GOVERNOR MODES

The governor mode is selected at the factory. The engine control unit (ECU) can provide either all-speed governing or min-max governing.

When using all-speed governing, the ECU controls the engine speed based on the throttle command and the selected speed. When using the min-max governing mode, the ECU provides the same minimum (slow idle) and maximum (fast idle) speed governing as with the

Fuel Shutoff Solenoid and Fuel Temperature Sensor

A—Fuel Temperature Sensor   B—Fuel Shutoff Solenoid

all-speed governor. However, between the minimum and maximum speeds, the throttle and engine speed inputs are used by the ECU to select an optimum fuel quantity — the throttle commands fuel quantity rather than engine speed in the min-max governor mode.

The engine application determines which combinations of slow idle, rated speed, and fast idle are programmed into the ECU. For further details, see the engine technical manual.

Continued on next page

SS40167,00000F2 -19-23JUL09-9/14

MAXIMUM FUEL QUANTITY CONTROL

The engine control unit (ECU) limits the maximum fuel delivery as a function of engine speed. This function is set at the factory to obtain the proper fuel limit curve (power curve).

Some electronic governors provide an optional feature that allows the power curve to be switched to any one of three factory-programmed power curves. Switching between the power curves can be accomplished by the ECU while the engine is running.

The normal power curve is selected if no input signal is present.

A derated power curve is usually selected by shorting the input signal to ground. A temperature switch is one method used to select the derated power curve.

The third curve is typically used for a Power Boost mode since operation using this curve can be limited by timers within the ECU. This power curve is selected by connecting the input to ground through a 2 kilo-ohm resistor. The ECU will limit Power Boost operation to the amount of time programmed in the ON timer. If the ON time limit is reached, the ECU will automatically switch back to the normal power curve. The OFF time must then be reached before the ECU will allow reselection of the Power Boost curve.

THROTTLE OPTIONS

There are three throttle options available for use with the electronically controlled fuel injection system. The selection of a throttle option is dependent on the application and is programmed into the ECU. These options are:

- Analog Throttle
- Three-State Throttle
- Pulse-Width-Modulated (PWM) Throttle

The **analog throttle** is commonly used with either all-speed or min-max governing. The analog throttle provides a continuously variable voltage input to the engine control unit (ECU). The voltage input normally passes through a potentiometer (variable resistor). Then

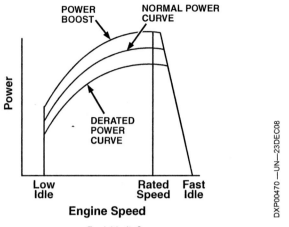

Fuel Limit Curve

the ECU converts this voltage into a percentage of full throttle command. The signal is sent to the injection pump actuator solenoid to control fuel delivery and engine speed accordingly.

A **three-state throttle** uses a simple switching arrangement to select one of three fixed speeds. Typical applications are generator sets where one or two fixed speeds are desired, or agricultural combines which use two fixed speeds.

The **pulse-width-modulated (PWM) throttle** can be used alone or with an analog throttle. The PWM throttle is a signal received from another electronic module (commonly a transmission controller) which uses a pulse width to indicate the desired percent of full throttle. "Pulse-Width-Modulated" means that the frequency of the signal remains the same, and the width of the pulse is changed to indicate a change in value.

When used in combination with an analog throttle, the PWM throttle input has priority over the analog throttle. If the ECU receives a PWM throttle signal, it will stop using the analog throttle input and will start using the PWM throttle to determine the throttle command. If the PWM throttle signal is turned off or is disconnected, the ECU will revert to using the analog throttle again.

Continued on next page   SS40167,00000F2 -19-23JUL09-10/14

TROUBLESHOOTING ELECTRONIC FUEL INJECTION

The engine control unit (ECU) contains electronic circuitry and a computer program that performs diagnostic functions when the ECU is in the governor mode (normal engine operating mode). In most cases, the ECU will display a diagnostic code to indicate a specific problem. The ECU can also flash a fault lamp.

When troubleshooting the fuel injection system, check if there are diagnostic codes being displayed. If no diagnostic codes are present, verify that a problem does not exist in the basic electrical system or in the fuel supply system before continuing troubleshooting of the fuel injection system. Some areas that should be checked include:

1. **Charging System** — Alternator output, condition of the batteries and battery cables.
2. **Power Distribution System** — Fuses or circuit breakers, key switch and any relays between the battery and fuel injection system, starter solenoid wiring connections.
3. **Fuel Source** — Fuel in the tank, quality of fuel, condition of fuel filters and fuel lines.

If the fault is in one of the above areas, fuel injection system diagnostic procedures will probably not locate the problem.

Checking Wiring and Connectors

When diagnosing electrical system problems, note the condition of the wiring and connectors since a high percentage of problems originate here.

Check for loose, dirty, damaged, or disconnected connectors. Inspect for wires that may have been pulled out of their connector terminal. Look for poorly positioned terminals and inspect for corroded wires and terminals.

Inspect wiring for possible shorts caused by wires rubbing against metal edges or other external parts.

Running Engine at Different Speeds

If the engine will start, operate the throttle control(s) at slow-to- medium rate between the slow idle and the fast idle stops with the engine running. If the engine speed "catches" or "drops," this may indicate a problem with the throttle adjustment, throttle sensor operation, or the wiring between the engine control unit (ECU) and the throttle sensor(s). Using this method of changing engine speed, you may be able to identify problems that only occur at certain speeds or throttle settings.

Self-Diagnosis and Backup Features

NOTE: *If the operator is able to keep the engine running during a fault condition, no damage to the engine should result. However, the problem should be fixed at the earliest convenience.*

The engine control unit (ECU) self-diagnoses as many system faults as practical. This includes determining if

Tachometer Display of Diagnostic Codes

any of the sensor input voltages are too high or too low, if the engine speed signals are valid, and if the pump control rack is responding properly.

The ECU also monitors its own operation for problems. In most cases, it will output a diagnostic code to indicate the specific problem that has been detected. If the ECU indicates a fault condition, it will also store the diagnostic code within its own memory. The diagnostic code messages will help isolate the problem and allow quicker repair.

A diagnostic reader is used to interpret and display diagnostic codes to the operator. Diagnostic codes are typically displayed as a number or a series of numbers.

A diagnostic reader can be as simple as a digital tachometer located in the vehicle's instrument panel or as sophisticated as a portable electronic governor testing device. The ECU can also flash a fault lamp that indicates the diagnostic status of the system.

Each diagnostic code represents a specific fault condition that may be caused by one or more conditions. If only one diagnostic condition is present, the code for that condition is transmitted every second and displayed continuously by the reader. If multiple conditions are present, the codes are transmitted one after another, once per second, until all conditions have been transmitted. The order in which the codes are transmitted has no meaning, and the display sequence is repeated continuously.

A diagnostic code is transmitted as long as the fault condition exists. If an intermittent condition is present, the code may be present only for a short time.

The ECU has the capability to save the diagnostic codes that have occurred. The codes are saved even when the engine key switch is turned off. This allows a service person to check to see if a diagnostic code has occurred in the past even though it is not presently occurring. This is helpful for intermittent failure conditions.

NOTE: *Record stored codes before disconnecting the battery. Once the battery is disconnected, stored codes will be erased.*

Continued on next page   SS40167,00000F2 -19-23JUL09-11/14

If the ECU detects a problem that can be diagnosed, it will switch to a safe mode of operation as a backup whenever possible, or it will stop the engine if control of the engine cannot be maintained.

In some fault conditions, little or no degradation in performance will be noticed. For example, the engine will continue to operate normally using the auxiliary speed sensor in the event of a primary speed sensor fault.

Consult the technical manual specific to your engine for the specifications and diagnostic procedures for that engine.

Diagnostic Codes

Following is a sample listing of diagnostic codes that can occur in an electronic governor system. Not all of these codes will be present in all engine applications. Refer to the engine technical manual for detailed information on diagnostic codes.

28 — Engine Control Unit (ECU) Failure
29 — Sensor Excitation Voltage Too High or Too Low

32 — Actuator Circuit Fault
33 — Actuator Solenoid Output Shorted High
34 — Rack Position Error
35 — Rack Position Voltage Too Low
36 — Rack Position Voltage Too High
37 — Fuel Temperature Input Voltage Too High
38 — Fuel Temperature Input Voltage Too Low
39 — Primary Speed Input Error
41 — Start Signal Missing
42 — Engine Over-Speed
43 — PWM Throttle Input Erratic
44 — Auxiliary Speed Input Error
47 — Derated Torque Curve Selected
51 — Electrical Noise on Analog Throttle Input
54 — Electrical Noise on Rack Position Voltage Input
55 — Electrical Noise on Fuel Temperature Input
56 — Electrical Noise on Fuel Limit Select Input
57 — Electrical Noise on Speed Regulation Select Input
58 — Electrical Noise on 3-State Throttle Input
59 — Electrical Noise on +5V Sensor Excitation

SS40167,00000F2 -19-23JUL09-12/14

USING ELECTRONIC GOVERNOR TESTER

The electronic governor tester is a portable diagnostic reader that provides a display of diagnostic codes, throttle position, or commanded fuel quantity. The electronic governor tester is used by connecting it to the diagnostic reader connector of the governor system wiring harness. The reader is powered through this connector so the engine key switch must be ON to obtain a display. Refer to the engine technical manual for detailed information on the diagnostic reader connector.

Diagnostic codes are obtained by pressing the DISPLAY CODES switch. Stored diagnostic codes can be recalled by pressing the DISPLAY CODES and then holding the RECALL CODES switch to display the codes. Stored diagnostic codes can be cleared by simultaneously pressing both the RECALL CODES and CLEAR CODES switches for at least 1 second.

Throttle position is obtained by pressing the % THROTTLE switch. The throttle position is given as a percentage of full throttle. The range for the percent throttle is 0–100%.

Commanded fuel quantity is obtained by pressing the % FUEL switch. Fuel quantity is given as a percentage

Electronic Governor Tester

DXP00472 —UN—23DEC08

of rated fuel delivery. The range for the percent fuel is 0–159%.

For detailed test procedures and specifications of a particular engine, consult the appropriate technical manual for that engine.

Continued on next page
SS40167,00000F2 -19-23JUL09-13/14

Diagnostic Voltages Connector

The diagnostic voltages connector, a part of the electronic governor system wiring harness, provides access to important governor system voltages that can be read using a digital multimeter. The following voltages should be present at the connector:

- Analog Throttle +5V
- Throttle Input Voltage
- Rack Position Voltage
- Sensor Common

The **analog throttle +5V** is the source voltage from the engine control unit (ECU) for the analog throttle. This voltage signal can be used to check on the condition of the ECU internal power supply system. The nominal voltage should be 4.8–5.2 volts. If the voltage is outside this range, the ECU may be faulty or there may be a short within the wiring harness. The absence of the +5V indicates a failure within the ECU or a problem in the wiring harness or throttle sensor.

The **throttle input voltage** is the input to the ECU from the throttle sensor. It is proportional to the throttle position and the analog +5V supply. This signal can be used to check throttle sensor adjustment and performance of the throttle sensor.

As the throttle is moved from slow idle to fast idle, the throttle input voltage should vary from low voltage to a high voltage. The voltage should change smoothly as the throttle is moved slowly through its range. If voltage readings do not change or change erratically rather than smoothly, the throttle sensor may be defective.

The **rack position voltage** signal is proportional to the injection pump rack position. The voltage reading can be used to determine if the governor control system is actually moving the rack to the proper range for a given operating condition. Consult the engine Technical Manual for the rack position voltages specific to the engine.

The **sensor common** is the system common or "ground" reference point for the throttle and rack position voltages.

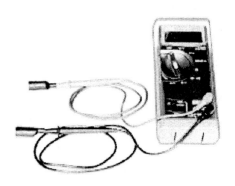

Digital Multimeter

The "ground" input should be connected to the common socket of the sensor when measuring these voltages.

Refer to the engine technical manual for detailed information on diagnostic voltage testing.

Summary of Electronic Governor Benefits

Some benefits of the electronic governor system are:

1. Fewer mechanical parts in governor to encounter wear.
2. Different governor modes are possible for specific operating conditions.
3. Precise fuel control based upon engine speed and fuel temperature.
4. Quick start-up with less smoke and faster warm-up.
5. Improved fuel economy and reduced exhaust smoke.
6. Consistent power and performance.
7. Engine performance is fine-tuned to meet a specific load requirement.
8. Engine Control Unit (ECU) is programmable to provide precise torque curve characteristics designed for each application.
9. Capable of communicating with other machine equipment such as transmission controllers.
10. Built-in self-diagnostics allow for quick and accurate repairs.

SS40167,00000F2 -19-23JUL09-14/14

Servicing Injection Pump

The following service and inspection procedures are only intended to be used for general information purposes for a typical injection pump. Service and inspection information pertaining to a specific injection pump is provided by the manufacturer of that pump.

A word of caution — **never spray a warm pump with cold water when the pump is running and never steam clean a pump** — severe damage or pump seizure can result.

Before removing an injection pump, clean the exterior of the pump and the area around it, using the appropriate method as outlined by the manufacturer of the pump.

When removing the injection lines from a pump, cap the fuel lines as well as all pump ports to prevent dirt from contaminating the lines and pump.

Keep in mind the following key points when servicing a pump:

1. Disassemble and inspect the pump only if there is clear indication that the pump is malfunctioning.
2. Troubleshoot the fuel system as a whole and pinpoint the failure before servicing a component. (For more information on troubleshooting a pump, see Troubleshooting in this chapter.)
3. Careless service could damage a pump — injection pumps are precision instruments.
4. Only a qualified mechanic with the proper tools should service an injection pump. Pump service and calibration should only be performed by an authorized repair shop.

Inspecting Injection Pump

1. Parts that stick could have a light buildup of varnish. If so, use a soft brush and lacquer thinner to clean the part.
2. Some matching parts such as rotors and hydraulic heads must be replaced as a set. The fit on some of these parts is so close that heat from your hands is enough to expand the parts to where they temporarily do not fit.
3. Replace all seals and gaskets regardless of their condition. Use special tools and lubricants when recommended.
4. Flush all parts with clean fuel during their reassembly.

Service Procedures

To perform quality service on an injection pump, the following items must be on hand or available:

1. A clean, well-lit work area, isolated from the repair shop.
2. Clean tools in good working condition. Many injection pumps require special tools for service and repair. Review the appropriate technical manual for your particular injection pump for the required special tools.

Using an Injection Pump Test Station

3. Calibrated test equipment as recommended by the manufacturer of the injection pump.
4. All appropriate service and repair technical publications for your particular pump.

Testing Injection Pump

The only method for testing an injection pump is with an injection pump test station. Because each test station is unique, the instructions for each test station must be followed accordingly.

After the injection pump is connected to the test station, run the pump for at least 10 minutes. This will warm the pump and bleed any air through loosened injection lines.

LEAKAGE TEST — With all connections tight, run the pump at full-load rated speed for several minutes. Clean the exterior of the pump and dry it off with compressed air. Then check the exterior of the pump for leaks. Make all necessary repairs.

VACUUM TEST — The vacuum test helps determine if air is being drawn into the pump. With the fuel supply cut off, the test gauge should read and hold a specified number of inches of mercury (Hg). If it doesn't, check for a possible air leak.

TRANSFER PUMP PRESSURE TEST — The transfer pump pressure is tested at various speeds. The correct pressures are listed in the specifications provided by the manufacturer of the pump.

Continued on next page   SS40167,00000F1 -19-05AUG09-1/2

FUEL DELIVERY TEST — Graduated glass tubes in the pump test station measure the fuel flow at various speeds. The amount of fuel injected into the glass tubes is recorded at various speeds and the average is compared to the factory specifications.

TORQUE CONTROL ADJUSTMENTS — If the pump is equipped with a torque adjustment, perform the adjustment while the pump is on the test station.

IDLE ADJUSTMENT — Slow and fast idle adjustments can be set to factory specifications on most test stations. Once all idle adjustments are complete, the pump is ready for installation.

Installing and Timing an Injection Pump

When installing an injection pump, make sure the pump and engine timing are synchronized. The specialized tools required to perform this procedure should be listed in the technical manual for each particular pump.

If an advance mechanism is used, it should be adjusted after the pump is installed. This helps prevent an engine from misfiring and helps prevent a loss of engine power while the engine is running.

SS40167,00000F1 -19-05AUG09-2/2

Mechanical Fuel Injection Nozzles

Distributor and in-line injection pump systems typically use mechanical fuel injection nozzles.

Injection nozzles perform three functions:

1. **Disperse fuel spray to mix fuel with air**
2. **Atomize fuel for efficient combustion**
3. **Complete nozzle shutoff to prevent fuel leakage**

Also, the injection nozzles in a multi-cylinder engine must provide an equal amount of fuel to each cylinder for smooth and balanced operation.

TYPES OF INJECTION NOZZLES

Nozzles are simple devices. They use a spring to oppose fuel pressure until the right instant for injecting fuel, when the nozzle valve opens. The opening pressure usually can be adjusted.

Most mechanical nozzles today are **closed** types — that is, the fuel pressure acts on only one side of the nozzle valve.

Closed-type nozzles are classified as follows:

- Inward-Opening Nozzles
  - Differential Pressure
  - Hole Type
  - Pintle Type
  - Pintaux Type
- Outward-Opening Nozzles
  - Poppet Type
  - Pintle Type
  - Multi-Orifice Type

Inward-Opening Nozzles (A)

For inward-opening nozzles, fuel pressure acts against the lower end of the valve and moves the valve inward. This allows the nozzle to release a spray of fuel. A spring at the upper end of the valve can be adjusted to vary the pressure required to open the valve.

Inward-opening nozzles include the following types:

- Differential Pressure Nozzle — is operated hydraulically. When fuel pressure to the nozzle rises to the set level, the nozzle valve opens against spring pressure and injects fuel. As pressure drops, the valve closes and injection stops. The valve shown at left in illustration operates like this. This nozzle is adjusted by spring tension for precise operation.
- Hole-Type Nozzle — provides some additional features not found in a differential pressure nozzle. The number

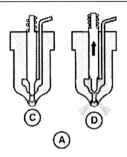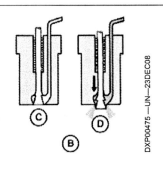

A—Inward-Opening Nozzle   C—Closed
B—Outward-Opening Nozzle   D—Open

and size of the spray holes can be varied for a precise spray pattern. This type of nozzle is best for an engine with open combustion chambers.
- Pintle-Type Nozzle — is a tapered valve that seats in a single orifice located in the valve body. The pintle-type nozzle is fairly simple, but produces a hollow spray and is not as versatile as a hole-type nozzle.
- Pintaux-Type Nozzle — is a variation on the pintle type nozzle. An extra hole is added to the head for spraying fuel at an angle during low engine speeds. This angled hole provides assistance when starting the engine, during cold weather for example. By angling the spray of fuel toward the hottest part of the combustion chamber, faster cylinder firing occurs.

Outward-Opening Nozzles (B)

For outward-opening nozzles, fuel pressure acts against the valve only in the direction of flow. Since combustion chamber pressures tend to close the valves, the opening pressure is relatively low. So fuel leak-off is not required.

Outward-opening nozzles include the following types:

- Poppet-Type Nozzle — opens outward and sprays a fine vapor but has little penetration for most diesels.
- Pintle-Type Nozzle — opens outward (illustration, right side). Fuel pressure acts on the tapered head to unseat the valve at the preset pressure. As the valve moves, the extra shoulder is exposed to fuel pressure, resulting in the valve opening rapidly for a spray of fuel, with no dribbling. This nozzle is simple in design and lightweight.
- Multi-Orifice Nozzle — is a variation of the pintle type nozzle. The pintle fits closely on its seat and during injection, fuel is sprayed from a ring of holes rather than past the pintle and then out of the head.

Continued on next page   SS40167,00000F6 -19-05AUG09-1/6

Typical Injection Nozzle which Opens Inward and Operates by Pressure Differences

A—Seal
B—Nozzle Body
C—Locating Clamp
D—Pressure Spring
E—Pressure Adjusting Screw
F—Fuel Leak-Off
G—Lift Adjusting Screw
H—Fuel Inlet
I— Nozzle Valve
J— Spray Tip

OPERATION OF A TYPICAL NOZZLE

PARTS OF A NOZZLE — The nozzle body (B) contains the moving parts — the nozzle valve (I) and the pressure spring (D).

Pressure to open a nozzle is controlled by the tension of the pressure spring (D). Tension on the pressure spring is adjusted by turning the pressure adjusting screw (E). (Other injectors use shims to adjust the pressure.)

How far a valve moves when injecting fuel is controlled by the lift adjusting screw (G). (A lift adjusting screw is not used on all injectors.)

HOW A NOZZLE WORKS — Metered fuel at high pressure from the pump enters the fuel inlet (H).

This fuel surrounds the nozzle valve (I) and forces the valve from its seat at a preset pressure. A measured amount of fuel is then sprayed from the nozzle tip into the engine combustion chamber at high velocity. After fuel is injected, pressure in the injector drops and the spring rapidly closes the valve.

A small amount of fuel then leaks past the nozzle valve to lubricate the working parts. The excess lubricating fuel is removed from the top of the nozzle at the fuel leak-off port and then returned to the tank.

SERVICING AND TESTING A TYPICAL NOZZLE

An injection nozzle is a precision-fitted part that operates rapidly in conditions of extreme heat and pressure. Nozzles are susceptible to failure.

Nozzle failure is typically caused by dirt clogging the orifices and passages. This sort of failure can be easily remedied by a thorough cleaning.

However, a nozzle does not have to stop injecting fuel completely to be considered a failure. For a list of nozzle failures and their possible cause, see the Nozzle Troubleshooting Chart.

REMOVING A NOZZLE — Before removing a nozzle, first clean the area around the nozzle to help eliminate any chance of further contamination. Then remove and cap off the fuel injection line and the leak-off line. Finally, remove the nozzle from the engine. (Most nozzles require a special tool for removal.)

CLEANING A NOZZLE — Soak the entire nozzle assembly in clean solvent or fuel after discarding the outer seals. Clean the spray tip and nozzle body with a **brass** wire brush.

IMPORTANT: **Never use emery cloth or a steel wire brush to clean a nozzle as the precision tip will be damaged.**

Continued on next page

SS40167,00000F6 -19-05AUG09-2/6

Nozzle Troubleshooting Chart	
Problem	**Possible Causes**
Nozzle opens at wrong pressure	Wrong spring pressure adjustment
	Broken spring
Nozzle will not open	Plugged orifices (spray tip only)
Poor spray pattern	Plugged or chipped orifices (spray tip only)
	Chipped or broken pintle end
	Deposits on pintle seat
	Chipped pintle seat
Poorly misting fuel	Plugged or chipped orifices (spray tip only)
	Valve not free
	Cracked tip
Valve operates erratically	Valve spring misaligned
	Spring broken
	Deposits on pintle seat
	Bent valve
	Distorted body
Valve will not operate	Valve spring misaligned
	Spring broken
	Varnish on valve
	Valve seat eroded or pitted
	Distorted body
Too much fuel leaks off	Valve guide wear
Too little fuel leaks off	Varnish on valve
	Not enough clearance between valve and guide

Continued on next page   SS40167,00000F6 -19-05AUG09-3/6

TESTING A NOZZLE

⚠ **CAUTION: The tip of an injection nozzle should always be directed away from you or other personnel in the work area. Fuel spray from an orifice can penetrate clothing and skin, causing serious personal injury. Enclosing a nozzle in a glass container, such as a beaker, is recommended.**

Before applying pressure with a nozzle tester, make sure all connections are tight and the fittings are not damaged. Fluid escaping from a small hole can be nearly invisible. Use a piece of cardboard or wood, rather than your hand, to locate a leak.

Fuel is ejected from a nozzle at extremely high pressure. This high pressure can penetrate clothing and skin, resulting in serious personal injury. If injured by a high-pressure stream of fluid, see a doctor immediately. Fluid injected into the skin must be surgically removed within a few hours or gangrene may result.

When testing a nozzle, always refer to the technical manual for that particular nozzle.

A nozzle tester is a high-pressure hand pump designed to force fuel through the nozzle.

Once the nozzle is attached to the tester and all fittings are checked for leaks, flush the nozzle by rapidly pumping the tester (with the gauge closed).

Test the nozzle for:

- Spray pattern and chatter
- Opening pressure
- Valve leakage

A word of caution when using this type of tester: The fuel comes out of the nozzle at extremely high pressure — which can penetrate clothing and skin and cause injury. Always keep the nozzle pointed away from you or, better yet, enclose it in a transparent beaker as shown.

After the nozzle is attached to the tester, flush out the nozzle by rapidly pumping the tester (with the gauge closed).

Perform the following tests:

- Spray Pattern Test
  - Test the spray pattern first. The fuel should be finely atomized and distributed evenly.
  - There should be no stream or large visible drops.
  - The spray pattern should match the one specified for that particular engine.
  - If the resulting spray pattern is not as specified, look for clogged or eroded tip orifices or a bent valve.

Nozzle Tester

- Opening Pressure Test
  - The pressure required to open or "crack" the nozzle is checked by pumping the tester steadily until the needle on the tester's gauge falls rapidly. Compare the pressure reading against the one specified in the technical manual for that particular nozzle.
  - If the pressure is too low, a weak or broken spring might be the cause or the spring pressure may need adjustment.
  - If the pressure is too high, the tip may be plugged or the valve may be binding in the valve guide.
- Leakage Test
  - As pressure builds up when working the tester, check for leakage using a piece of cardboard or wood. All nozzle connections and fittings must be tight.
- Summary: Testing a Nozzle
  - If all tests prove that the nozzle is working as specified, do not disassemble the nozzle any further.
  - If the test shows a failure, the nozzle must be disassembled, inspected, cleaned, and repaired as necessary.

SERVICING THE NOZZLES

Introduction

- All fuel nozzles require careful handling. Refer to the nozzle or engine technical manual for the precise service procedures for your particular nozzle.
- Each nozzle requires a special tool kit to perform service. Never service a nozzle without the appropriate tool kit.
- When working on several nozzles, DO NOT mix nozzle parts.

Continued on next page SS40167,00000F6 -19-05AUG09-4/6

Cleaning Nozzle Parts

Most nozzle servicing is just cleaning off carbon and dirt deposits. A good soaking in clean solvent will remove much of the residue. Usually, a **brass** wire brush will clean off the remainder. **DO NOT use a steel wire brush.** It can scratch smooth surfaces or erode the spray orifices.

Special tools from the nozzle kit are often needed to clean internal seats, orifices, and passages. For example, the tiny spray orifices in some nozzle tips are very critical in size and require special gauged cleaning wires.

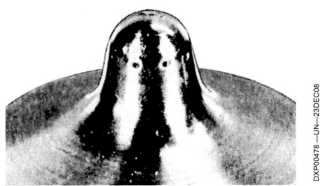

Nozzle Orifices Eroded by Use of a Steel Wire Brush for Cleaning

SS40167,00000F6 -19-05AUG09-5/6

Inspecting Nozzle Parts

Normal wear on valve parts can be ignored unless it is causing the valve to fail.

Inspect the parts for abnormal wear, chipped edges, scratches, misalignment, and breakage. The Nozzle Troubleshooting Chart will provide ideas on where to look for problems.

Using a magnifying glass during the inspection will help you locate problems.

Repairing a Nozzle

If cleaning does not resolve the problem, inspect each part very carefully before replacing it. For example:

1. If a valve is sticking and it is not bent, a little polish or lapping compound around the valve guide might be all that is necessary to correct the problem.
2. Other parts may also be repaired by lapping their surfaces to remove tiny scratches or burrs. Always be careful when lapping any part to avoid excess wear on the part.

Parts that are chipped, broken, or bent should be replaced. If the damage is extensive, replace the entire nozzle.

NOTE: Some nozzle parts are sold in matched sets. In this case, replace both parts if either is damaged.

DXP00479 —UN—23DEC08

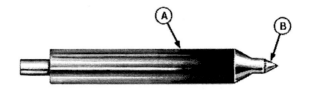

Nozzle Valve Damaged Beyond Repair

A—Lapped Surface Scored   **B—Tip Worn**

Reassembling a Nozzle

Flush each part of the nozzle in clean fuel before the nozzle is reassembled. Once it is reassembled, retest the nozzle to verify that it is working as specified.

Spring pressure adjustments are generally performed before the nozzles are retested.

If the retest still indicates a malfunction, the nozzle will have to be further serviced or replaced.

SS40167,00000F6 -19-05AUG09-6/6

Electronic Unit Injector System

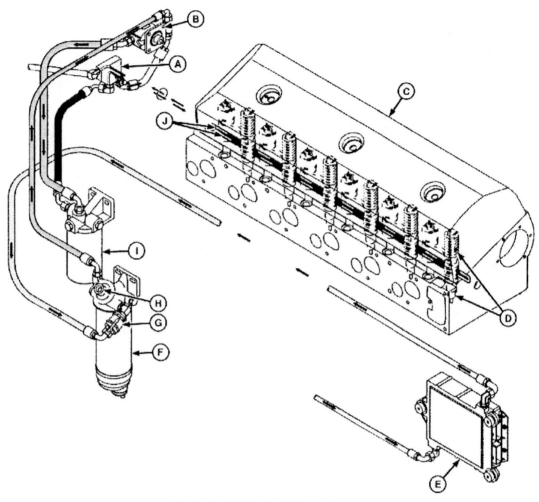

Electronic Unit Injector (EUI) Fuel System

A—Fuel Manifold
B—Fuel Transfer Pump
C—Engine Cylinder Head
D—Unit Injector
E—Engine Control Unit (ECU)
F—Primary Fuel Filter/Water Separator
G—Check Valve
H—Primer Pump
I— Secondary Fuel Filter
J— Fuel Rails

The electronic unit injector (EUI) uses a combination pump and nozzle for each engine cylinder.

An EUI is mechanically actuated by revolving cams or hydraulically actuated by high-pressure engine oil. Injection timing control and fuel quantity metering are electronically controlled by an engine control unit (ECU).

The electronic unit injector system includes these notable features:

1. **Precise control of timing and fuel** — The beginning and end of an injection cycle are electronically controlled.
2. **One unit injector per cylinder** — Provides electronic control of each cylinder.

3. **High injection pressures** — Provides a more efficient combustion and lower exhaust emissions.
4. **Compact design** — Eliminates external injection pump and external steel lines.

The main parts of an EUI system are:

- Fuel Transfer Pump
- Fuel Manifold
- Fuel Rails
- Secondary Fuel Filter
- Primer Pump
- Check Valve
- Primary Fuel Filter/Water Separator
- Engine Cylinder Head
- Unit Injector
- Engine Control Unit (ECU)

SS40167,00000F8 -19-21JUL09-1/1

Electronic Unit Injector (EUI) Fuel System Sub-Systems

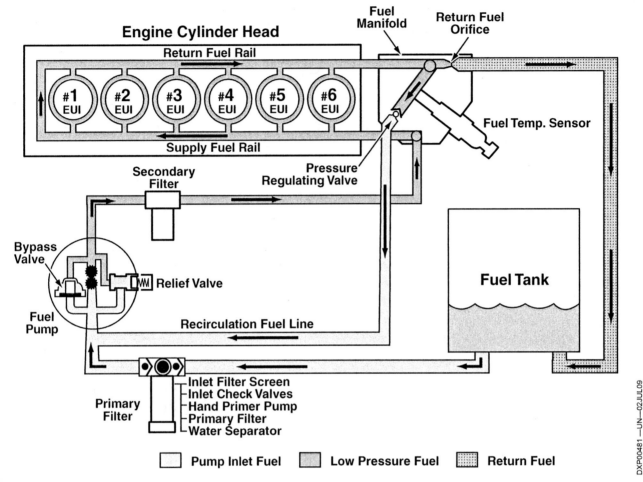

Low-Pressure Fuel Supply System

The typical electronic unit injector (EUI) fuel system consists of two sub-systems:

- **Low-Pressure Fuel Supply System**
- **High-Pressure Fuel Injection System**

LOW-PRESSURE FUEL SUPPLY SYSTEM

The purpose of a low-pressure fuel system is to provide the unit injectors with fuel at a uniform pressure and temperature.

The fuel transfer pump pulls fuel from the tank through a check valve located on the primary fuel filter. A primer pump is located on the primary filter. This pump can be operated by hand to pull fuel through the system after a filter has been changed or when the fuel lines have been disconnected. Some systems use an electric primer instead of a hand primer to purge the system of air.

Fuel flows from the primary filter to the gear-type transfer pump. The transfer pump sends the fuel at low pressure to the secondary fuel filter. Fuel flows from the outlet of the secondary filter to the fuel manifold.

The fuel manifold contains a fuel temperature sensor, pressure regulating valve, return-to-tank orifice, and passages to route fuel to and from the fuel rails in the engine cylinder head. The supply fuel rail and the return fuel rail are passages through the entire length of the cylinder head. (In some systems, these passages are combined.)

Fuel flows from the fuel manifold into the fuel supply rail in the cylinder head. The fuel supply rail routes fuel to each electronic unit injector (EUI).

Excess fuel not needed by the unit injectors flows into the return fuel rail. Fuel in the return rail flows from the cylinder head into the fuel manifold. The fuel manifold will direct a portion of the return fuel to the fuel tank. The remainder of the fuel will flow back to the inlet of the fuel transfer pump through the recirculation fuel line.

This type of system is called a recirculating fuel system. It provides fuel at a uniform temperature at the injector inlet, resulting in reduced power variations due to inlet fuel temperature variations.

Continued on next page

SS40167,00000F9 -19-21JUL09-1/4

Fuel Transfer Pump

The fuel transfer pump ensures fuel is supplied to the unit injectors at all times. Fuel is drawn into a primary fuel filter and pumped through a secondary filter to ensure only clean fuel reaches the unit injectors.

The transfer pump is a gear-type pump. It contains a pressure relief valve and a bypass valve. The relief valve protects the system from excessive pressure. The bypass valve is a one-way check valve located in the pump outlet. It allows fuel to bypass the pump gears during primer pump operation.

Fuel Manifold

The fuel manifold directs the low pressure fuel from the transfer pump into the fuel supply rail in the cylinder head. The fuel supply rail routes fuel to each unit injector. The fuel manifold contains a pressure regulator valve that maintains the fuel pressure in the supply rail at a set value. Any excess fuel is returned to the fuel manifold.

At this point fuel is routed in two directions. A portion of the fuel flows through an orifice and is returned to the fuel tank. The remainder flows through the pressure regulating valve and returned to the inlet of the transfer pump.

Continued on next page

SS40167,00000F9 -19-21JUL09-2/4

HIGH-PRESSURE FUEL INJECTION SYSTEM

The electronic unit injector (EUI) consists of three sections:

- **Pumping Section (A)**
- **Actuator Section (C)**
- **Injector Section (F)**

The pumping section (A) of the EUI consists of a plunger and a return spring. The pumping action is created by the up-and-down movement of the plunger. The plunger may be actuated by either mechanical force or hydraulic force.

Most EUI systems control the plunger movement mechanically by the rotation of a camshaft and the rocking action of a rocker arm.

The pumping action begins when the camshaft lobe pushes on the rocker arm to cause the plunger to move down. The return spring moves the plunger up as the camshaft rotates and relaxes the force on the rocker arm.

The mechanically actuated plunger is operated at a constant stroke; that is, it moves the same distance each time the cam actuates it. To vary the amount of fuel delivered per stroke, a solenoid-operated spill valve (D) controls the start and end (duration) of injection.

By controlling the duration of injection, the spill valve is able to regulate the quantity of fuel delivered to the engine to satisfy varying load conditions. The solenoid (B) is controlled by the electronic engine control unit (ECU).

Some EUI systems control plunger movement hydraulically using high-pressure engine oil. The pressurized oil applies force to a piston that causes the plunger to move down.

In this system, fuel delivery is entirely dependent on the movement of the plunger. Injection timing, quantity of fuel delivered, and injection pressure are controlled by regulating the pressurized oil supply driving the plunger piston.

The pressurized oil supply is regulated by a solenoid, which in turn is controlled by the ECU. When the solenoid is energized, pressurized oil is directed to the plunger piston. The hydraulic force moves the plunger downward and injection begins.

When the solenoid is de-energized, oil pressure is quickly relieved, and plunger movement immediately stops to end the injection cycle.

By varying the hydraulic input pressure, the injection pressure can be controlled within a specific range. This system has the capability to control plunger movement and fuel delivery independent of the engine speed.

For most EUI systems, the actuator section (C) consists of a solenoid (B) and a spill valve (D). The spill valve controls the start of injection, the quantity of fuel delivered

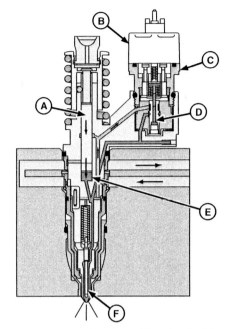

Electronic Unit Injector (Pumping Cycle)

A—Pumping Section
B—Solenoid
C—Actuator Section
D—Spill Valve
E—High Pressure Fuel
F—Injector Section

to the engine, and the end of injection. The spill valve is actuated by a solenoid controlled by the ECU.

When the ECU shuts off current to the solenoid, spring pressure moves the spill valve downward into the open position. This connects the drilled passages in the unit injector body, allowing fuel to flow from the fuel supply rail to the unit injector plunger cylinder, which fills as the plunger moves up.

When the ECU energizes the solenoid, the spill valve moves upward into the closed position. The land on the spill valve contacts the shoulder on the spill valve housing. This closes the path for fuel to escape from the plunger cylinder.

Downward movement of the plunger will cause the fuel pressure in the plunger cylinder to rapidly rise. The rising fuel pressure will cause the injector needle to move up and injection will begin.

The spill valve return spring is compressed when the spill valve moves up. This spring-loads the spill valve so it opens rapidly when the solenoid is de-energized. Fuel pressure in the plunger cylinder immediately drops and the injector needle will close instantly to halt fuel injection.

The injector section (F) is located below the pumping section (A) and consists of a conventional nozzle and needle valve.

Continued on next page

SS40167,00000F9 -19-21JUL09-3/4

The downward (pumping) movement of the plunger causes the fuel pressure to rise in the fuel chamber surrounding the injector needle. The rising fuel pressure overcomes the downward force of the needle spring, causing the needle to move up and injection begins.

As the plunger continues to move down, fuel pressure rises rapidly and injection continues until the actuator section opens the spill port. High-pressure fuel then flows around the spill valve and into the fuel supply rail, allowing a rapid decrease in fuel pressure at the injection needle. Spring pressure then forces the needle onto its seat to halt injection.

SS40167,00000F9 -19-21JUL09-4/4

Electronic Unit Injector (EUI) Operation

To understand how an electronic unit injector (EUI) works, let's trace the fuel through an injector during a complete cycle.

Fill Cycle

The pumping action of an EUI is created by the up-and-down movement of the plunger (F). The downward movement of the plunger is the result of mechanical force being applied to the top of the plunger. The large return spring will then move the plunger up as the downward force on the plunger is relaxed.

During the fill cycle, the plunger (F) is moving up. The engine control unit (ECU) has stopped the electrical signal to the solenoid, which in turn allows spring pressure to move the spill valve (C) to the open position.

The unit injector body has drilled passages to route the fuel through the unit injector.

Low-pressure fuel from the fuel supply rail (D) enters passage A (A) of the unit injector. The fuel flows past the open spill valve into passage B (B), routing the fuel into the plunger cylinder (E), which fills as the plunger moves up.

Fuel also flows to the nozzle, but the fuel pressure is not sufficient to open the needle so no injection takes place.

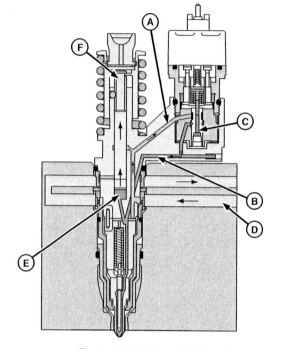

Electronic Unit Injector (Fill Cycle)

A—Passage A
B—Passage B
C—Spill Valve Open
D—Fuel Supply Rail
E—Plunger Cylinder
F—Plunger Moving Up

Continued on next page   SS40167,00000FA -19-21JUL09-1/4

Vent Cycle

As the plunger (A) nears the top of the fill cycle stroke, the vent port (C) opens. This allows trapped fuel and air to flow out of the vent port to the fuel return rail.

A—Plunger at Top of Stroke
B—Fuel Supply Rail
C—Vent Port (Open)

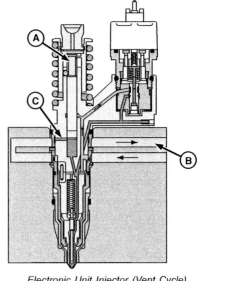

Electronic Unit Injector (Vent Cycle)

SS40167,00000FA -19-21JUL09-2/4

Pumping Cycle

The pumping cycle begins when the camshaft lobe pushes on the rocker arm to cause the plunger (F) to start moving down. During the initial downward movement of the plunger, fuel and air can flow out the vent port (E) until the plunger closes the vent port.

Once the vent port is closed, fuel will be forced out of the plunger cylinder through passage (B). Since the spill valve (C) is open, fuel can flow around the spill valve, into passage (A), and back to the fuel supply rail (D). This flow continues until the spill valve closes.

A—Passage
B—Passage
C—Spill Valve (Open)
D—Fuel Supply Rail
E—Vent Port (Closed)
F—Plunger Moving Down

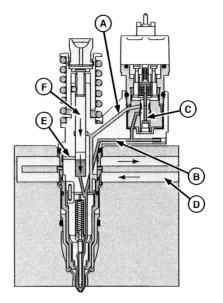

Electronic Unit Injector (Pumping Cycle)

Continued on next page

SS40167,00000FA -19-21JUL09-3/4

Injection Cycle

The injection cycle will start when the engine control unit (ECU) energizes the EUI solenoid (B). This occurs during the downward (pumping) stroke of the plunger.

The energized solenoid moves the spill valve (D) up. The land on the spill valve contacts the shoulder on the spill valve housing. This closes the path for the fuel to escape from the plunger cylinder through passage B to passage A. With the spill valve closed, the downward movement of the plunger causes the pressure of the trapped fuel to rise rapidly.

The rising fuel pressure exerts a hydraulic force on the lower end of the injector needle. When this force overcomes the downward force of the injector needle spring, the needle moves up and injection begins. As the plunger continues to move down, the fuel pressure remains high and injection continues until the ECU de-energizes the solenoid.

When the ECU turns the solenoid off, spring pressure rapidly moves the spill valve to the open position. High-pressure fuel (E) then flows around the spill valve (D) into fuel passage A and back to the fuel supply rail, allowing the fuel pressure to drop rapidly. Spring pressure forces the injector needle down onto its seat and injection stops.

The EUI has the ability to produce injection pressures in excess of 20,000 psi (138 000 kPa).

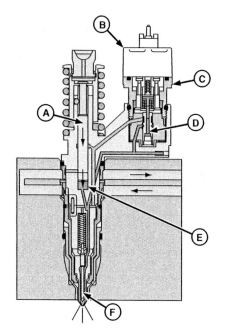

Electronic Unit Injector (EUI) (Injection Cycle)

A—Pumping Section   D—Spill Valve
B—EUI Solenoid   E—High Pressure Fuel
C—Actuator Section   F—Injector Section

SS40167,00000FA -19-21JUL09-4/4

Servicing Electronic Unit Injectors

The electronic unit injectors (EUI) are precision-fitted parts operating rapidly under extreme conditions of heat and pressure. The harsh conditions can cause the EUIs to malfunction. Thorough fuel system diagnostics will determine if an EUI is functioning properly. For details, refer to the engine technical manual for your particular EUI.

Removing Unit Injector

First clean the engine. Remove the rocker arm cover and the rocker arm shaft. Disconnect fuel lines and drain the fuel from the fuel rails in the engine cylinder head. Disconnect the injector solenoid wiring leads.

Remove the injector hold-down clamp cap screw. Pry upward on the hold-down clamp and remove the unit injector and clamp from the engine cylinder head. Label the injector for installation in the same cylinder location as removed.

Most unit injectors have no field serviceable components except the external O-ring seals. Normally, unit injectors are returned to the manufacturer for testing and rebuilding.

Refer to the engine technical manual for instructions specific to the engine.

Installing Unit Injector

Clean the injector bore in the cylinder head. Install new O-rings on the EUI body. Lubricate the O-rings with clean engine oil or petroleum jelly.

Electronic Unit Injector (Removed)

Install the EUI into the same cylinder as removed. Push on the top of the injector plunger with the palm of your hand to properly seat the O-rings and center the injector between the valve springs. Do not use excessive force. If the EUI fails to seat properly in the cylinder head, remove it and determine the cause.

Tighten the injector hold-down clamp cap screws to the torque specified in the engine technical manual.

Connect the injector solenoid wire leads to the solenoid studs. Install the valve-bridges, push tubes, and rocker arm assembly. Adjust the valve stem-to-bridge clearance and the injector preload. Consult the engine technical manual for details.

Continued on next page   SS40167,00000FB -19-23JUL09-1/2

Adjusting Unit Injector (Preload)

Follow the manufacturer's procedure for the unit injector preload adjustment. The following is a typical electronic unit injector (EUI) preload adjustment procedure for a six-cylinder diesel engine:

IMPORTANT: The crankshaft and camshaft must be properly synchronized to accurately adjust the EUI preload.

1. Rotate the engine flywheel in normal running direction until the piston in the No. 1 or first cylinder is at Top Dead Center (TDC) of its compression stroke. The intake and exhaust rocker arms on No. 1 cylinder should be loose.

2. Special timing pins can be inserted into timing slots in the camshaft and crankshaft on some engines to ensure that the camshaft and crankshaft are properly timed.

3. To relieve tension on the injector plunger, loosen the lock nuts (B) and injector rocker arm adjusting screws (A) for No. 3, No. 5, and No. 6 cylinders.

4. Slowly tighten the adjusting screws until the rocker arm roller just makes contact with the camshaft lobe.

5. Tighten the adjusting screws an additional 1/2 turn (180 degrees) to preload the injector. Tighten the

Electronic Unit Injector Adjusting Screw

A—Adjusting Screw　　　B—Lock Nut

adjusting screw lock nut while holding the adjusting screw stationary.

6. Rotate crankshaft one full revolution (360 degrees). Engine will now be at TDC of No. 6 cylinder's compression stroke.

7. Set the injector preload on No. 1, No. 2, and No. 4 cylinders in the same manner as described for No. 3, No. 5, and No. 6 cylinders.

SS40167,00000FB -19-23JUL09-2/2

Electronic Injector/High-Pressure Common Rail Fuel System

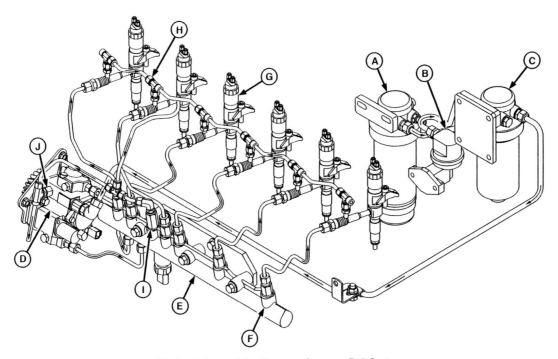

Electronic Injector/High-Pressure Common Rail System

A—Pre-Filter
B—Fuel Transfer Pump
C—Final Filter
D—High-Pressure Fuel Pump
E—High-Pressure Common Rail
F—Flow Damper
G—Electronic Injector
H—Fuel Leak-Off Line
I— Pressure Limiter
J— Overflow Orifice

The high-pressure common rail system utilizes an electronic injector for each cylinder head, with the operation of each injector controlled by a two-way electromagnetic valve (TWV). Because each TWV is controlled by the engine control unit (ECU) that is constantly monitoring the various engine parameters, each valve is controlled to deliver a precise amount of fuel at the exact moment necessary for smooth and efficient engine operation.

Main Parts of the System

The main parts of the system are shown in the illustration.

How the System Works

The fuel transfer pump (B) draws fuel from the tank through a pre-filter (A) and then delivers it to the final filter (C). From the final filter, fuel is transferred to a high-pressure pump (D). The high-pressure pump raises the pressure of the fuel to the required pressure for injection.

After fuel passes through the high-pressure pump, pressurized fuel is routed to a high-pressure common

rail (HPCR) (E), which evenly distributes fuel to each electronic injector (G). The engine control unit opens and closes the two-way electromagnetic valve (TWV) located on each injector to control the volume of fuel, the timing of delivery, and the rate of delivery for each injector.

To maintain a constant, consistent pressure of fuel at each injector, the HPCR utilizes flow dampers (F). If fuel pressure exceeds the established limit of the damper, a pressure limiter (I) will open the damper to relieve the pressure. From the damper, any relieved fuel passes through an overflow orifice (J) and is returned to the fuel tank by way of a fuel return line. (On some applications, this fuel passes through a cooler before returning to the tank.)

The open and close cycle of each two-way electromagnetic valve is dependent on the various parameters monitored by the engine control unit (ECU). The amount of time a damper is open is dependent on the fuel pressure monitored by the pressure limiter. (For more information, see Engine Control Unit (ECU) in this chapter.

Continued on next page   SS40167,00000FC -19-21JUL09-1/6

100709
PN=284

High-Pressure Fuel Pump

MAIN PARTS OF THE PUMP

The main parts of a mechanically controlled high-pressure fuel pump are shown in the illustration.

HOW THE PUMP WORKS

Filtered fuel enters the high-pressure pump through a fuel inlet (A). After passing through the inlet, fuel is filtered once again before being passed along to an internal transfer pump. At this stage, fuel is used to lubricate the pump crankcase or is sent on to the pump control valve (E).

A high-pressure pump has two chambers — one on top of the pump, the other on the bottom of the pump. Fuel in each chamber is pressurized when the pump camshaft rotates. Excess fuel drains from the pump at the overflow orifice (B) and is redirected to the fuel tank by way of a fuel return line.

Some high-pressure fuel pumps utilize a temperature sensor (D) that allows the engine control unit (ECU) to monitor the temperature of the fuel. Because fuel temperature affects the efficiency of an engine, the ECU

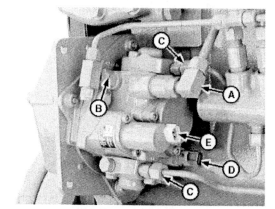

High-Pressure Fuel Pump

A—Fuel Inlet
B—Overflow Orifice
C—Fuel Outlets
D—Fuel Temperature Sensor
E—Pump Control Valve

considers fuel temperature when calculating how much fuel to inject into a cylinder.

SS40167,00000FC -19-21JUL09-2/6

High-Pressure Common Rail

MAIN PARTS OF THE COMMON RAIL

The main parts of a common rail are shown in the illustration.

HOW THE COMMON RAIL WORKS

Fuel from the high-pressure pump is delivered to the high-pressure common rail (HPCR) by way of two high-pressure delivery lines (A). From the HPCR, fuel is distributed to the fuel injectors by a series of common-rail delivery lines (E).

A pressure sensor (B) monitors fuel pressure in the rail. And the engine control unit (ECU) uses this pressure data to establish pump control valve timing on the fuel pump. If an abnormally high pressure is detected, the pressure limiter (C) will release fuel to relieve the pressure. The released fuel will then be redirected back to the fuel tank.

Flow dampers (D) protect the injectors from fuel leaks or injector failures, by stopping fuel flow to the failed injector.

High-Pressure Common Rail

A—Pump Delivery Lines
B—Pressure Sensor
C—Pressure Limiter
D—Flow Dampers
E—Common-Rail Delivery Lines

Continued on next page

SS40167,00000FC -19-21JUL09-3/6

Electronic Fuel Injector

MAIN PARTS OF THE ELECTRONIC FUEL INJECTOR

The main parts of an electronic injector are shown in the illustration.

HOW THE ELECTRONIC INJECTOR WORKS

An electronic fuel injector is simply a gate that controls the flow of fuel from the high-pressure common rail to the engine cylinder associated with the injector.

The open and close cycle of an injector is controlled by the engine control unit (ECU) as it monitors the various operating parameters of the engine. The time duration the injector is open to allow fuel to flow is directly proportional to the amount of time the ECU holds the injector valve open.

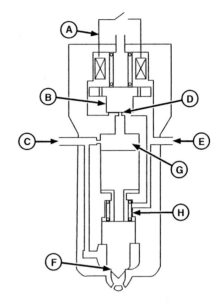

A—Two-Way Valve (TWV)
B—Solenoid Valve
C—Fuel Inlet
D—Orifice Seat
E—Fuel Leak-Off Port
F—Nozzle
G—Control Chamber
H—Valve Spring

Electronic Fuel Injector

SS40167,00000FC -19-21JUL09-4/6

Fuel from the high-pressure common rail (HPCR) enters the injector at the fuel inlet (C) as shown. Fuel injection begins when electrical current from the ECU is applied to the TWV (A).

By energizing the TWV, the resulting electromagnetic field pulls up the solenoid valve (B), causing the orifice seat (D) to open. Fuel from the control chamber (G) then flows through the orifice seat, where it makes its way to the fuel leak-off port (E). From the leak-off port, fuel is returned to the fuel tank by way of a fuel leak-off line.

As fuel exits the control chamber, negative hydraulic pressure is created, resulting in the hydraulic piston moving away from the injector nozzle (F). Because hydraulic pressure against the shoulder of the nozzle is greater than the pressure of the valve spring (H) that holds the nozzle tight against the tip of the injector, the nozzle is forced up, allowing fuel to spray into the cylinder.

The spray of fuel stops once the ECU cuts electrical current to the two-way valve. The solenoid valve retracts, fuel fills the control chamber, and the valve spring (along with hydraulic fuel pressure) pushes the needle tight against the tip of the injector to stop the spray of fuel to the cylinder.

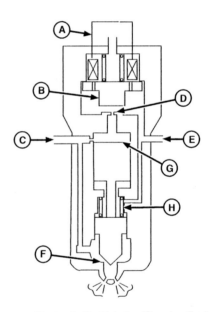

Electronic Fuel Injector (Spraying Fuel)

A—Two-Way Valve (TWV)
B—Solenoid Valve
C—Fuel Inlet
D—Orifice Seat
E—Fuel Leak-Off Port
F—Nozzle
G—Control Chamber
H—Valve Spring

Continued on next page

SS40167,00000FC -19-21JUL09-5/6

SERVICING ELECTRONIC INJECTORS

For details on removing, installing, and servicing electronic injectors, refer to the engine technical manual for your particular injector.

After removing injectors, always store them in a clean location. Cover injector bores to prevent contaminating the fuel systems. Do not allow oils to come in contact with fuel passages or sealing surfaces. Always use new seals when reinstalling injectors.

Light deposits can be cleaned from the injector bore using a fine thread cleaning brush. Injectors can be cleaned using diesel fuel and a brass wire brush to remove carbon deposits. The injector orifice cannot be cleaned. If it is plugged, replace the injector.

Electronic Fuel Injector

SS40167,00000FC -19-21JUL09-6/6

Electronic Control System for EUI and EI Fuel Systems

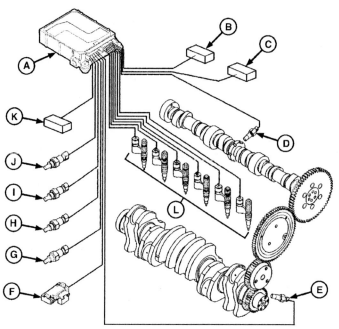

Electronic Unit Injector (EUI) Control System

A—Engine Control Unit (ECU)	D—Cam Position Sensor	G—Manifold Air Temperature (MAT)	J— Coolant Temperature
B—Monitor Output	E—Crank Position Sensor	H—Water in Fuel	K—Throttle Position Sensor
C—Diagnostics	F—Manifold Air Pressure (MAP)	I— Fuel Temperature	L—Electronic Fuel Injectors

For an electronic unit injector (EUI) system or an electronic injector (EI) system, the engine control unit (ECU) (A) serves as the engine governor by controlling all injector functions. The ECU regulates fuel delivery according to a given set of engine conditions, in precise amounts, and at a precise time in relation to position of the piston. In order to achieve this, the following parameters are monitored by the ECU:

• Engine Coolant Temperature
• Fuel Temperature (I)
• Water in Fuel (H)
• Manifold Air Temperature (MAT) (G)

• Manifold Air Pressure (MAP) (F)
• Throttle Position
• Engine Speed
• Camshaft Position
• Crankshaft Position

The ECU constantly receives input signals from the engine sensors and compares those signals to information stored in the memory of the ECU. After analyzing the input signals, the ECU determines when to start injection and the amount of fuel to be delivered. The ECU then sends output signals to the electronic injectors to precisely control the timing and fuel delivery.

SS40167,00000FD -19-21JUL09-1/1

DXP00495 —UN—23DEC08

Engine Control Unit (ECU)

The engine control unit (ECU) is a self-contained electronic device that is software controlled. The ECU performs the following functions:

- Monitors all engine operating parameters and makes the necessary adjustments to ensure the engine runs as smoothly and efficiently as possible.
- Converts the electrical signals (analog voltage signals) from various sensors into digital data.
- Controls electronic fuel injectors for optimum fuel delivery and engine efficiency.
- Provides all-speed governing.
- Performs self-diagnostic testing on the control system.
- Stores error codes.

The ECU consists of the following subsystems:

- Analog/Digital Converters
- Central Processing Unit (CPU)
- Memory

The **Analog/Digital Converter** converts the analog voltage readings from the various sensors into digital data for the central processing unit (CPU).

The **Central Processing Unit (CPU)** compares signals from the various sensors around the engine, to the preprogrammed values stored in the ECU. Using this information, the CPU determines the precise amount of fuel to inject in each cylinder and when it should be injected. The CPU also controls system diagnostics.

The ECU contains three types of **Memory**:

- Random-Access Memory (RAM) — Temporarily stores data from the sensors and the results of various calculations. Information in RAM is lost when battery voltage to the ECU is removed.
- Read-Only Memory (ROM) — Contains programmed information that can only be read, not changed. ROM information is retained when battery voltage is removed.
- Programmable Read-Only Memory (PROM) — Contains information programmed at the factory such as data specific to the engine. Information in PROM is retained when battery voltage is removed.

For more information on electronic fuel injection systems, see the FOS manual on Electronic and Electrical Systems.

ENGINE CONTROL UNIT SELF-DIAGNOSTICS

The engine control unit (ECU) has the ability to diagnose problems internally and in the electronic control system.

This includes determining if any of the sensor input voltages are too high or too low, if the camshaft and crankshaft position sensor inputs are valid, and if the unit injector solenoids are responding properly.

If the ECU detects a problem with the electronic control system, a diagnostic trouble code (DTC) specific to the failed system will be stored in the ECU's memory. There are two types of DTCs: Active and Inactive.

- **Active trouble codes** — Indicate that a failure is occurring. These types of failures are know as "hard" failures.
- **Inactive trouble codes** — Indicate that a failure has occurred in the past, but is not occurring at the moment. This type of trouble code can be caused by an intermittent failure, such as a bad connection or a wire briefly shorting to ground.

If the ECU detects a problem that can be diagnosed, whenever possible it will switch to a safe mode of operation to back up the system.

DIAGNOSTIC PROCEDURE

Make sure all engine mechanical and electrical systems not related to the electronic control system are operating properly.

A diagnostic reader is used to analyze the diagnostic output signal of the ECU and display the associated diagnostic codes to the operator as a number or a series of numbers. A diagnostic reader may simply be a digital tachometer in the vehicle's instrument panel or a specialized electronically controlled diagnostic scan tool, such as a portable laptop computer. The ECU can also flash a fault lamp (warning light) to indicate the diagnostic status of the system.

Read and record the diagnostic trouble code (DTC). Consult the engine technical manual for a description of the DTCs that can occur in the electronic control system.

NOTE: If more than one DTC is displayed, refer to the chart corresponding to the lowest numbered DTC received and work through that particular problem first.

After any repairs are made, recheck the system to make sure each DTC has been eliminated.

SS40167,00000FF -19-21JUL09-1/1

Electronic Control System Operation

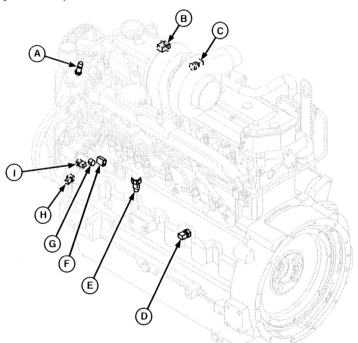

Electronic Control System

A—Crank Position Sensor
B—Water in Fuel (WIF) Sensor Harness
C—Fuel Temperature Sensor
D—Oil Pressure Sensor
E—Fuel Rail Pressure Sensor
F—Pump Position Sensor
G—Pump Control Valve (PCV)
H—Engine Coolant Temperature (ECT) Sensor
I— Manifold Air Temperature (MAT) Sensor

NOTE: Some of the sensors shown are optional items and are NOT used on all applications.

The electronic control system serves as an engine governor by controlling the electronic injectors (EI) so that fuel is delivered according to a given set of engine conditions, precise amounts, and at the precise time in relation to piston position. In order to achieve this, the control system performs the following functions:

- Constantly monitors engine operation conditions.
- Precisely determines piston position.
- Delivers optimum amount of fuel for a given set of operating conditions.
- Delivers fuel at optimum piston position.
- Provides multiple control modes.
- Performs self-diagnosis.

MONITORING ENGINE PARAMETERS

In order for the electronic control system to deliver fuel according to a given set of operating conditions and, on some applications to provide engine protection, the following engine parameters are monitored by the ECU:

- Engine Coolant Temperature (ECT)
- Fuel Temperature
- Manifold Air Temperature (MAT)

- Fuel Rail Pressure
- Oil Pressure
- Crank Position
- Pump Position
- Pump Control Valve (PCV)
- Water in Fuel (WIF)

Engine Coolant Temperature (ECT) Sensor

The ECU monitors engine coolant temperature for:

- Engine protection purposes.
- Starting fuel quantity determination.
  - The ECU will adjust the amount of fuel delivered during start-up based on the initial ECT readings.
- Idle speed determination.
  - In order to speed engine warm-up, the ECU will increase idle speed after start-up if low coolant temperature is measured.

Fuel Temperature Sensor

The ECU uses this sensor input to calculate fuel density and adjust fuel delivery accordingly. The ECU also uses the fuel temperature sensor for engine protection purposes.

Manifold Air Temperature (MAT) Sensor

Continued on next page SS40167,00000FE -19-21JUL09-1/2

The MAT sensor measures intake air temperature to help the ECU calculate for correct fueling. This is an optional sensor that is not included on all applications. The ECU also monitors manifold air temperature for engine protection.

Fuel Rail Pressure Sensor

The ECU monitors fuel pressure to control the amount and timing of fuel transferred from the high pressure fuel pump to the high-pressure common rail (HPCR).

Oil Pressure Sensor

The ECU monitors oil pressure for engine protection.

Crank Position Sensor

The crank sensor is an inductive-type pickup sensor that detects teeth on the oil pump drive gear. The ECU uses the crank position input to determine engine speed and precise piston position in relation to the firing order. The oil pump drive gear is composed of 72 evenly-space teeth with two of the teeth ground off. These two missing teeth help the ECU determine when No. 1 cylinder is at top-dead-center (TDC).

Pump Position Sensor

The pump position sensor is an inductive-type pickup sensor that detects notches on the auxiliary gear of the high-pressure fuel pump camshaft. The auxiliary gear consists of six evenly spaced notches with one additional offset notch. This offset notch helps the ECU determine when cylinder number one is approaching top-dead-center (TDC).

Pump Control Valve (PCV) Sensor

The ECU sends an electronic signal to the PCV to regulate the delivery of fuel to the high-pressure common rail (HPCR). When the PCV is energized, fuel is allowed to discharge from the fuel outlet located on the high-pressure fuel pump, to the HPCR.

Water-in-Fuel (WIF) Sensor

When water is detected in the fuel, a signal is sent to the ECU. The WIF sensor uses the resistance of the fuel and water in the fuel system, along with the principle that water is a better conductor than fuel. If water is present voltage will be lower. The ECU monitors this for engine protection.

SS40167,00000FE -19-21JUL09-2/2

Combustion Chambers

As described earlier, an injection nozzle atomizes fuel with air to create a fine particle spray that is injected into the combustion chamber. However, it is the design of the combustion chamber that plays the key role in providing the proper fuel-to-air ratio necessary for efficient combustion.

OPEN COMBUSTION CHAMBER

The most commonly-found combustion chamber is an "open" combustion chamber.

The open combustion chamber is designed to mix atomized fuel with the fresh air that is drawn into the combustion chamber during the intake stroke. Because the top of the piston is domed and the area of the cylinder head just above the piston is dished, swirling air turbulence is created to mix the fuel and air more thoroughly during the intake stroke, to provide a more efficient combustion stroke later on.

OTHER TYPES OF COMBUSTION CHAMBERS

In some small, high-speed diesel engines, the combustion stroke is slow and inefficient. For these engines, a more efficient combustion chamber is necessary to compensate for this lag and inefficiency. Three combustion chambers capable of making this compensation are:

* Turbulence Chamber
* Pre-Combustion Chamber
* Energy Cell Chamber

Turbulence Chamber

Fuel is injected into a small spherical chamber connected to the cylinder. The shape of this chamber produces a highly turbulent condition. As the piston starts the compression stroke, air is forced into the turbulence chamber, to set up for a rotary motion. Near the top of the piston stroke, fuel is injected into the swirling air for a more thorough mixing of the fuel and air.

Pre-Combustion Chamber

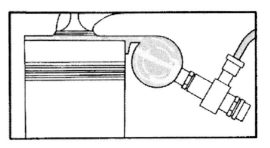

Pre-Combustion Chamber

A pre-combustion chamber is similar to a turbulence chamber. However, the combustion stroke begins by burning a rich fuel mixture in a pre-combustion chamber. As the combustion stroke begins, burning fuel is forced from the pre-combustion chamber into the combustion chamber. The velocity at which this transfer occurs results in the burning fuel mixing with the remaining fresh air in the combustion chamber.

SS40167,0000100 -19-05AUG09-1/2

Energy Cell Chamber

During the compression stroke, the piston forces air into a small cell (chamber) separate from the main combustion chamber. Near the end of the compression stroke, the fuel nozzle sprays atomized fuel across the top of the piston, toward the energy cell and ignites.

Approximately half of the fuel is burned in the main chamber, while the remaining fuel is forced into the energy cell. The remaining fuel then ignites in the cell and blows out into the main chamber. The turbulence created by the combustion in the energy cell, mixes and ignites any remaining unburned fuel in the main chamber.

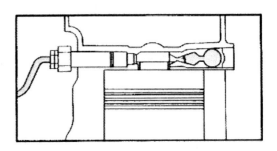

Energy Cell Chamber

SS40167,0000100 -19-05AUG09-2/2

General Problems of Diesel Combustion

The primary goal of the diesel fuel system is to deliver fuel to the engine cylinder for combustion.

However, good combustion depends on four conditions:

- **Fuel must be of good quality and correct type.**
- **Fuel must be finely atomized.**
- **There must be high temperatures.**
- **Fuel and air must be completely mixed.**

Let's look at a few of these briefly.

CORRECT TYPE OF FUEL

Diesel fuel is graded for various types of engine service. Always follow the operator's manual for the engine when selecting fuel.

Other factors are sulfur content, water and sediment contents, pour point, and cloud point.

The cetane number rates the ignition quality of the diesel fuel — at a given pressure and temperature that the fuel will ignite and burn.

ATOMIZING FUEL

The injection nozzle atomizes the fuel which helps the fuel mix with the air. However, the nozzle also helps to make a **vapor** of the fuel.

Vaporizing the fuel is important. **Remember, only a gas or a vapor will burn**. In this case, the heat generated during compression vaporizes the fuel. But, because so little time is allowed for combustion, the fuel particles must be as small as possible so that they will vaporize fast enough to permit a complete combustion.

This does not mean that fuel becomes a gas and this gas mixes with the air and then ignites. Actually, each fuel particle absorbs heat and throws off vapor. This vapor then mixes with the nearest air molecule and ignites. All the combined fuel vapor-air particles do not ignite at the same time. However, the speed of this process is so quick it appears they do ignite at the same time.

HIGH TEMPERATURE REQUIRED FOR COMBUSTION

We learned in Chapter 1 that heat generated by compressed air does the same job in a diesel, as a spark does in a gasoline engine. The amount of heat is determined by the compression ratio of the engine, the size of the cylinders, and the efficiency of the cooling system. Diesel engines with small cylinders have a higher compression ratio as similar-size gasoline engines; and the higher the compression, the more heat available to initiate combustion.

In summary: Higher diesel compression equals more heat, which equals combustion without a spark.

SS40167,0000101 -19-21JUL09-1/1

Starting a Cold Diesel Engine

Diesel fuel does not efficiently ignite in a cold engine. There are two main types of cold start aids that can be used to increase cylinder combustion chamber heat. They are:

- Cold weather starting fluid added to the intake manifold.
- Thermal heating elements (glow plugs) that warm parts of the engine.

Always follow the engine operator's manual when a starting aid is used.

Starting fluid is not affected by low temperatures because it has a low combustion point. When it is injected into the intake manifold and compressed in the cylinder head, starting fluid ignites without the use of a thermal heating element. Starting fluid may also be sprayed into the air cleaner element from an aerosol can.

⚠ **CAUTION: Do not use thermal starting aids (glow plugs) in the cylinder head or the intake manifold when starting fluid is used on a cold diesel engine. Glow plugs in the turbulence chamber, intake manifold heating coils, and other thermal aids can cause the starting fluid to ignite prematurely. This can result in personal injury and damage to the engine.**

If you have a starting fluid can mounted in the fuel system, spray starting fluid into the engine only while cranking with the starter. Too much starting fluid can cause an uncontrolled explosion in cylinder chambers and damage the engine. Also, starting fluid must never come in contact with a thermal starting aid.

Thermal starting aids include:

- **Glow plugs**
- **Lubricant and coolant heating coils**
- **Intake air heater coils**

Thermal starting aids can be used individually or in a combination. Thermal starting aids are coils of wire that carry electric current much like an electric blanket. Heat is created by the electrical resistance of the coil. This heat is transferred to the air intake, coolant, or lubricant reservoirs.

The most common thermal starting aid is the glow plug (A). It differs from other thermal aids because it is mounted in the turbulence chamber of the cylinder head as shown. The injection nozzle also is in the chamber.

The glow plug is turned on to preheat the cylinder chamber. It is then turned off just before the starter motor

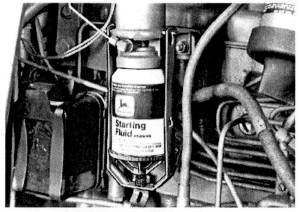

Injecting Starting Fluid for Cold Weather

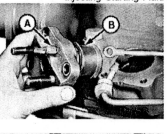
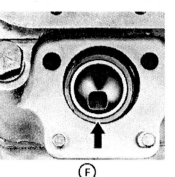
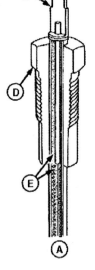

Cold Weather Glow Plug

A—Glow Plug	D—Body
B—Upper Turbulence Chamber	E—Heating Element
C—Terminal	F—Lower Turbulence Chamber

is engaged. After the initial ignition, sufficient enough to keep the engine running, heat is generated by diesel fuel combustion and cylinder movement.

If starting fluid is sprayed onto the glow plug when it is on or hot, the starting fluid could explode. Starting fluid contacting heating elements in the air intake also is dangerous. **Do not use starting fluid and air intake aids simultaneously.**

SS40167,0000102 -19-21JUL09-1/1

Bleeding Air from a Diesel Fuel System

Whenever the diesel fuel system has been opened for service (fuel lines disconnected or filters removed), it will be necessary to bleed air from the system. The reason for this is that air will compress while a liquid (fuel) will not. If air is present in the fuel lines, the injection pump plunger is not capable of producing sufficient pressure and volume on its pumping stroke (due to the compressibility of the air) to lift the needle in the injector off its seat to purge the air.

Therefore, the air must be purged from the system so that only non-compressible diesel fuel is in the fuel system.

Failure to bleed air from a fuel system may result in the engine not starting, engine misfiring, or a loss of engine power.

The fuel system may be bled at one of several locations. Consult the engine technical manual or operator's manual for the recommended bleeding locations and procedure.

Here is a general guide for bleeding air from a diesel fuel system:

1. Fill the fuel tank with the correct type of diesel fuel.

2. Open the fuel shutoff valve at the tank. Some applications use a fuel shutoff solenoid, which must be energized to allow fuel to flow through the system. Turn the engine key switch to the ON position to energize the solenoid.

3. Loosen the bleed plug (A) on the fuel filter base. On engines with series filters, the air can be bled from both filters by loosening the bleed plug on the second stage filter.

4. Actuate the hand primer (B) on the fuel supply pump until a smooth flow of fuel, free of bubbles, flows out of the filter bleed plug hole.

5. Gently stroke the hand primer down and at the same time close the bleed plug.

6. When bleeding is completed, continue operating the hand primer until slight pressure is felt. The pressure indicates that fuel has filled the gallery in the injection pump.

   If the engine fails to start after bleeding air from the filters, you will also have to bleed air from the injection lines.

Fuel Filter Bleed Plug

DXP00502 —UN—23DEC08

Fuel Pump Primer Lever

A—Air Plug Bleed   C—Primer Lever
B—Hand Primer

⚠ CAUTION: To prevent injury from high-pressure fluid, loosen injection line connectors only one turn to avoid excessive spray.

7. Hold the injection line nuts with a wrench and loosen the injection lines at the injection nozzles. If equipped with a fuel shutoff solenoid, turn the engine key switch to the ON position. Move the hand throttle to the slow idle position. Crank the engine with the starter until fuel without foam flows from the loose fuel line connections.

   Tighten the injection line connections carefully. Avoid bending the line connection. Tighten the lines only until snug and free of leaks.

8. Start the engine and run it at fast idle for at least 3 to 5 minutes.

SS40167,0000103 -19-21JUL09-1/1

Troubleshooting a Diesel Fuel System

Since we have already tested and serviced the individual components of a diesel fuel system, here we will provide general troubleshooting information for a diesel fuel system.

Listed below are causes for typical failures for a diesel fuel system. For troubleshooting a complete engine, refer to the charts in Diagnosis and Testing of Engines in Chapter 13.

Engine Will Not Start, Starts Hard, or Misfires

1. No fuel in tank; leaky or clogged supply line or filter; water in fuel.
2. Fuel transfer pump not functioning.
3. Air lock in injection pump.
4. Throttle linkage to pump loose or broken.
5. Drive shaft or key sheared; pump plungers or distributor seized.
6. Pump delivery valves not seating properly due to dirt or broken seals.
7. Pump out of time.
8. Nozzles not functioning because of stuck valves or plugged orifices.

Engine Will Not Idle Smoothly

1. Governor idling adjustment not set properly or throttle linkage worn.

2. Pump rack or control sticky or stuck; stuck plunger.
3. Nozzle opening pressure incorrect; stuck nozzle valves.
4. Pump out of calibration; loose control sleeves; leaky delivery valves; pump timing incorrect.
5. Dirty fuel filter.

Engine Smokes or Knocks

1. Pump out of time.
2. Dirty or fouled nozzles; valves stuck open; opening pressure too low.
3. Fuel stop setting incorrect so that fuel delivery is excessive.
4. Other possible causes:
   - Air cleaner dirty.
   - Broken engine valves.
   - Scored engine pistons.
   - Turbocharger speed reduced by leaky manifold gaskets.

Engine Lacks Power

1. Pump timing retarded.
2. Pump plungers or distributor worn.
3. Faulty nozzles.
4. Governor out of adjustment.
5. Plugged air or fuel filters.

SS40167,0000104 -19-21JUL09-1/1

Test Yourself

QUESTIONS

1. (True or False?) Diesel combustion takes place when fuel and hot air are ignited by a spark.

2. Why is the distributor injection pump system well suited for a multi-cylinder engine?

3. How is the distributor injection pump lubricated?

4. (Fill in the blanks.) Injection nozzles are simple mechanisms; they use _____ pressure to oppose _____ pressure.

5. What kind of a brush should be used to clean deposits from injector nozzles?

6. How does fuel flow through two series filters as compared to two parallel filters?

7. Which type of filters clean fuel the best?

8. Name the two most common types of electric fuel gauges.

9. (Fill in the blanks.) The electronic unit injector (EUI) uses a combination _____ and _____ for each engine cylinder.

10. Name the three major sections of the electronic unit injector (EUI).

11. (True or False?) Fuel injection will stop when the spill valve of the electronic unit injector (EUI) moves to the closed position.

12. (True or False?) The electronically controlled fuel injection system uses an electronic actuator to replace the mechanical governor located on the rear of the in-line injection pump.

13. (True or False?) The engine control unit (ECU) or "black box" contains electronic circuitry and computer software that together perform governor and diagnostic functions.

14. Name two types of diagnostic readers that may be used to read the diagnostic output signal from the engine control unit (ECU).

15. (True or False?) A thermostatic fuel gauge uses a single bimetal blade.

16. (True or False?) On a high-pressure common rail fuel system, flow dampers are used to maintain constant and consistent fuel pressure at each injector.

17. (Fill in the blanks.) The amount of fuel injected into the engine is directly proportional to the _____ that the ECU holds the _____ valve open.

18. (True or False?) On a high-pressure common rail fuel system, dampers leak off excess fuel to the fuel tank if an abnormally high fuel pressure is detected.

19. (True or False?) On a high-pressure common rail fuel system, based on the input received from the ECT sensor, the ECU will adjust engine idle speed to cause the engine to warm up faster.

SS40167,0000105 -19-21JUL09-1/1

Intake and Exhaust Systems — Introduction

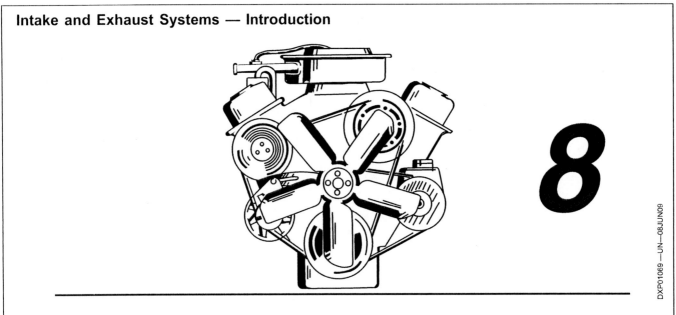

DP79986,0000045 -19-04JUN09-1/1

Air Intake Systems

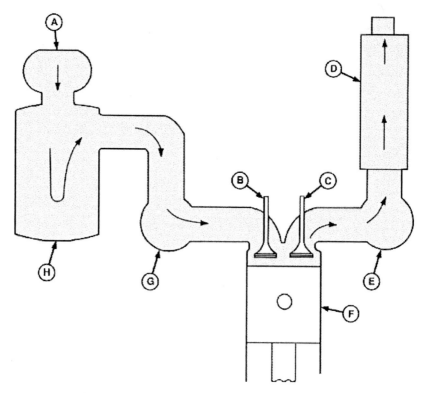

A—Pre-Cleaner
B—Intake Valve
C—Exhaust Valve
D—Muffler
E—Exhaust Manifold
F—Cylinder
G—Intake Manifold
H—Air Cleaner

The intake system carries the air-fuel mixture into the engine, while the exhaust system carries exhaust gases away.

The air intake system supplies the engine with the proper amount of clean air, at the right temperature, with the correct amount of fuel that will be mixed for best combustion.

A malfunction of any part of the air intake system can cause severe damage to the engine very quickly. Airborne dirt and dust particles that find their way into the combustion chamber are turned into extremely hard abrasive particles by the heat of combustion. These abrasive particles account for the majority of the wear that occurs to the pistons, rings, cylinders, and bearings.

There are two basic types of air intake systems:

• Cylinder Feed
• Crankcase Feed

CYLINDER FEED — Fuel is mixed with air as it passes through the carburetor (spark-ignition engines). Then the air-fuel mixture passes into the cylinder through the intake valve.

Air only passes through the intake manifold and intake valve into the cylinder of the diesel engine. Then fuel is sprayed directly into the engine combustion chamber where it mixes with the compressed air.

CRANKCASE FEED — Some two-stroke cycle engines (see Chapter 1) draw the air-fuel mixture into the crankcase before routing it, under pressure, into the cylinder.

The intake system consists of:

• **Air Cleaner, Pre-cleaner (if used)**
• **Supercharger or Turbocharger (if used)**
• **Carburetor Air Inlet (spark-ignition engines)**
• **Intake Manifold**
• **Intake Valve**

AIR CLEANER filters dust and dirt from the air passing through it en route to the intake manifold.

PRE-CLEANER prevents larger particles from reaching the air cleaner and plugging it.

SUPERCHARGER or TURBOCHARGER increases horsepower by packing more air or air-fuel mixture into the engine cylinders than the engine could take in by normal atmospheric pressure.

CARBURETOR mixes fuel with incoming air in the proper proportion for combustion in spark-ignited engines.

INTAKE MANIFOLD transports the air-fuel mixture (air only on diesel engines) to the engine cylinders.

Continued on next page

SS40167,0000108 -19-22JUL09-1/3

DXP00503 —UN—23DEC08

100709
PN=299

INTAKE VALVE admits air to diesel engines and air-fuel mixture to spark-ignition engines. See Chapter 3 for engine information.

Continued on next page   SS40167,0000108 -19-22JUL09-2/3

Operation of Air Intake System

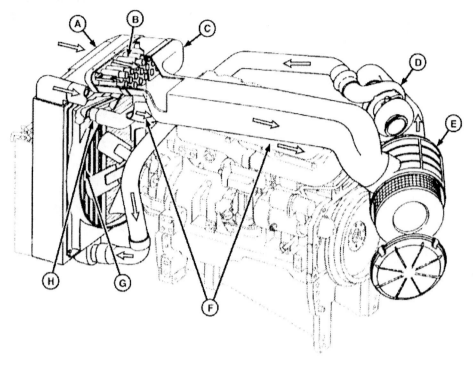

A—Air Intake Duct
B—Tubes

C—Pre-Cleaner
D—Turbocharger

E—Air Cleaner
F—Intake Manifold

G—Air-To-Air Aftercooler
H—Fan Shroud Aspirator

The air intake system supplies clean air for the engine combustion process. Clean air is a necessity for long engine life. If airborne dirt or dust particles are allowed to enter the combustion chamber, premature engine wear and damage will result.

The air intake system shown in the illustration is typical of one used on a modern high-output diesel engine. In this system, air enters through a grille screen located at the front of the air intake duct and passes into a pre-cleaner. Tubes are housed in the pre-cleaner compartment directly behind the air duct.

Air flows through the pre-cleaner tubes and passes over angled fins in the tubes that cause the air to swirl. Heavy dust particles are forced toward the outside of the tubes by centrifugal force where they are expelled with a portion of the air through a tube wall opening.

The dirty air is removed from the pre-cleaner compartment through a hose to the fan shroud aspirator, which creates a partial vacuum in the pre-cleaner compartment. About 94% of the dust, two microns or larger, is separated from the incoming air at the pre-cleaner.

The pre-cleaned air continues through the tubes into the air duct, passing into the air cleaner housing. Filtering is done as the air passes through a paper element contained in the cleaner housing. The air then passes through a secondary paper element and flows into the turbocharger compressor inlet.

The turbocharger is basically an air pump that compresses the air in the cylinder, permitting a more optimized air-fuel ratio and complete combustion. The compressed air leaving the turbocharger may flow directly to the intake manifold, or it may be directed to a aftercooler before entering the intake manifold.

The air intake system shown in the illustration directs the pressurized air from the turbocharger outlet to an air-to-air aftercooler heat exchanger mounted ahead of the engine fan. The compressed air flows through a series of tubes in the aftercooler, while a stream of air created by the fan flows around the tubes.

The cooler air flowing around the aftercooler tubes reduces the temperature of the compressed air from the turbocharger before it enters the engine intake manifold. This cooler, compressed air increases fuel efficiency and minimizes exhaust emissions.

SS40167,0000108 -19-22JUL09-3/3

Air Cleaners

Clean air is essential to satisfactory performance and long engine life. The air cleaner must be able to remove fine materials such as dust and blown sand as well as chaff or lint from the air.

A cleaner must also have a reservoir large enough to hold material taken out of the air so operation over a reasonable period of time is possible before cleaning and servicing is necessary.

A buildup of dust and dirt in the air cleaner passages will eventually choke off the air supply, causing incomplete combustion and heavy carbon deposits on valves and pistons.

Multiple air cleaner installations are sometimes used where engines are operated under extremely dusty air conditions or where two small air cleaners must be used in place of a single large cleaner.

The most common types of air cleaners are:

- Pre-cleaners
- Dry-type cleaners
- Dry element cleaners
- Viscous-impingement cleaners
- Spiral-rotor cleaners
- Oil bath cleaners

Pre-Cleaners

Pre-cleaners are usually installed at the end of a pipe extended upward into the air from the air cleaner inlet. On some applications, the pre-cleaner may be positioned at the front of the vehicle. This places it in an area relatively free of dust.

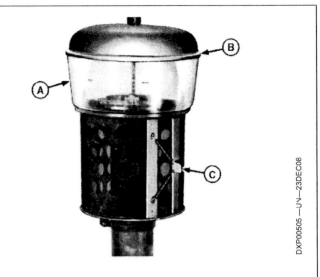

A—Collector Bowl   C—Pre-Screener
B—Pre-Cleaner

Pre-cleaners are simple devices that remove larger particles of dirt or other foreign matter from the air before it enters the main air cleaner. This relieves much of the load on the main cleaner and allows longer intervals between servicing.

Most pre-cleaners have a pre-screener, which prevents lint, chaff, and leaves from entering the air intake.

SS40167,000010A -19-15JUL09-1/18

Dry-Type Air Cleaners

Dry-type air cleaners are attached directly to the carburetor or manifold. They are only used on engines in which the demand for air is small.

Dry-type air cleaners or filters clean the air by passing it through layers of cloth or felt. It is most effective in removing larger dirt particles from the air and is used mostly on small engines.

A—Dry-Type Air Cleaner

Continued on next page   SS40167,000010A -19-15JUL09-2/18

Dry Element Air Cleaners

Two major types of dry element cleaners are used at the present time.

Dry element air cleaners are built for two-stage cleaning:

1. Pre-cleaning
2. Filtering

This first stage (pre-cleaning) directs the air into the cleaner at high speed so that it sets up centrifugal rotation (cyclone action) around the filter element.

The cleaner shown on the right in the illustration conducts the air past tilted fins which start the centrifugal action. When the air reaches the end of the cleaner housing, the dirt passes through a slot in the top of the cleaner and enters the dust cup.

This pre-cleaning action removes 80 to 90% of the dirt particles and greatly reduces the load on the filter.

The partially cleaned air then passes through the holes in the metal jacket surrounding the pleated-paper filter. Filtering is done as the air passes through the paper filter. It filters out almost all of the remaining small particles. This is the second stage cleaning.

Some heavy-duty cleaners use a spiral rotor device for pre-cleaning the air. Others have a small safety element built into the unit in case the main element fouls.

If the air cleaner has a dust cup, it should be emptied daily. If an automatic dust unloader is used in place of a

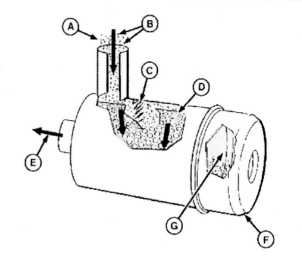

A—Air Inlet
B—Dirt Particles
C—Tilted Fins
D—Filter Element

E—Cleaned Air Outlet
F—Dust Cup
G—Slot

cup, it is usually recommended that it be checked at least once daily to make certain it doesn't become clogged. The dust unloader is a rubber duck-bill device that is held closed by engine suction while the engine is running. When it stops, the weight of the accumulated dirt helps open the flaps so the dirt can drop out.

SS40167,000010A -19-15JUL09-3/18

Viscous-Impingement Air Cleaners

In viscous-impingement air cleaners, air flows through a maze of metal wool, wire, or screens saturated with oil.

This type was used for automotive applications where dust conditions are relatively light. The lower portion of the unit acts as a hollow chamber to quiet the air intake noise.

Viscous-impingement type air cleaners are now occasionally used only on small engines. They utilize an oil-impregnated element for filtering.

A—Oil-Saturated Material
B—Metal Wool or Screen

C—Outlet
D—Air Inlet

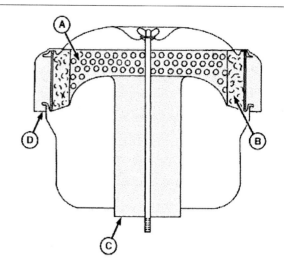

Continued on next page

SS40167,000010A -19-15JUL09-4/18

Spiral-Rotor Air Cleaners

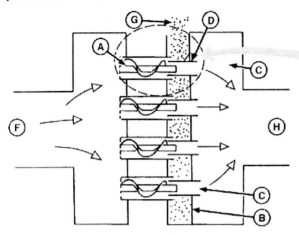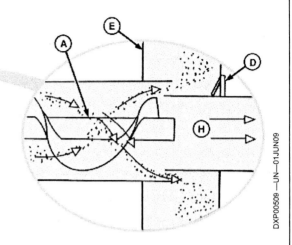

A—Spiral Rotors
B—Compartment
C—Small Tubes
D—Valve

E—Tube to Exhaust
F—Air Inlet
G—Dirt Out
H—Clean Air Outlet

The spiral rotor air cleaner consists of a number of spiral rotors installed in tubes in the air inlet. These tubes protrude into a compartment.

Smaller tubes extend into the larger tubes. The smaller tubes lead to the main air inlet.

A valve allows air to flow out of the main air cleaner inlet compartment but not into it.

The tube connects the compartment to the muffler assembly. The muffler provides a constant vacuum or suction.

Air passing through the tubes is spun by the rotors, causing the dirt particles to be thrown against the walls of the larger tubes and drawn off by the exhaust vacuum.

Clean air passes on through the smaller tubes to the compartment and the main air outlet.

SS40167,000010A -19-15JUL09-5/18

Oil Bath Air Cleaners

Oil bath air cleaners use a combination of a relatively porous filter element and a reservoir of engine oil to separate dust from the incoming air.

Light-duty oil bath air cleaners draw air through the air inlet surrounding the air cleaner body. The velocity of the incoming air increases as it is drawn through the narrow air inlet between the filter cover and the oil sump.

The incoming air reverses when it strikes the surface of the oil, causing most of the heavy dirt particles to become trapped by the oil and settle out in the sump.

The air then passes upward through the screen element where the lighter dust particles and suspended oil are removed. The oil with the suspended dust particles drains back into the sump.

A—Oil Sump
B—Air Inlet
C—Dirty Air Strikes Oil Surface,
    Dirt Particles Stick to Oil
D—Air Outlet

Continued on next page

SS40167,000010A -19-15JUL09-6/18

Medium-duty oil bath cleaners draw air into a center tube where it strikes the surface of oil in a partly filled cup. The impact causes a mixture of air and oil spray to be carried up into the element of baffles and wire mesh.

The separating element breaks up the dust-laden air with the fine dust particles trapped by the oil film. The particles are then washed down as the oil later drains back into the cup. Clean air continues through the element and on to the engine.

The oil wash keeps the filter element fairly clean.

The major services are to keep the oil cup filled to the proper level with the correct weight of oil, and to replace the oil when it gets dirty or thickens, reducing its ability to clean particles from the air.

However, at least once each year, the air cleaner should be disassembled and the filter element cleaned in solvent.

NOTE: *Today's detergent oils may hold dirt in suspension. For this reason, the need for oil cup service cannot be determined by the layer of dirt that has settled into the bottom of the cup. Instead, see if the oil has thickened due to dirt.*

Some medium-duty air cleaners are fitted with an oil trap. This trap catches the oil that otherwise would be thrown out of the cleaner when the engine backfires.

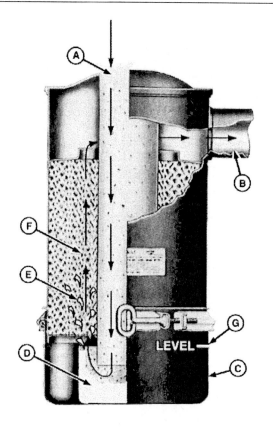

A—Dirty Air In
B—Clean Air to Engine
C—Oil Cup
D—Oil Bath
E—Oil Spray Drawn Up, Cleans Particles from Air
F—Filtering Element
G—Level

Continued on next page

SS40167,000010A -19-15JUL09-7/18

Heavy-duty oil bath air cleaners operate the same as medium-duty ones. They are used on larger engines, especially diesels, because these engines use more air.

Air Cleaner Efficiency

Oil bath air cleaners are not efficient at speeds less than engine rated speed. These cleaners depend primarily on air velocity to separate the dust particles from the air. Since air velocity is reduced at slow engine speeds, the cleaning efficiency of the oil bath cleaner is also reduced.

To operate efficiently, the oil bath air cleaner usually requires service more often than the modern dry element type air cleaner. Properly maintaining the oil bath cleaner is more labor intensive than for the dry-type cleaner, and the dirty oil must be disposed of properly.

Because of these negative factors, oil bath air cleaners are not commonly used on modern engines.

When properly maintained, dry element type air cleaners are extremely efficient at all engine speeds. They are used almost exclusively on modern engines.

Air Silencers

Air silencers eliminate noise caused by the pulsations of incoming air created by the engine. A screen between the air cleaner and engine keeps out foreign matter.

An open space at the air outlet end of the air cleaner housing also acts as an intake air silencer.

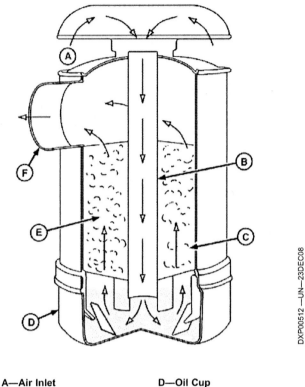

DXP00512 —UN—23DEC08

A—Air Inlet
B—Air Tube
C—Inner Oil Cup
D—Oil Cup
E—Main Element
F—Air Outlet

Continued on next page   SS40167,000010A -19-15JUL09-8/18

Air Cleaner Maintenance

The function of the air cleaner is to remove impurities from the air but at the same time allow sufficient volume of air to enter the engine to ensure complete combustion of the fuel.

The air cleaner will only fulfill this function if it is correctly and regularly maintained. A poorly maintained air cleaner will mean loss of power, excessive fuel consumption, and a reduction in engine life.

Generally, there is no specific service interval for the dry air cleaner. Always follow the recommendations of the engine or vehicle manufacturer for servicing the dry filter.

Some engines are equipped with an air cleaner restriction indicator that monitors the air intake vacuum at the filter outlet. When the indicator signal tells of a restriction, clean or replace the filter. If not equipped with a restriction indicator, clean or replace the element at recommended intervals — more often during unusually dusty conditions.

Oil bath cleaners must be properly maintained or the oil cup will become filled with sludge, preventing the screens and elements from cleaning the air properly.

Restricted air flow to the engine will cause the same effect as choking the engine, resulting in poor engine performance and may allow unclean air to enter.

Restricted air flow through the air cleaner will eventually cause incomplete combustion, increased carbon formation, and crankcase dilution.

Leaks in the connecting pipes, loose hose connections, or damaged gaskets that permit dust-laden air to enter the engine can defeat the efficiency of the air cleaner.

Dirty air going directly into an engine cylinder is abrasive and will cause premature wear of moving parts.

Use the following maintenance rules for good air cleaner operation:

1. Keep air cleaner-to-engine connections tight.

2. Keep air cleaner properly assembled so all joints are air tight.

3. Make careful periodic inspections of the entire air intake system. Repair damaged parts at once. Remember that a partial vacuum is created in the air intake system from the air cleaner to the engine (naturally aspirated engines) or turbocharger when the engine is running. Enough dust-laden air can be sucked into the engine through an almost invisible crack or a loose connection over a period of time to severely damage the engine.

A—Hose Clamp

4. Inspect the pre-cleaner, if used, for dirt accumulation and clean if needed.

5. Inspect the air cleaner frequently under dusty conditions. Dust accumulation on the element gradually restricts the air flow and decreases engine efficiency.

6. Service oil bath cleaners often enough to prevent oil from becoming thick with sludge.

7. Use correct quality of oil. Keep oil at the proper level in the cup. **Do not overfill.**

IMPORTANT: Oil from an overfilled cup can be drawn into an engine. This oil acts as fuel in a diesel engine, over which there is no control and can cause the engine to run away (overspeed), resulting in severe damage.

Air Cleaner Service

No hard and fast rule can be given for servicing an air cleaner since this depends upon the type of cleaner, air condition, and type of application.

Normal service intervals are designated by the manufacturer, but frequent inspection can tell whether or not the service schedule is adequate for the conditions under which the engine is operated.

An air cleaner operating in severe dust conditions will require more frequent service than one operating in clean air.

Always refer to the operator's manual for service information.

Continued on next page   SS40167,000010A -19-15JUL09-9/18

100709
PN=307

Pre-Cleaner and Pre-Screener Service

Remove the pre-screener and blow or brush off any accumulation of lint, chaff, or other foreign matter.

Some pre-cleaners use a removable collector bowl. Remove and thoroughly clean it.

If too much dirt is allowed to collect, the pre-cleaner becomes clogged and a greater load is placed on the main cleaner.

A—Collector Bowl   B—Pre-Screener

SS40167,000010A -19-15JUL09-10/18

Dry Element Type Air Cleaner Service

When the engine smokes too much or loses power, there may be a restriction in the dry air cleaner.

Many engines have an air filter restriction indicator that reflects the intake air vacuum at the outlet side of the air filter. The restriction indicator may be a self-contained type, which shows some type of mechanical red flag when restriction reaches a predetermined level, or it may be of the vacuum-actuated warning light type. The warning light is usually remotely mounted where it may be easily observed by the vehicle or engine operator.

Clean the air filter element at the following times:

1. Units with restriction indicator: Clean the filter element at recommended intervals and whenever the indicator signal tells of a restriction.

2. Units without indicator: Clean the filter element at recommended intervals — more often during dusty or unusual operation.

A—Filter Element   C—Baffle
B—Restriction Indicator   D—Dust Cup

Continued on next page
SS40167,000010A -19-15JUL09-11/18

Primary Element: Remove the primary filter element.

NOTE: The dry element air cleaner may consist of an outer (primary) element and an inner (safety) element. Normally the inner element should not be removed unless it is to be replaced. DO NOT attempt to clean the inner element.

Thoroughly clean all dirt from inside the filter canister taking care not to damage the secondary filter element. If dust is present inside the outer element, it is damaged and must be replaced. Replace the inner element also at this time.

⚠ **CAUTION: Reduce compressed air pressure to less than 30 psi (206 kPa) when using for cleaning purposes. Wear proper protective safety equipment including eye protection.**

Use a dry element compressed air cleaning gun to clean the element. Direct clean, dry air up and down the pleats, blowing from inside to outside.

IMPORTANT: BE CAREFUL NOT TO RUPTURE THE ELEMENT. Do not try to clean the outside of the filter element with an air hose.

To clean with water: Some dry elements can be cleaned using water. First blow out dirt with compressed air as described above. Then attach a garden hose to the cleaning gun and flush remainder of dirt from inside to outside of element. Allow the element to dry thoroughly.

Always follow the instructions in the operator's manual for the particular air cleaner being serviced.

Oily or Sooty Elements: Blow dust from element with compressed air or flush with clean water. Soak the element in a solution of lukewarm water and commercial filter element cleaner or an equivalent non-sudsing detergent. Let the element soak at least 15 minutes, then agitate gently to flush out dirt.

Rinse the element thoroughly from the inside with clean water. Use the element cleaning gun or a free-running hose. Keep the water pressure under 40 psi (280 kPa) to avoid damaging the filter element.

Shake excess water from the element and allow to dry completely before using. This usually takes from one to three days. Do not oven dry or use drying agents. Protect element from freezing until dry.

Keep a spare element to use while the washed one is drying. The spare element should be sealed in a plastic bag and stored in a dry place to protect against dust and damage. The element must be dry before storing in the plastic bag.

IMPORTANT: Do not operate the engine with a wet filter element. The moist paper element will severely restrict the intake air flow. The

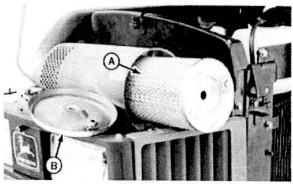
Removing Primary Element

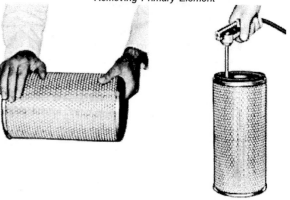
Cleaning the Dry Element

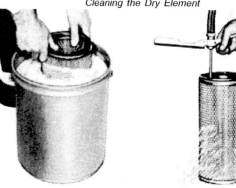
Washable Dry Element

A—Primary Element   B—Cover

resulting high vacuum created in the air intake when the engine is running may cause the element to collapse and rupture. Then the damaged element will not be able to protect the engine from dust-laden air.

Never wash a dry element in fuel oil, gasoline, or solvent. Never use compressed air to dry the element; otherwise the element may be damaged.

Continued on next page   SS40167,000010A -19-15JUL09-12/18

After cleaning the element, hold a bright light inside the element and carefully inspect it for holes. Discard any element that shows the slightest rupture, indicated by light visible through the hole.

Inspect the filter element rubber gasket for damage. Replace the element if the gasket is damaged or missing.

Be sure the outer metal screen is not dented. Vibration would quickly wear a hole in the element.

IMPORTANT: Replace the filter element:

1. **If damaged**
2. **When attempts to clean it fail**
3. **After the recommended maximum number of cleanings**
4. **After a recommended service period (such as one year)**

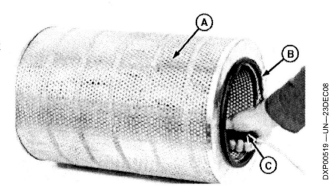

A—Metal Screen   C—Light
B—Gasket

SS40167,000010A -19-15JUL09-13/18

Clean out dirt in the air cleaner body with a clean, damp cloth. Remove and clean the rubber dust unloading valve, if used. Inspect the valve for cracks or deterioration and replace if necessary. Always install the valve with the slot facing the direction of travel.

Install the element in the air cleaner body, gasket end first. Be sure the gasket is in place.

IMPORTANT: Do not operate the engine without the air cleaner element or rubber dust unloading valve installed.

Draw the cover tight on the cleaner body.

A—Dust Unloading Valve

SS40167,000010A -19-15JUL09-14/18

Servicing Safety Element (If Used)

Some engines have an inner safety element in case the primary element ruptures and fails. Normally, these elements are not cleaned, only replaced once each season.

NOTE: Remove the safety element only if it is to be replaced. DO NOT attempt to clean the inner element.

However, carefully check the condition of the safety element during service. If it is dirty, the primary element has probably failed and must be replaced. In this case, the safety element also must be replaced.

A—Safety Element

Continued on next page   SS40167,000010A -19-15JUL09-15/18

Dry Air Cleaner Service

Remove the filter element and shake vigorously to remove most of the dust. Use compressed air or a vacuum cleaner to complete the job.

Wash the element with water and detergent if it is excessively dirty. Allow the filter to dry before using.

Replace the element if it can not be cleaned, has been damaged, has been washed more than six times, or has been in service for more than one year.

IMPORTANT: Never wash the dry element in fuel oil, gasoline, or solvents. Do not oil the element.

Viscous-Impingement Type Air Cleaner Service

Remove the filter element and clean by soaking in solvent. Swish the element up and down in the solvent to thoroughly clean the element. Then dry it with compressed air.

After cleaning and drying the element, immerse it in engine oil and hang it up to allow any excess oil to drip out.

Wipe out the inlet tube with a lint-free cloth.

Install the element, making sure all connections are air-tight.

Light-Duty Oil Bath Air Cleaner Service

Remove the cleaner assembly from the air inlet pipe. The upper section of the cleaner contains the metal wool or screen filtering element. The lower section is made up of the oil cup and inlet tube.

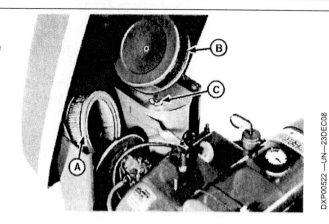

A—Air Filter
B—Cover
C—Wing Nut

IMPORTANT: Dispose of the waste oil properly. Do not pour oil onto the ground, down a drain, or into any water source.

Clean the filter element by soaking it in solvent or diesel fuel. Flush the loosened dirt out with solvent.

Clean the oil cup with solvent and wipe out the inlet tube with a clean, lint-free cloth. **Never** use cotton waste to wipe the center tube.

Fill oil cup to level line with clean engine oil. Assemble and install on engine.

Continued on next page

SS40167,000010A -19-15JUL09-16/18

Medium-Duty Oil Bath Air Cleaner Service

Remove the oil cup, dispose of the oil properly, remove the sediment, and thoroughly clean the cup.

Inspect the bottom surface of the fixed element for a collection of lint, trash, or other foreign matter. If any of these are present, the cleaner should be removed and cleaned.

To clean, soak the element in solvent to loosen accumulated dirt. Flush thoroughly by running solvent through the element from the air inlet end. Allow excess solvent to drip out. Blow out with compressed air.

IMPORTANT: Never attempt to clean the element with a steam cleaner. The force of the steam cannot be maintained throughout the element and will only force the dirt to the center of the element.

Wipe out the center tube with a clean, lint-free cloth.

Inspect the inside of the air cleaner-to-manifold pipe for accumulation of oil and dirt. Remove the pipe and clean it, if necessary.

Install the air intake pipe and air cleaner on the engine. Attach the oil trap to cleaner.

Fill the oil cup to level line with the same weight and quality of oil as used in the engine. Install the cup on the

A—Clamp

cleaner. Be sure it is tightly secured in position with a leak-proof joint.

Also make sure all joints are air-tight.

Continued on next page   SS40167,000010A -19-15JUL09-17/18

Heavy-Duty Oil Bath Air Cleaner Service

Remove and clean oil cup and tray. Refer to "Medium-Duty Oil Bath Air Cleaner Service."

If the filter element is dirty enough to restrict air flow, remove the air cleaner and soak it in solvent or run solvent through it in the opposite direction that air normally passes.

Reinstall the air cleaner. Fill the oil cup to proper level with oil of the same weight and quality as used in the engine. Install the cup and tray on the air cleaner.

Check all tubes and joints to make sure they are tight.

Since air cleaners vary in construction, always follow the instructions in the operator's manual for the particular cleaner being serviced.

In some heavy-duty air cleaners, a collector screen attached to the inlet tube will be visible. Remove by loosening the wing nuts and rotating the tray to unlock it from the tube. In other models, the screen rests on the lip of the inner cup and is not secured by wing nuts.

This screen catches chaff and other foreign matter. Clean the screen each time the air cleaner is serviced.

Some screens may be separated for dirt and lint removal.

Wash screens in solvent and blow out with compressed air.

When a clean screen is held up to the light, an even pattern of light should be visible. If not, repeat the cleaning operation or replace the screen.

NOTE: *The fixed elements of heavy-duty air cleaners are self-cleaning. However, it may be necessary to*

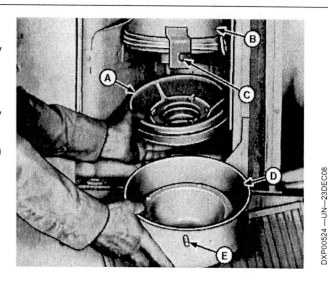

A—Tray
B—Air Cleaner
C—Slot
D—Oil Cup
E—Pin

remove and clean these elements periodically. See the operator's manual for complete information.

Oil Trap

An oil trap prevents oil from blowing out when an ignition engine backfires.

Remove and clean the oil trap each time the air cleaner is serviced.

SS40167,000010A -19-15JUL09-18/18

Intake Systems

Intake Manifolds

The illustration shows a typical intake manifold for a gasoline engine. The air-fuel mixture is carried from the carburetor to the engine intake valves by the manifold.

Heat is supplied to the intake manifold either by having the exhaust manifold touch the intake manifold at various spots or by having exhaust passages built into the intake manifold.

The heat transfer from the exhaust manifold assists in vaporizing the intake fuel after it leaves the carburetor. This is especially important when operating in cool ambient temperatures in order to improve driveability and reduce exhaust emissions.

LP-gas and natural gas fuels vaporize at normal temperatures, so a mixing valve is commonly used instead of a carburetor. Even though heat transfer from the exhaust manifold is not required to vaporize the fuel, the same intake manifold system may be used for gaseous fuel systems as for gasoline systems.

Diesel engine intake manifolds are similar to spark-ignition engine manifolds except that they do not require heat and are usually located on the side of the engine opposite the exhaust manifold.

Some turbocharged diesel engines have an aftercooler heat exchanger located in the intake manifold that cools the compressed air from the turbocharger before it enters the engine.

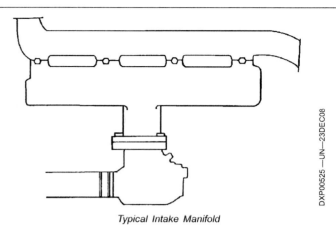

Typical Intake Manifold

Integral Intake Manifolds

The intake manifold is an integral part of the cylinder head on some engines.

The advantages of this are:

1. The intake passages are surrounded by the engine coolant, and the incoming air is warmed before entering the engine cylinders.

2. There are fewer joints, requiring fewer gaskets, and so there is less danger of leakage.

However, external intake manifolds have the advantage of being easier to remove when repairs are necessary.

Continued on next page   SS40167,000010B -19-15JUL09-1/5

100709
PN=314

Crankcase Intake

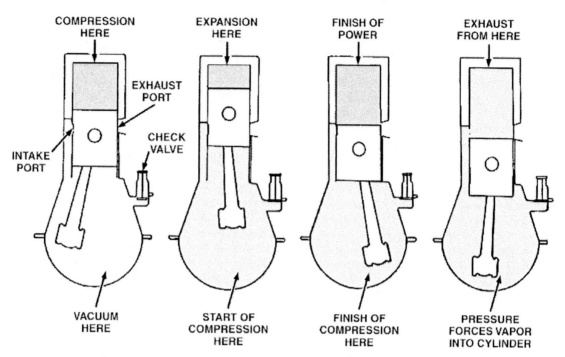

Two-stroke cycle engines (see Chapter 1) ordinarily provide intake through the crankcase. The carburetor is located directly on the crankcase.

On the most simple engines, the air-fuel mixture is drawn into the engine each time the piston moves upward on the compression stroke.

A check valve prevents the vapor from being forced back out the carburetor on the power stroke.

As the piston moves downward on the power stroke, pressure is built up in the crankcase until the intake port and exhaust port are uncovered.

The pressurized vapor flows into the cylinder, forcing the burned gases out the exhaust port.

The action of the crankcase is two stage: **vacuum** to pull in air-fuel; and **compression** to push the mixture in the cylinder.

Continued on next page

SS40167,000010B -19-15JUL09-2/5

Two-Stroke Cycle Engine Blower

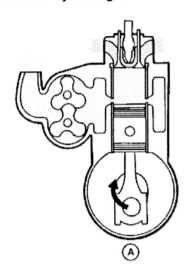
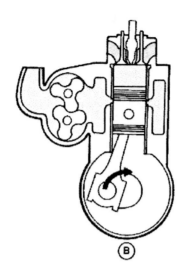
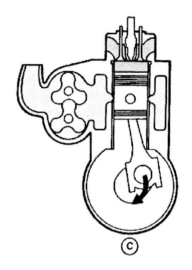

A—Intake and Exhaust   B—Compression   C—Power

Another type of two-stroke cycle engine is one that uses a Roots-type blower or supercharger.

The Roots is a positive-displacement blower that compresses air by trapping it at the inlet port between the rotor lobes and the housing, as the impellers turn. When the trapped air reaches the outlet port, it is forced out of the blower into the engine. The same amount of air is

trapped between the lobes with each turn, regardless of impeller speed.

The air-fuel mixture is forced directly into the cylinder each time the piston moves downward far enough to uncover the intake port. The exhaust valve releases exhaust gases during the exhaust portion of the power/exhaust stroke. The crankcase is never pressurized on this type of engine.

SS40167,000010B -19-15JUL09-3/5

Manifold Heat Control Valve

Several types of thermostatic heat control valves are used to direct exhaust gases against some area of the intake manifold when the engine is cold, and deflect gases away from the intake manifold when the engine has warmed up.

The thermostatic bimetallic coil spring holds the valve open until exhaust heat causes it to expand. Force from the expanding spring closes the valve and shuts off heat to the intake manifold.

Acids from the burned fuel, together with heat from the gases, tend to seize the valve. Use a special heat-control lubricating oil to keep the valve free at all times.

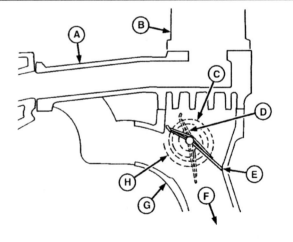

A—Intake Manifold
B—Carburetor
C—Thermostatic Bimetallic Coil Spring
D—Cold Position
E—Hot Position
F—To Muffler
G—Exhaust Manifold
H—Heat Causes Spring to Rotate

Continued on next page   SS40167,000010B -19-15JUL09-4/5

Diesel Intake Manifolds

Diesel engines do not require heat transfer from the exhaust manifold to assist in keeping intake fuel vaporized, since the fuel is injected directly into the cylinder as required.

Heat transfer to the intake air is detrimental to the efficient operation of the diesel engine since heat causes the air to expand and become less dense. The result is that less air will enter the engine.

For this reason, diesel intake and exhaust manifolds are usually separated to prevent heat transfer. The intake manifold is commonly located on the side of the engine opposite the exhaust manifold.

Some turbocharged diesel engines have an aftercooler that cools the heated air from the turbocharger before it enters the engine. The aftercooler is a heat exchanger that uses either engine coolant or the air flow created by the engine fan to reduce the temperature of the intake air.

Aftercoolers that use the engine coolant are located within the intake manifold. They cool the intake air as it enters the manifold.

Aftercoolers that use engine air flow to cool the intake air are normally located ahead of the engine fan. Both aftercoolers cool the intake air before it enters the manifold.

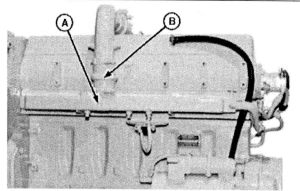

Typical Diesel Engine Intake Manifold with Aftercooler

A—Intake Manifold/After-Cooler   B—Air Inlet

Aftercooler heat exchangers are described in more detail later in this chapter.

Service of Intake Manifolds

Check the intake manifold for restrictions and air leaks. Inspect intake passages of integral manifolds by removing the cylinder head. Replace warped manifolds.

SS40167,000010B -19-15JUL09-5/5

Exhaust Systems

The exhaust system collects exhaust gases from the engine and carries them away. This system:

1. Removes heat
2. Muffles engine sound
3. Carries away burned and unburned gases

The exhaust system consists of:

- **Exhaust Valve**
- **Exhaust Manifold**
- **Turbocharger (if used)**
- **Muffler**

EXHAUST VALVES seal the burning gases within the cylinder until most of the energy has been expended, then open so the cylinder can clear before the next air-fuel charge is admitted. See Chapter 3.

EXHAUST MANIFOLDS receive burned gases from each cylinder and carry them away from the engine. Some heat from the exhaust manifold is used on gasoline engines to maintain the intake manifold at the proper temperatures.

TURBOCHARGERS use engine exhaust gases to drive the compressor wheel of the turbocharger.

MUFFLERS carry away exhaust gases and heat, and quiet engine exhaust noise.

Exhaust Manifold

Exhaust manifolds collect the spent gases from the engine cylinders and conduct them to the muffler or to some other location away from the engine.

A—Exhaust Manifold

The exhaust ports in the engine and the passages in the exhaust manifold are large to allow free flow and expansion of the escaping gases. This is important as it permits better scavenging of the engine cylinders.

If any burned gases are left in the cylinders following the exhaust stroke, the amount of air-fuel mixture that can be taken in on the next intake stroke is limited. This reduces engine power and increases fuel consumption.

Exhaust Manifold Service

Keep inner surfaces of the passages free of carbon buildup. Remove any buildup by scraping, by using carbon solvent, or with a combination of both.

SS40167,000010C -19-15JUL09-1/1

Mufflers

There are two common types of mufflers:

- **Straight-Through**
- **Reverse-Flow**

STRAIGHT-THROUGH mufflers consist of a perforated inner pipe enclosed by an outer pipe roughly three times larger in diameter. The space between the pipes is sometimes filled with a sound-absorbing and heat-resistant material.

REVERSE-FLOW mufflers are hollow chambers using short lengths of pipe and baffles to force the exhaust gases to travel a maze-type path before being discharged.

The muffler acts as an expansion chamber, reducing the noise of the spent gases.

It also serves as a spark arrester to eliminate fire hazards when operating near combustible material.

Muffler Service

Exhaust systems are designed to provide the least amount of restriction. Excessive restrictions cause back-pressure, resulting in incomplete cylinder scavenging. This, in turn, causes loss of power and increased fuel consumption.

It has been estimated that for every 2 psi (14 kPa) of back-pressure, there is a loss of about 4 hp (3 kW).

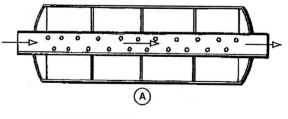

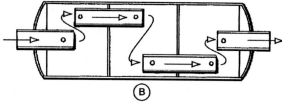

A—Straight-Through Muffler   B—Reverse-Flow Muffler

⚠ **CAUTION: One of the products of combustion is carbon monoxide. This is a deadly, odorless, poisonous gas. Provide good ventilation anytime the engine is operating.**

The entire system must also be free of leaks. When they do occur, repair immediately. This is particularly true when the engine is in a machine equipped with a cab or other enclosure.

SS40167,000010D -19-15JUL09-1/1

Turbochargers and Superchargers

The power delivered by an internal combustion engine is determined by the amount of fuel and air that can be packed into each cylinder. The more air-fuel mixture, the more power the engine develops.

Turbochargers and superchargers are types of air pumps that are designed to increase the amount of air delivered to each of the engine's cylinders.

Turbochargers are driven by the waste exhaust gases from the engine's exhaust system. The exhaust-driven turbine drives a centrifugal compressor wheel. The compressor is usually located between the air cleaner and the engine intake manifold, while the turbine is located between the exhaust manifold and the muffler.

The compressor wheel rotates at high speed inside a circular housing, pulling air in at its center, and then accelerating it to a high velocity by throwing it off the tips of the impeller. The high-velocity air flows radially outward through the housing, where the velocity is slowed and converted to air pressure.

The turbocharger, being light-weight and small in size, is most commonly used on four-stroke cycle engines to improve engine power and efficiency.

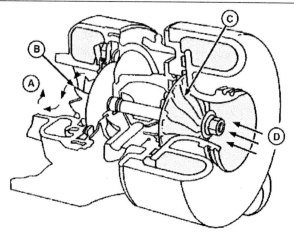

Exhaust-Driven Turbocharger

A—Exhaust Air   C—Compressor Wheel
B—Exhaust Turbine   D—Air Intake

Continued on next page   SS40167,000010E -19-05AUG09-1/22

Superchargers are mechanically driven by the engine crankshaft. They are positive-displacement air compressors. The rotation of the rotors compresses the incoming air against the walls of the blower housing. The compressed air is then forced out through the outlet opening of the unit into the engine cylinders. Thus, it improves the engine's efficiency and performance by compressing and packing more air into the cylinders.

The Roots-type supercharger is used primarily on two-stroke cycle engines to force air into the cylinder and exhaust gases out (called scavenging). Some four-stroke cycle engines use a supercharger to force air into the cylinders under pressure, thereby improving efficiency and performance.

The primary function of the turbocharger and supercharger is, by compressing the air, to force more air into the engine cylinders. This allows the engine to efficiently burn more fuel, thereby producing more horsepower.

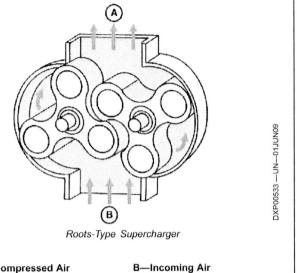

Roots-Type Supercharger

A—Compressed Air   **B—Incoming Air**

SS40167,000010E -19-05AUG09-2/22

Operation of Turbocharger

All of the engine exhaust gases pass through the turbine housing. The flow of these gases, acting on the turbine wheel, causes it to turn. After passing through the turbine, the exhaust gases are routed to the atmosphere.

The turbine also functions as a spark arrester. For example, it is recognized by the U.S. Department of Agriculture as providing a spark arrester function adequate for forestry operations.

A—Compressed Air   **D—Exhaust**
B—Incoming Air   **E—Exhaust Gases**
C—Turbocharger   **F—Engine Cylinder**

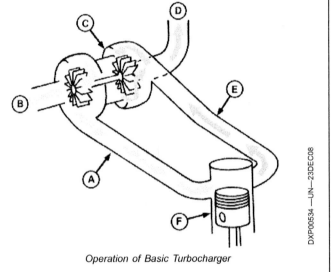

Operation of Basic Turbocharger

Continued on next page   SS40167,000010E -19-05AUG09-3/22

The compressor is directly connected to the turbine by a shaft. The only power loss from the turbine to the compressor is the slight friction of the journal bearings.

Air is drawn in through a filtered air intake system, compressed by the wheel, and discharged into the engine intake manifold.

A—Floating Bearings   D—Shaft
B—Thrust Collar   E—Turbine
C—Compressor

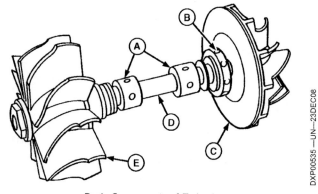

Basic Components of Turbocharger

SS40167,000010E -19-05AUG09-4/22

The extra air provided by the turbocharger allows more fuel to be burned, which increases horsepower output. Lack of air is one factor limiting the engine horsepower of naturally aspirated engines.

As engine speed increases, the length of time the intake valves are open decreases, giving the air less time to fill the cylinders. On an engine running at 2500 rpm, the intake valves are open less than 0.017 second. The air drawn into a naturally aspirated engine cylinder is at less than atmospheric pressure. A turbocharger packs the air into the cylinder at greater than atmospheric pressure.

The flow of exhaust gas from each cylinder occurs intermittently as the exhaust valve opens. This results in fluctuating gas pressures (pulse energy) at the turbine inlet. With a conventional turbine housing, only a small amount of the pulse energy is used.

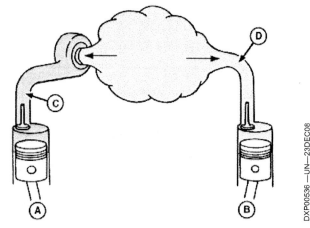

Air Intake Comparison

A—Turbocharged Engine   C—Greater than Atmospheric
B—Naturally Aspirated Engine   Pressure
     D—Less than Atmospheric
       Pressure

Continued on next page SS40167,000010E -19-05AUG09-5/22

100709

To better utilize these impulses, one design has an internal division in the turbine housing and the exhaust manifold, which directs these exhaust gases to the turbine wheel. There is a separate passage for each half of the engine cylinder exhaust.

On a six-cylinder engine, there is a separate passage for the front three cylinders and another passage for the rear three cylinders.

By using a fully divided exhaust system combined with a dual scroll turbine housing, the result is a highly effective nozzle velocity. This produces higher turbine speeds and manifold pressures than can be obtained with an undivided exhaust system.

The turbocharger offers a distinct advantage to an engine operating at high altitudes. The turbocharger automatically compensates for the normal loss of air density and power as the altitude increases.

With a naturally aspirated engine, horsepower drops off 3% per 1000 ft (300 m) because of the 3% decrease in air density. If fuel delivery is not reduced, exhaust and fuel dilution increases with altitude.

With a turbocharged engine, an increase in altitude also increases the pressure drop across the turbine. Inlet turbine pressure remains the same, but the outlet pressure decreases as the altitude increases. Turbine speed also increases as the pressure differential increases. The compressor wheel turns faster, providing approximately the same inlet manifold pressure as at sea level, even though the incoming air is less dense.

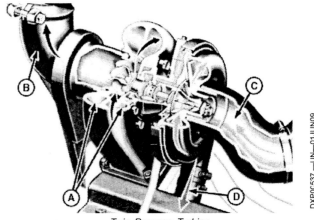

Twin Passage Turbine

A—Twin Passages
B—Exhaust Out
C—Incoming Air
D—Compressed Air

However, there are limitations to the actual amount of altitude compensation a turbocharged engine has. This is primarily determined by the amount of turbocharger boost and the turbocharger-to-engine match.

All turbochargers operate at a very high speed. This can range from 40,000 to 130,000 rpm or more.

Continued on next page
SS40167,000010E -19-05AUG09-6/22

Turbocharger Waste Gate

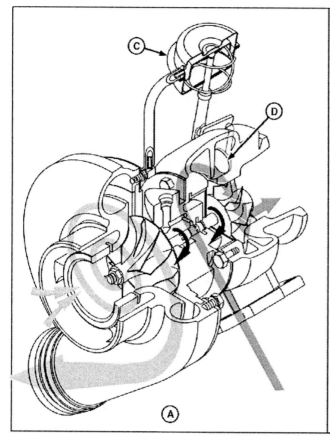

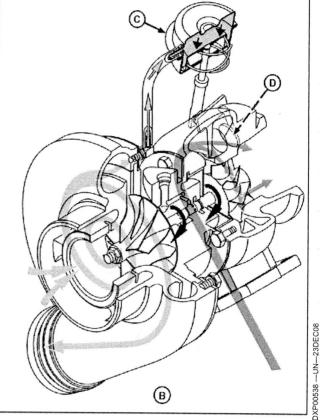

Waste Gate

A—Bypass Closed   B—Bypass Open   C—Diaphragm   D—Waste Gate Actuator Bypass Valve

Some applications of engines have a waste gate actuator bypass valve to help control turbine speed and boost pressure during high-engine rpm operation. This device is integral to the turbine housing and is diaphragm activated or, in some cases, electronically controlled.

The waste gate actuator is precisely calibrated and opens a valve to bypass the exhaust gas around the turbine wheel. This limits the shaft speed, which in turn controls boost pressure. The valve allows the system to develop peak charge-air pressures for maximum engine boost response while eliminating the chance of excessive manifold pressure (over-boost) at high speeds or heavy loads.

Continued on next page

SS40167,000010E -19-05AUG09-7/22

Altitude Compensator

The appearance, construction, and operation of the altitude compensator is the same as that of a turbocharger. However, the purpose is different.

The purpose of a turbocharger is to increase the power output of an engine by supplying compressed air to the engine intake manifold so increased fuel can be utilized for combustion.

The purpose of the altitude compensator is to maintain consistent power output and efficiency of an engine operating at all altitudes. This is done by supplying compressed air to the engine intake manifold at a pressure about equal to that at sea level.

There is no increase of fuel for combustion and consequently no increase in basic horsepower of the engine. However, the extra air provided by the altitude compensator normally increases combustion efficiency, which generally will improve fuel economy and reduce smoke level.

Altitude Compensator

Why Use Turbochargers?

There are five basic reasons for using a turbocharger:

1. **To increase horsepower output of a given displacement engine.** Where the engine compartment of a machine is of a given size, a turbocharged engine can be used to provide increased horsepower without having to enlarge the engine compartment for a larger displacement engine.

2. **To reduce weight.** Turbocharged engines have more horsepower per pound than non-turbocharged engines.

3. **To keep down costs.** Initial cost of turbocharged engines, on a dollar per horsepower basis, is less than for a naturally aspirated (N.A.) engine, and the differential increases with the rate of turbocharging. It all adds up to more horsepower per dollar.

4. **To maintain power at higher altitudes.** The altitude compensator also falls in this category, giving vital machine productivity at high altitudes.

5. **To reduce smoke.** Turbocharging can be an effective way to reduce exhaust density by providing excess air. However, using a turbocharger does not ensure this, as many other components also affect exhaust density and these must be properly designed and matched to provide an acceptable smoke level.

Continued on next page   SS40167,000010E -19-05AUG09-8/22

Operation of Superchargers

Roots-type superchargers are positive-displacement air compressors — each revolution of the rotors displaces a given amount of air. They resemble a gear-type oil pump in design.

Incoming air enters the blower housing due to rotation of the rotors. The blower compresses the air by trapping it at the inlet port between the rotor lobes and the walls of the housing as the rotors turn. When the compressed air reaches the outlet port, it is forced out of the blower into the engine. Since the same amount of air is trapped between the lobes and the housing with each turn, the volume of air that the blower delivers to the engine is directly proportional to the blower speed.

This type of blower is usually chain, belt, or gear driven.

A Note on Supercharger Kits

Installing a supercharger on most engines will increase horsepower by about 30%.

But adding a supercharger will also cause:

• A substantial increase in peak cylinder pressure
• Much higher cylinder temperatures
• Increased air intake volume
• Increased wear and stress on the engine and drive components
• Increased oil consumption

Before mounting a supercharger on an engine, make certain that it also has been equipped with the following:

• Heavy-duty engine and drive components
• High capacity cooling system
• High capacity lubrication system

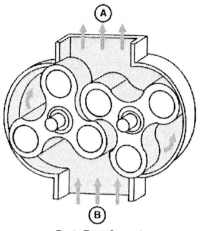

Roots-Type Supercharger

A—Compressed Air   B—Incoming Air

• High capacity air cleaner

Superchargers for Two-Stroke Cycle Engines

An air pump of some type must be used on two-stroke cycle engines to force air into the cylinder and exhaust gases out (called scavenging). (On four-stroke cycle engines, this is done by the exhaust and intake stroke.)

There are three methods of providing this scavenging through a direct engine coupling. They are:

1. Crankcase scavenging
2. Power-piston scavenging
3. Pump or blower scavenging

SS40167,000010E -19-05AUG09-9/22

Crankcase scavenging is used when the air-fuel vapor enters the engine through the crankcase.

Each downward movement of the piston compresses the vapor within the crankcase until the intake port or valve opens.

The compressed vapor then escapes into the cylinder at a pressure nearly equal to atmospheric pressure.

A—Compression/Power Stroke   E—Air-Fuel Inlet
B—Intake/Exhaust Stroke   F—Crankcase
C—Piston   G—Intake Port
D—Exhaust Port

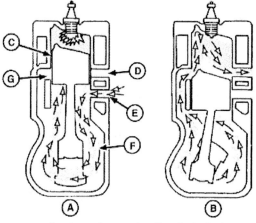

Crankcase Scavenging (Two-Cycle Engine)

Continued on next page   SS40167,000010E -19-05AUG09-10/22

Power-piston scavenging uses a separate piston and cylinder, driven by the engine crankshaft, to push the vapor into the cylinder as the intake port or valve opens.

With **blower scavenging**, a positive-displacement rotary blower, driven by the engine, compresses the air-fuel vapor into an air chamber surrounding the intake ports.

This type of blower has an advantage over an engine-driven centrifugal blower because it delivers practically the same amount of air per revolution regardless of speed or pressure.

A—Scavenging Pump
B—Air-Fuel Vapor
C—Power Cylinder
D—Crankshaft Drives Pump
E—Exhaust Port
F—Inlet
G—Roots-Type Blower

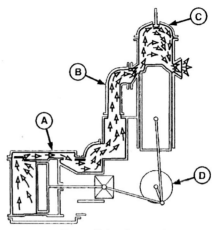

Power-Piston Scavenging

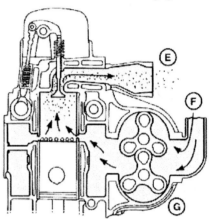

Positive-Displacement Blower

Continued on next page　　　　SS40167,000010E -19-05AUG09-11/22

Scavenging Pump and Turbocharger

A turbocharger may be used in addition to the regular scavenging system. In this case the air drawn into the scavenging pump or blower is compressed to scavenging pressure in the normal manner and then passed to the turbocharger, where it is raised to supercharged pressure.

At light loads when there is little energy available to drive the turbocharger, the mechanically driven blower alone puts the scavenging air into the cylinders.

At increased loads, the turbocharger speeds up and takes in so much air that its inlet pressure drops to atmospheric level, causing the blower check valve to open.

At this engine speed, the blower becomes unloaded (saving engine power) and the turbocharger enters the load range where it alone can provide scavenging and supercharging.

Under the most favorable conditions, the engine starting air contains enough energy to start the turbocharger and also supply enough combustion air to burn the fuel.

However, usual practice today is to equip the turbocharger with some method for supplying additional scavenging air while the engine is being started and while it is running at slow speed.

Two of the methods are:

1. Mechanical — At starting and slow speeds, both the blower and turbocharger are driven from the crankshaft. When speed picks up, the drive coupling disconnects and the turbocharger operates on exhaust gas only.

2. Jet Air Starting — Used only for starting when high-pressure air is available. Air is blown through jets into the turbocharger turbine or compressor. Air passing through the compressor passes on into the engine to assist in scavenging.

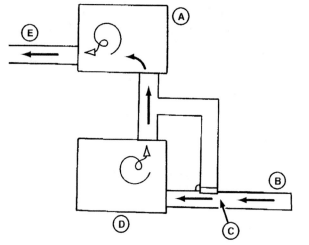

Scavenging Pump and Turbocharger

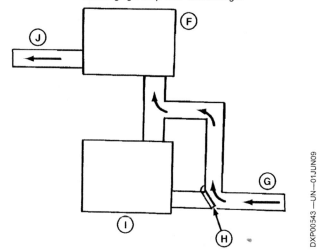

Scavenging Pump and Turbocharger

A—Turbocharger (Idle)
B—Air Inlet
C—Valve (Closed)
D—Blower (Running)
E—Pressurized Air to Engine
F—Turbocharger (Running)
G—Air Inlet
H—Valve (Open)
I— Blower (Idle)
J— To Engine

Continued on next page   SS40167,000010E -19-05AUG09-12/22

Aftercoolers

When the turbocharger compresses the engine intake air, the air becomes heated (due to compression) and expands. When the heated air expands, it becomes less dense. The result is that part of the purpose of the turbocharger is defeated; that is, due to heat expansion, less air is forced into the engine.

To overcome this condition, some turbocharged engines are equipped with an aftercooler that is installed between the turbocharger and the engine intake manifold.

The aftercooler reduces the temperature of the compressed air by 80 to 90°F (44 to 50°C). This makes the air denser, allowing more to be packed into the combustion chambers.

The result is:

1. More power
    Sufficient air is provided to burn the fuel, resulting in higher horsepower.
2. Greater economy
    The fuel is burned more completely, giving more power from a given amount of fuel.
3. Quieter combustion
    By controlling warm air for air-fuel mixing, there is a smoother pressure rise in the engine cylinder.

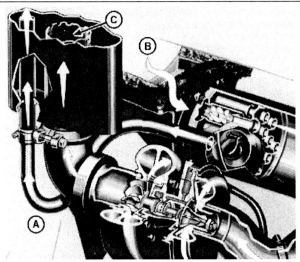

Intake-Exhaust System on Turbocharged, Aftercooled Engine

A—Muffler   C—Exhaust Out
B—Air In

SS40167,000010E -19-05AUG09-13/22

The aftercooler is nothing more than a heat exchanger. There are two methods of accomplishing the exchange of heat in the aftercooler. They are:

1. Air-to-Liquid
2. Air-to-Air

The **air-to-liquid aftercooler** uses engine coolant to reduce the temperature of the intake air. The intake air flows over a series of tubes through which engine coolant is circulated. Heat from the intake air is transferred to the engine coolant and carried out of the aftercooler to the coolant radiator.

The air-to-liquid aftercooler is not as efficient in removing heat from the intake air as the air-to-air aftercooler. The air-to- liquid aftercoolers can obtain air temperatures only as low as the coolant temperatures in the engine radiator.

Since the air-to-liquid aftercooler is exposed to the engine coolant, it is susceptible to the same problems as the coolant radiator: rust, corrosion, plugged coolant passages, and leakage.

The air-to-liquid aftercooler is reliable, however, it is very important that engine coolant does not leak into the intake

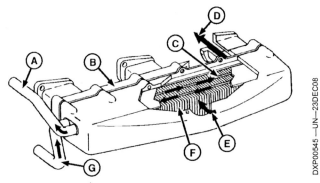

Air-to-Liquid Aftercooler

A—Coolant Outlet   E—Heated Air Flow
B—Intake Manifold   F—Aftercooler
C—Coolant Tubes   G—Coolant Inlet
D—Cooled Air Flow

manifold. Serious engine damage may result if engine coolant enters the cylinders.

Continued on next page   SS40167,000010E -19-05AUG09-14/22

Air-to-Air Aftercooler

A—Air Out
B—Turbocharger

C—Air Cleaner
D—Intake Manifold

E—Air In
F—Air-to-Air After Cooler

The **air-to-air aftercooler** uses the air flow created by the engine fan to reduce the temperature of the intake air. The air-to-air heat exchanger unit is located ahead of the engine fan. The compressed air from the turbocharger flows through a series of tubes in the heat exchanger as a stream of cooler air is forced over the tubes by the engine fan. Heat from the intake air is transferred to the air flowing around the tubes, providing the necessary cooling.

The air-to-air aftercooler is a very simple and efficient heat exchanger system.

Continued on next page   SS40167,000010E -19-05AUG09-15/22

Testing and Repairing the Aftercooler

Inspect the aftercooler for overall condition. The fins should be relatively straight, and there should be no evidence of cracks in the aftercooler tubes.

Test the aftercooler for leaks by plugging either the inlet or outlet. Apply compressed air (A) to the other opening while the unit is submerged under water. Use 20–25 psi (140–170 kPa) air pressure for testing. Watch for air bubbles escaping from the unit, which mean leaks.

A minor leak that is accessible may be repaired. However, if the condition of the core is questionable, replace the aftercooler.

IMPORTANT: Coolant leakage from the air-to-liquid aftercooler may cause severe engine damage.

Special Operating Instructions for Turbocharged Engine

Follow these special operating instructions when operating a turbocharged engine.

1. After starting the engine, do not accelerate or apply load until there is positive indication of oil pressure.

2. After starting during cold weather, allow the engine to run 5 minutes at half throttle to ensure oil pressure at the turbocharger before putting the engine under load.

3. Before stopping the engine, allow it to run at near slow idle speed for a few minutes to allow internal engine temperature to normalize. Failure to do this can damage the turbocharger (as well as the engine) due to distortion and "coking" of the oil in the passages.

4. Should the engine stall when operating at normal operating temperature, restart it immediately to

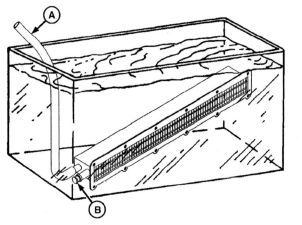

Testing Aftercooler for Leaks

A—Compressed Air   B—Plug

prevent temperature "soaking" of the turbocharger. The turbocharger gets very hot during operation. If stopped or stalled, oil in the center section may "coke," causing oil passages to clog.

IMPORTANT: When transporting an idle turbocharged engine with the exhaust outlet exposed, cover the exhaust outlet to prevent entrance of foreign matter and possible rotation of the turbocharger. Rotation of the turbocharger could damage the rotor bearings due to lack of lubrication.

On other occasions, when the engine is not operating, cover the exhaust outlet to prevent infiltration of water or other foreign matter.

Continued on next page   SS40167,000010E -19-05AUG09-16/22

Lubrication of Turbochargers

The turbocharger center housing contains lubrication passages through which engine oil, under pressure, is directed to the journal bearings and thrust washers. Oil is sealed from the compressor and turbine by a piston ring at both ends of the bearing housing.

If the unit has floating sleeve-type bearings, they provide oil clearance between the bearing outside diameter and housing bore as well as oil clearance between the bearing inside diameter and the shaft outside diameter. When the turbocharger is operating, this allows the bearing to turn as the shaft rotates.

All clearances in the turbocharger are controlled by closely maintained machine tolerances of detailed parts. Since all parts of the rotating assembly are protected by a film of oil, no metal contact occurs. Consequently bearing and shaft wear is minimal if a constant supply of clean oil is supplied to the unit.

The oil not only lubricates the turbocharger's spinning shaft and bearings, it also carries away heat generated by friction of the bearings and from the hot exhaust gases. When oil flow stops or is reduced, heat is immediately transferred from the hot turbine wheel to the bearings. The bearings are also heating up because of the increased friction due to the lack of oil. This causes the turbocharger shaft temperature to increase rapidly.

If the oil flow does not increase and the process continues, the bearings will fail. Once the bearings fail, the seals, shaft, turbine wheel, and compressor wheel can also be damaged.

Some turbochargers make use of the engine coolant in addition to the lubricating oil to help cool the turbocharger. The engine coolant circulates through passages in the center housing and carries heat away from the bearings and shaft.

Periodic Inspection of Turbochargers

1. Inspect the mounting and connections of the turbocharger to be certain they are secure and there is no leakage of oil or air.

2. Check the engine crankcase to be sure there is no restriction to oil flow.

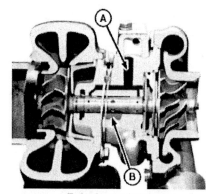

Turbocharger Lubrication

A—Pressure Oil   B—Discharge Oil

3. Operate the engine at approximate rated output and listen for unusual turbocharger noise. If a shrill whine (other than normal) is heard, stop the engine immediately. The whine means that the bearings are about to fail. Remove the turbocharger for inspection.

NOTE: Do not confuse the whine heard during "run down," as the engine stops, with a bearing failure during operation.

Other unusual turbocharger noises could mean improper clearance between the turbine wheel and housing. If such noises are heard, remove the turbocharger for inspection. See the engine technical manual.

4. Check the turbocharger for unusual vibration while the engine is operating at rated output. If necessary, remove the turbocharger for inspection.

5. Check the engine under load conditions. Excessive exhaust smoke indicates incorrect air-fuel mixture. This could be due to engine overload or turbocharger malfunction.

6. Inspect and service the air cleaner according to instructions in the operator's manual.

Continued on next page   SS40167,000010E -19-05AUG09-17/22

Pre-Lubricating Turbocharger

It is important that a new or remanufactured turbocharger be properly lubricated before operating the engine. Significant damage may result to the bearings if a turbocharger operates for even a short period of time without sufficient lubrication.

IMPORTANT: DO NOT spin the rotor assembly with compressed air. Damage to the bearings can occur by high-speed spinning of rotor without lubrication.

A new or remanufactured turbocharger should be pre-lubricated with clean engine oil before it is installed on the engine. Fill the oil return (drain) port in the center housing with engine oil and spin the rotating assembly by hand to properly lubricate the bearings.

Install the turbocharger as instructed in the engine technical manual. Crank the engine with the starting motor (but do not start) for several seconds to allow engine oil to reach the turbocharger bearings. DO NOT

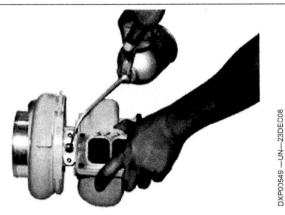

Pre-Lubricating the Turbocharger Bearings

crank the engine for longer than 30 seconds at a time to avoid damage to the starting motor.

Start and run the engine at slow idle speed while checking the oil inlet and air piping connections for leaks.

Continued on next page   SS40167,000010E -19-05AUG09-18/22

Testing of Turbochargers

These diagnostic procedures will allow you to determine the condition of the turbocharger. If the turbocharger has failed, determine the cause and correct the problem to prevent a repeat failure.

Mount the turbocharger in a vise.

1. Rotate the shaft, using both hands, to check rotation and clearance. The shaft should turn freely; however, there may be a slight amount of drag. If the shaft binds, internal problems are indicated. Consult the engine technical manual for disassembly procedure.

2. Next, pull up on the compressor end of the shaft and press down on the turbine end while rotating the shaft. Neither the compressor wheel nor the turbine wheel should contact the housing at any point.

NOTE: *There will be some "play" because the bearings inside the center housing are free-floating.*

3. Check shaft end play by moving the shaft back and forth by hand while rotating. There will be some end play, but not to the extent that the wheels contact the housings. Too much end play indicates wear of the rotating parts. Too little end play indicates carbon buildup on the rotating parts.

4. The final test will give an indication of the condition of the axial bearing within the center housing and the rotating assembly.

    Mount a dial indicator so that the indicator tip rests on the end of the shaft. Preload the indicator tip and zero the dial on the indicator. Move the shaft back and forth by hand. Observe and record total dial indicator movement.

    Refer to the engine technical manual for end play specifications. If end play is not within specified range, repair or replace the turbocharger.

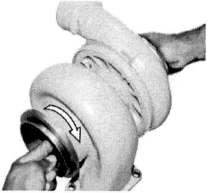

Checking Shaft Rotation

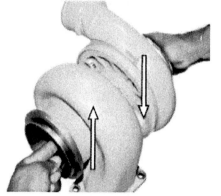

Checking Shaft Radial Play

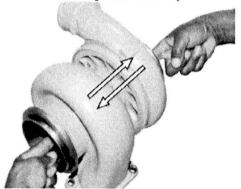

Checking Shaft Axial End Play

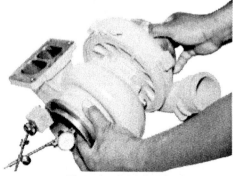

Measuring Axial End Play

Continued on next page   SS40167,000010E -19-05AUG09-19/22

100709
PN=333

Inspection of Damaged Turbocharger Parts

If an impeller or turbine wheel ever sustains damage as shown on this page, replace it. Also be sure to eliminate the cause of the failure.

FOREIGN MATERIAL IN INTAKE SYSTEM

Appearance:

1. Leading edges of compressor wheel badly nicked (A).
2. Blade tips usually intact (B).
3. Compressor wheel may or may not have contacted compressor housing (C).
4. Imbalance of rotating assembly may cause bearing extrusion and heavy seal wear, (not shown).
5. Inlet section of compressor housing will be rough and pitted, (not shown).

Probable Causes:

1. Leak in connections of joints.
2. Loose material left in duct.
3. Welding slag not removed from duct.

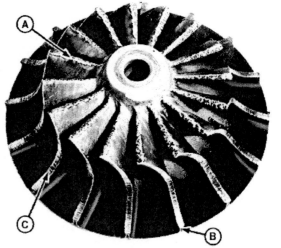

Foreign Object Damage to Intake System Impeller

4. Ice formation in duct from water pulled into air ducting.
5. Loose wire particles from air cleaner.

SS40167,000010E -19-05AUG09-20/22

SAND BLASTING OF IMPELLER

Appearance:

1. Nicked leading edges (A).
2. Blade tips thinned from peening (B).
3. Blade contour deeply eroded (C).

NOTE: Sand blasting problems may or may not introduce imbalance and cause bearing failure.

Probable Causes:

1. Loose connections at joints of air cleaner or ducting.
2. Dry-type paper element split.
3. "Channeling" in oil bath air cleaner.

Impeller Eroded by Sand Blasting Problem

Continued on next page

SS40167,000010E -19-05AUG09-21/22

FOREIGN MATERIAL IN EXHAUST SYSTEM

Appearance:

1. Blade tips of turbine wheel chewed and battered (A).
2. In extreme cases, imbalance will destroy seals and bearings (B).
3. Shaft journals in good condition (C).

NOTE: Thrust bearing may or may not be damaged. Pounding and heat may cause shaft failure.

Probable Causes:

1. Piston ring breakage.
2. Engine valve breakage.
3. Loose material left in manifold or broken injector tips.

See the engine technical manual for detailed inspection procedure and disassembly instructions.

Refer to the FOS manual **Identification of Parts Failures** for additional information on the cause of turbocharger failures.

IMPORTANT: If the turbocharger failed because of foreign material entering the air intake system, be sure to examine the system and clean as required to prevent a repeat failure.

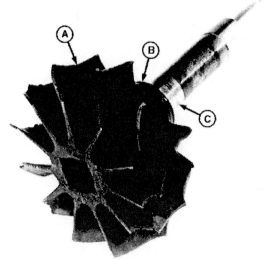

Foreign Object Damage to Turbine Wheel

Troubleshooting of Turbocharger

As has been explained, most turbocharger failures start in some other area of the engine. The turbocharger is a relatively simple device and, if properly serviced, will operate with little or no attention.

Turbocharger Troubleshooting Chart		
Problem	**Possible Cause**	**Solution**
Noisy operation or vibration.	Bearings not being lubricated.	Supply required oil pressure. Clean or replace oil line. If trouble persists, overhaul the turbocharger.
	Leak in engine intake or exhaust manifold.	Tighten loose connections or replace manifold gaskets.
Engine will not deliver rated power.	Clogged manifold system.	Clear all ducting.
	Foreign matter lodged in compressor, impeller or turbine.	Disassemble and clean.
	Excessive buildup in compressor.	Thoroughly clean compressor assembly. Clean air cleaner and check for leaks.
	Leak in engine intake or exhaust manifold.	Tighten loose connections or replace manifold gaskets.
	Rotating assembly bearing seizure.	Overhaul or replace turbocharger.
Oil seal leakage.	Failure of seal.	Overhaul turbocharger.
	Restriction in air cleaner or air intake creating suction.	Remove the restriction.

SS40167,000010E -19-05AUG09-22/22

Testing the Air Intake System

When the air flow into the engine is restricted, there is more vacuum or suction in the cylinders. This can cause oil to be drawn in around the valve stems and piston rings and so increase oil consumption. Restricted air flow can also cause an excessively rich air-fuel mixture, engine overheating, decreased power output, poor fuel economy, and increased exhaust emissions.

The test given here will tell if there is a restriction in the air intake system.

Intake Vacuum Test (Diesel Engines)

Test as follows:

1. Warm up the engine.

2. On engines without air restriction indicators, remove plug from intake manifold and connect the vacuum gauge to the intake manifold.

3. On engines with air filter restriction indicators, remove the indicator, install a pipe tee fitting and reinstall the indicator. Connect the vacuum gauge to the tee fitting.

4. Set engine speed at fast idle and note the reading on the gauge. **Too high a reading means that there is a restriction in the air intake system.** Check the engine technical manual for correct specifications.

5. On engines with air filter restriction indicators, the operation of the indicator can be checked as follows: Use a board or metal plate to very slowly cover the air intake opening while the engine is operating at fast idle speed. Note the action of the indicator in relation to the reading on the vacuum gauge. If the indicator does not operate properly, replace it.

Manifold Depression Test (Spark-Ignition Engines)

Use a vacuum gauge calibrated in inches of mercury to perform this test.

1. Connect the vacuum gauge to the intake manifold.

2. Warm up the engine and operate it at idle speed.

Connecting the Vacuum Gauge

DXP00557 —UN—23DEC08

3. Note the reading on the vacuum gauge. Check the engine technical manual for exact specifications.

4. Interpret the gauge reading as follows:

- **If the reading is steady and low,** loss of power in all cylinders is indicated. Possible causes are late ignition, incorrect valve timing, or loss of compression at valves or piston rings. A leaky carburetor gasket will also cause a low reading.
- **If the needle fluctuates steadily,** a partial or complete loss of power in one or more cylinders is indicated. This can be due to an ignition defect, a loss of compression due to stuck piston rings, or a leaky cylinder head gasket.
- **Intermittent needle fluctuation** indicates occasional loss of power due to an ignition defect or a sticking valve.
- **Slow needle fluctuation** is usually caused by improper carburetor idle mixture adjustment.
- **A gradual drop in the gauge reading** at idle engine speed indicates back-pressure in the exhaust system due to a restriction.

Be very careful in analyzing abnormal readings since the gauge readings can indicate more than one thing.

SS40167,000010F -19-15JUL09-1/1

Test Yourself

QUESTIONS

1. What are the two basic types of air intake systems?

2. Why is it important that a supply of clean air is provided for the engine combustion process?

3. What is the primary function of the air cleaner?

4. If dust is present on the inside of the dry-type filter element, what condition is indicated?

5. What are the three parts of the exhaust system?

6. What three things does the exhaust system do?

7. (Fill in the blank.) Turbochargers and superchargers are types of _____ that are designed to increase the amount of air delivered to each of the engine's cylinders. Which one delivers the same amount of air per revolution at all engine speeds?

8. How does the drive system for turbochargers and superchargers differ?

9. What is the function of an aftercooler?

10. (Fill in the blank.) Waste gates are used to help control turbine speed and boost pressure during _____ operation.

SS40167,0000110 -19-15JUL09-1/1

Lubrication Systems — Introduction

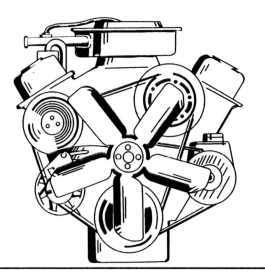

The lubrication system does these jobs for the engine:

- Reduces friction between moving parts
- Absorbs and dissipates heat
- Seals the piston rings and cylinder walls
- Cleans and flushes moving parts
- Helps deaden the noise of the engine

With lubricating oil, the system is able to perform all these jobs at once: without it, the engine would soon wear out and seize up.

A—Absorbs Heat
B—Seals Piston Rings
C—Cleans Parts
D—Deadens Noise
E—Reduces Friction and Wear

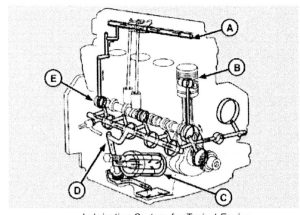

Lubrication System for Typical Engine

SS40167,0000111 -19-15JUL09-1/1

Types of Lubrication Systems

Engine lubrication systems may be classified as:

- **Circulating Splash**
- **Internal Force-Feed and Splash**
- **Full Internal Force-Feed**

The type of system used depends largely upon the size and design of the engine.

Let's discuss each system in detail.

Circulating Splash System

In the circulating splash system, an oil pump supplies oil to a splash pan located under the crankshaft. As the connecting rods revolve, scoops on the ends of the rods dip into troughs in the splash pan, creating the oil splash.

The splashing oil lubricates the moving parts nearby. Other parts are lubricated by oil splash which builds up in collecting troughs and is gravity fed through channels or lines.

The upper parts of the cylinders, pistons, and pins are lubricated more by oil mist than by the oil splash itself. This mist is created as the connecting rods spin.

The circulating splash system must have:

1. Proper oil level in the pan
2. Suitable oil for good splashing

There must be enough oil in the troughs of the splash pan for the connecting rods to splash. The oil pump must be working properly to provide this oil.

Because the oil must splash and flow freely, heavy oil will not work. Use only oil of the viscosity recommended by the engine manufacturer.

Internal Force-Feed and Splash System

In the internal force-feed and splash system, the pump forces oil directly to a main oil gallery in the engine block rather than to a splash pan.

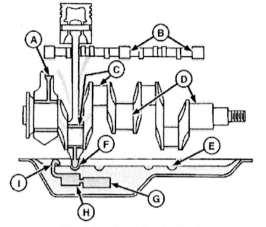

Circulating Splash Lubrication System

A—Oil Collection Trough	F—Oil Scoop
B—Camshaft Bearings	G—Oil Strainer
C—Rod Bearings	H—Oil Pump
D—Main Bearings	I— Oil Supply to Splash Pan
E—Splash Pan Troughs	

From the main oil gallery, the oil is forced through passages to the main bearings, connecting rod bearings, camshaft bearings, rocker arm shaft, filter, and pressure sending unit.

The oil escaping from the bearings creates a mist that also lubricates the upper cylinder walls, pistons, and pins.

Pressure of the lubricating oil can usually be adjusted in these systems.

Continued on next page

SS40167,0000112 -19-15JUL09-1/2

Full Internal Force-Feed System

The full internal force-feed system goes one step farther than the system above. Oil is forced not only to the crankshaft bearings, rocker arm shaft, filter, and pressure sending unit, but also to the piston pin bearings.

The piston pin bearings are lubricated through drilled passages in the connecting rods. The cylinder walls and pistons are lubricated by oil escaping from the piston pin bearings or the connecting rod bearings.

The force-feed system has one prime need: good oil pressure. This system is used (with a full-flow filter) in most agricultural and industrial machines.

As the force-feed system requires oil to be pumped a long distance and through many passages, sufficient oil pressure is required. The required oil pressure for most engines is 25–40 psi (170–275 kPa), but it may go as high as 65 psi (450 kPa).

A—Camshaft Bearings
B—Tappet Lever Shaft
C—Piston Pin Bearing
D—Main Oil Gallery
E—Oil Pump and Filter
F—Crankshaft Main Bearings

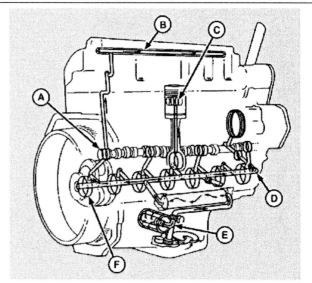

Full Internal Force-Feed Lubrication System

SS40167,0000112 -19-15JUL09-2/2

Oil Pumps

Pumps for engine lubrication are usually one of two types:

• **External Gear Pump**
• **Rotor Pump**

The external gear type is the most common.

Normally the oil pump is mechanically driven by the engine. Usually, the external gear pump is driven from the camshaft, while the rotor type is driven from the crankshaft. In some large engines, an electric motor is used to drive an auxiliary pump.

A—Oil Filter
B—Engine Oil Cooler
C—Piston Pin and Bushing
D—Camshaft Bushings
E—Main Bearings
F—Connecting Rod Bearings
G—Bypass Valve
H—Oil Pressure Regulating Valve

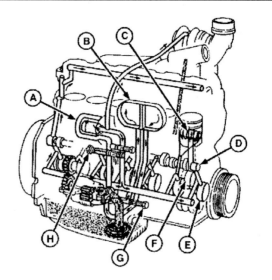

Crankshaft-Driven Oil Pump

Continued on next page

SS40167,0000113 -19-15JUL09-1/4

The **external gear pump** has two gears in mesh, closely fitted inside a housing. The drive shaft drives one gear, which in turn drives the other gear. The machined surfaces of the outer housing are used to seal the gears.

As the gears rotate and come out of mesh, they trap inlet oil between the gear teeth and the housing. The trapped oil is carried around to the outlet chamber. As the gears mesh again, they form a seal, which prevents oil from backing up to the inlet. The oil is forced out at the outlet and sent through the system.

A—Inlet
B—Internal Seal Formed Here

C—Outlet
D—Internal Seal Formed Here

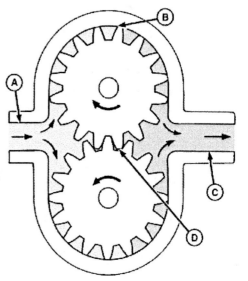

External Gear Oil Pump.

SS40167,0000113 -19-15JUL09-2/4

The **rotor pump**, which is a variation on the internal gear pump, is also relatively simple in design. An inner rotor turns inside a fixed rotor ring.

In operation, the inner rotor is driven inside the rotor ring. The inner rotor has one less lobe than the outer ring at any one time. This allows the other lobes to slide over the outer lobes, making a seal to prevent backup of oil. As the lobes slide up and over the lobes on the outer ring, oil is drawn in. As the lobes fall into the ring's cavities, oil is squeezed out.

A—Inlet
B—Rotor Ring

C—Outlet
D—Inner Rotor

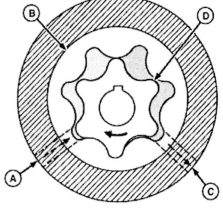

Rotor Oil Pump

Continued on next page   SS40167,0000113 -19-15JUL09-3/4

Servicing Oil Pumps

When servicing an oil pump, be careful not to damage the mounting surfaces of either the housing or cover. Many pumps do not use a gasket, and these machined surfaces form the only seal.

Check the machined surfaces of the pump housing and cover for warping. They must be perfectly flat if they are to seal.

In an external gear pump, most wear will occur between the teeth of the gears.

In a rotor pump, most wear will occur between the lobes on the inner rotor and the lobes on the rotor ring.

Use a micrometer to measure the widths of these parts. Check the results against specifications for the pump.

Also measure the pump drive shaft and compare it with the specifications.

With the gears mounted in the housing, use a feeler gauge to measure the clearance between the gears and the housing. If the clearance is more than specifications, replace the pump.

Place a straight edge across the top of the pump housing (to represent the cover) and measure the clearance between the gears and the straight edge. If the clearance is excessive, replace the pump.

Almost all pumps use a screen over the inlet to strain out foreign material. Where possible, remove the screen from the inlet pipe and clean both with a solvent. Use compressed air to dry the parts.

Inspect all bushings in the housings and replace those with excessive wear. If a gear set is replaced, always replace the gears as a matched pair. Never replace only one gear.

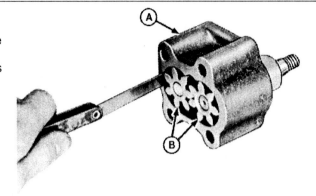

Measuring Clearance Between Pump Gears and Housing

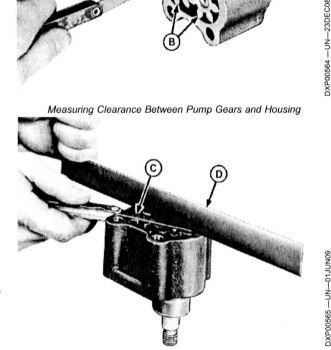

Measuring Clearance Height of Pump Gears

A—Pump Housing
B—Pump Gears
C—Feeler Gauge
D—Straight Edge

SS40167,0000113 -19-15JUL09-4/4

Oil Filters and Filtration Systems

Oil contamination reduces engine life more than any other factor. To help combat this, oil filters are designed into all modern lubrication systems.

There are two basic types of filters — surface and depth. They are used in two basic types of filtering systems — bypass and full flow. Some large engines use a combination bypass and full flow filtering system.

Let's look first at the two types of filters, then at the two types of filtering systems.

Types of Filters

Filters are classified as either surface filters or depth filters, depending on the way they remove dirt from the oil.

A—Surface Filter
B—Depth Filter

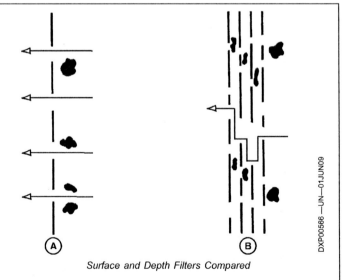

Surface and Depth Filters Compared

Continued on next page
SS40167,0000115 -19-15JUL09-1/8

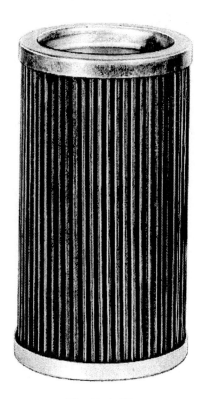

Wire Mesh Filter

Metal Edge Filter

SURFACE FILTERS have a single surface that catches and removes dirt particles larger than the holes in the filter. Dirt is strained or sheared from the oil and stopped outside the filter as oil passes through the holes in a straight path. Many of the large particles will fall to the bottom of the reservoir or filter container, but eventually enough particles will wedge in the holes of the filter to prevent further filtration. Then the filter must be cleaned or replaced.

A surface filter may be made of fine wire mesh, stacked metal or paper disks, metal ribbon wound edgewise to form a cylinder, cellulose material molded to the shape of the filter, or accordion-pleated paper.

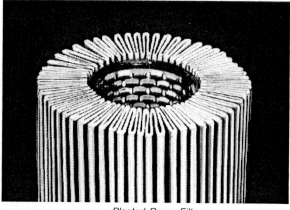

Pleated Paper Filter

Continued on next page

SS40167,0000115 -19-15JUL09-2/8

DEPTH FILTERS, in contrast to the surface type, use a large volume of filter material to make the oil move in many different directions before it finally gets into the lubrication system. The filter made of cotton waste shown in the illustration is an example of a depth filter.

Depth filters can be classified as either absorbent or adsorbent, depending on the way they remove dirt.

Absorbent filters operate mechanically like a sponge soaking up water. Oil passes through a large mass of porous material such as cotton waste, wood pulp, wool yarn, paper, or quartz, leaving dirt trapped in the filter. This type of filter will remove particles suspended in the oil and some water-soluble impurities.

Adbsorbent filters operate the same way as absorbent filters but also are chemically treated to attract and remove contaminants. This filter may be made of charcoal, chemically treated paper, or fuller's earth. It will remove contaminating particles, water-soluble impurities and, because of its chemical treatment, will also remove contamination caused by oxidation and deterioration. Adsorbent filters may also remove desirable additives from the oil and for this reason are not often used in lubrication systems.

Depth Filter, Cotton Waste Type

SS40167,0000115 -19-15JUL09-3/8

Degrees of Filtration

In addition to the type of filter, the degree of filtration is also important to a lubrication system.

It is the degree of filtration that tells just how small a particle the filter will remove. The most common measurement used to determine degree of filtration is a micron. One-micron is approximately 0.00004 in. or 400 millionths of an in. (0.001 mm). To get an idea of how big a micron really is, 25,000 particles of this size would have to be laid side by side to total just 1 in. (25.4 mm).

The smallest particle that can normally be seen with an unaided eye is about 40 microns (0.04 mm), so much of the dirt that is filtered out of lubrication systems is invisible.

Some filters, such as those made of wire mesh, may allow particles as big as 150 microns (0.152 mm) to pass. Although they do not provide as fine a cleaning action as some other types of filters, wire mesh offers less resistance to oil flow and is often used on pump inlet lines to prevent the possibility of starvation.

A filter will pass small solids and stop larger ones, but the actual amount of filtering done is difficult to determine. Because the material that is stopped by the filter is not taken away continuously, the size of the openings in the filter will usually decrease with use.

Two or more particles smaller than the holes in the filter may become wedged in the hole. The result is that the

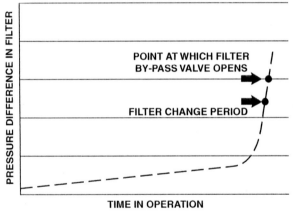

Life of a Filter Element

hole will now remove dirt particles far smaller than it would when new. As more dirt wedges in the filter, the holes become smaller and eventually plug.

Illustration shows the gradual reduction in the size of the filter pores until, near the end of the filter life, the pressure difference between the inside and outside of the filter rises sharply. At this point, the filter stops operating and should be replaced.

Now let's look at some types of filtration systems.

Continued on next page

SS40167,0000115 -19-15JUL09-4/8

Bypass Filtration System

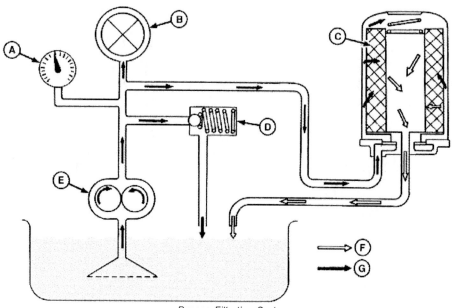

Bypass Filtration System

A—Pressure Gauge
B—Engine Bearings
C—Filter
D—Pressure Regulating Valve
E—Oil Pump
F—Filtered Oil
G—Unfiltered Oil

In the bypass filtration system, there are two separate oil flows, one to the engine bearings and one to the filter.

In this system, 5 to 10% of the oil delivered by the oil pump is routed or bypassed to the filter instead of the engine bearings. After filtering, the oil is returned to the crankcase. This system is sometimes called a "partial flow" system as only part of the supply oil is filtered at one time.

The volume of oil bypassed through the filter is initially controlled by a restriction in the filter outlet. However, as the filter passages become clogged, the volume of oil through the filter is reduced. This, of course, reduces the volume of filtered oil being returned to the crankcase. Because of the two separate oil flows, the oil pressure at the bearings is constant, regardless of the condition of the filter.

This means that the filter and the oil must be changed regularly to prevent loss of filtering.

Shunt System

The shunt filtration system is a variation of the bypass system. In this case, there is only one oil flow to the filter. Within the filter case, part of the oil goes through the filter to the bearings while the remainder is shunted directly to the bearings.

Initially, the volume of filtered oil is greater than the shunted oil. But as the filter becomes clogged, less filtered oil reaches the bearings, until finally no oil is filtered. However, oil still reaches the bearings even though it is not filtered.

Continued on next page   SS40167,0000115 -19-15JUL09-5/8

Full-Flow Filtration System

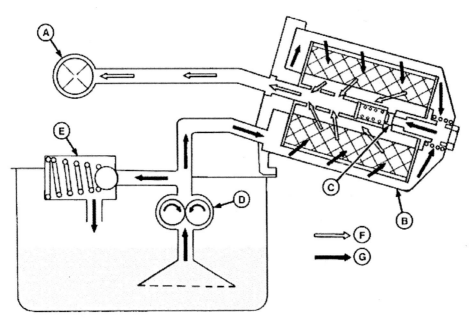

Full Flow Filtration System

A—Engine Bearings
B—Filter
C—Relief Valve
D—Oil Pump
E—Pressure Regulating Valve
F—Filtered Oil
G—Unfiltered Oil

In the full-flow filtration system, there is only one oil flow — from the oil pump to the filter and then to the engine bearings. A pressure gauge and pressure-regulating valve are used just as in the bypass system.

Notice the relief valve within the filter. When the filter is new, there is little pressure drop through the filter.

However, if the filter gets clogged, the resulting pressure will open the relief valve to allow unfiltered oil to bypass to the bearings. This relief valve may be located in the base of some filter housings.

Continued on next page   SS40167,0000115 -19-15JUL09-6/8

Flow Through Dual Filters and Oil Cooler

Full-flow systems with dual oil filters and oil cooler have an oil flow similar to that shown in the illustration.

Oil is pumped to the oil cooler and the oil cooler bypass valve. If the oil cooler is unrestricted, the oil flows through the oil cooler to the dual oil filters. Cool oil to the filters is also directed to the head of the oil filter bypass valve, the spring end of the oil cooler bypass valve, and the spring end of the pressure-regulating valve.

Cooled oil that passes through the filter is routed to the spring end of the oil filter bypass valve, the pressure regulating valve and the engine oil gallery.

The oil cooler bypass valve opens when the cooler is restricted to the point of creating enough pressure on the head of the valve to open it. When open, oil flows through the valve and bypasses the cooling element by flowing through an opening at one end of the cooler to the filters.

If the filters are restricted, pressure oil at the head of the filter bypass valve causes it to open. This allows unfiltered oil to flow to the oil gallery for lubrication.

When oil pressure in the oil gallery exceeds a predetermined pressure, the pressure-regulating valve spool opens, allowing cooled unfiltered oil to go to the

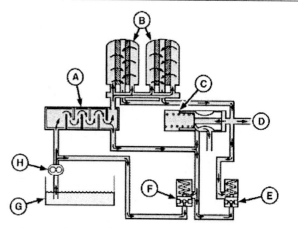

Oil Flow Through a System with Dual Filters and an Oil Cooler

A—Oil Cooler
B—Dual Oil Filters
C—Pressure-Regulating Valve
D—To Oil Gallery
E—Oil Filter Bypass Valve
F—Oil Cooler Bypass Valve
G—Oil Pan
H—Oil Pump

oil pan. The pressure regulating valve has orifices that provide dampening for the spool.

SS40167,0000115 -19-15JUL09-7/8

Servicing of Oil Filters

The filter may be enclosed in a disposable housing with a permanent base.

Service for these filters is simple — replace it at the regular intervals specified by the engine manufacturer.

If the filter element is contained in a removable housing, use a solvent to clean the housing and the mounting base. Dry thoroughly with compressed air. Likewise, clean and dry the relief valve if it is removable.

Edge-type metal filters should be cleaned with solvent and a soft brush. Hard varnish-like deposits may be removed by soaking the filter overnight in a strong solution of lye water. Flush the filter thoroughly after soaking to remove the loosened particles.

When servicing any filter, use new gaskets and seal rings. Tighten the housing firmly, but not too tight. Run the engine until oil pressure registers and check for leaks. Leaks are caused by the housing being too loose or the gasket installed improperly.

Often the crankcase oil is changed at the same time as the filter is serviced. This ensures both a clean oil supply and filter for the lubricating system.

IMPORTANT: Many full-flow filters have a special bypass valve. Replace them only with a genuine duplicate of the original filter.

Service Engine Oil Filters at Regular Intervals

A—Sealing Ring

B—Replaceable Filter Element

SS40167,0000115 -19-15JUL09-8/8

Lubricating Valves

Valves have two uses in the lubrication system:

- **To regulate oil pressure**
- **To bypass oil at filters and oil coolers**

The operation of the valve is the same in either case.

Basically, the valve has a poppet held in place over an opening by a spring.

As shown in the illustration, the valve is **closed** when the spring tension is greater than the oil pressure at the inlet. The spring tension holds the poppet securely in position.

The valve **opens** when pressure at the oil inlet exceeds that of the spring. This pushes the poppet off the inlet hole and oil flows through the valve.

Now let's see how this valve is used.

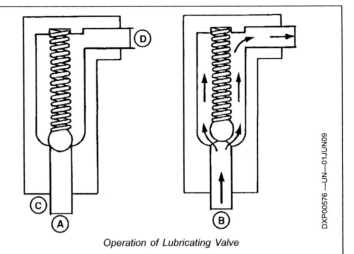

Operation of Lubricating Valve

A—Closed   C—Inlet
B—Open   D—Outlet

Oil Pressure Regulating Valves

The oil pressure-regulating valve maintains the correct pressure in the lubrication system regardless of the engine speed or the temperature of the oil. Most regulating valves are adjustable.

Most oil pumps can push out more oil than needed for engine lubrication. At idling speeds, or on older engines, this is not a problem. But what happens to the excess oil when the engine is new and operating properly?

This is where the pressure-regulating valve does its job. When the oil pressure exceeds the valve setting, the valve opens and returns the excess oil to the crankcase.

The regulating valve is usually connected to the main oil gallery through passages in the block. However, it may be a part of the oil pump.

SS40167,0000116 -19-15JUL09-1/6

Checking and Adjusting Engine Oil Pressure

Before checking the oil pressure, always check the condition of the oil filter. A dirty filter will limit the flow of filtered oil.

Many engines have tapped holes in the block as oil pressure test points. On other engines, the oil pressure sending unit hole is used for testing.

Install a master pressure gauge to the engine oil gallery port as shown in the illustration. Start and run the engine at fast idle. When the engine is up to operating temperature, record the pressure reading on the gauge. Compare this reading with the specifications for the engine.

A—Master Pressure Gauge   B—Engine Oil Gallery Port

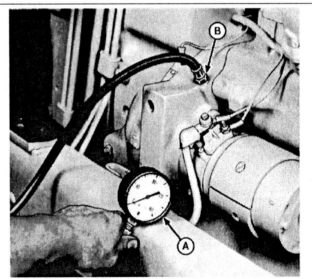

Checking the Engine Oil Pressure

Continued on next page
SS40167,0000116 -19-15JUL09-2/6

For adjusting pressure, shims and washers are used in some regulating valves.

To raise pressure, shims are added behind the spring to increase the tension of the spring. This in turn increases the setting at which the valve will open.

To lower pressure, shims are removed to decrease the tension of the spring and allow the valve to open at a lower setting.

Some regulating valves are adjusted by an adjusting screw. Normally, turn the screw in to increase the spring tension and the pressure setting. Turn the screw out to decrease the pressure setting.

Causes of Too-Low Engine Oil Pressure
1. Oil level in crankcase too low
2. Oil in crankcase too thin
3. Worn engine bearings
4. Worn oil pump
5. Filter or pump leaks
6. Regulating valve spring failed
7. Regulating valve needs adjustment
For solutions, see Chapter 13, "Diagnosing and Testing."

Causes of Too-High Engine Oil Pressure
1. Oil in crankcase too heavy
2. Defective pressure gauge or sender
3. Stuck regulating valve
4. Regulating valve needs adjustment
For solutions, see Chapter 13, "Diagnosing and Testing."

A—Valve Head
B—Aluminum Washer
C—Valve Plug
D—Shim
E—Spring
F—Adjusting Screw

Adjusting Engine Oil Pressure (Shim-Adjusted Valves)

Adjusting Engine Oil Pressure (Screw-Adjusted Valves)

SS40167,0000116 -19-15JUL09-3/6

Filter Bypass Valves

Every filter in a full-flow lubrication system must have a bypass valve. This valve bypasses oil around the filter when the filter becomes clogged.

In the full-flow system, all of the oil delivered by the pump passes either through the filter to the bearings or through the pressure-regulating valve to the crankcase. Let's see what would happen if a bypass valve were not provided.

A—Bypass Valve Inside Filter
    Housing
B—Bypass Valve in Mounting
    Pad of Filter
C—Oil Inlet
D—Oil Outlet
E—Spring
F—Valve Poppet

Filter Bypass Valves (Two Locations)

Continued on next page   SS40167,0000116 -19-15JUL09-4/6

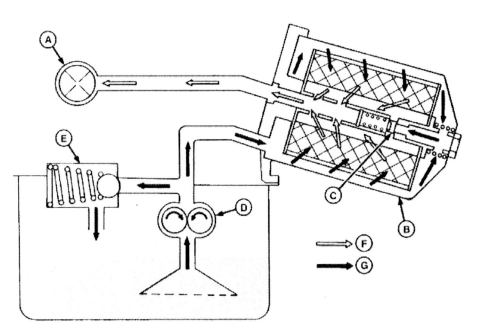

Full-Flow Filtration System

A—Engine Bearings
B—Filter
C—Relief Valve
D—Oil Pump
E—Pressure-Regulating Valve
F—Filtered Oil
G—Unfiltered Oil

As the filter becomes completely clogged, pressure would build up on the inlet side of the filter. This would cause the pressure-regulating valve to open completely, allowing all of the oil to return directly to the crankcase. Result — a burned-up engine.

Continued on next page

SS40167,0000116 -19-15JUL09-5/6

The bypass valve, then, is a safety device to ensure that the oil, filtered or dirty, will get to the bearings. The valve is usually set to open before the filter becomes completely clogged.

The illustration shows the filter bypass valve in two locations — inside the filter (A) and in the filter-mounting pad (B).

Many of the "spin-on" filter elements have the bypass built into the element.

When replacing a filter, **be sure to use only the recommended filter.** Another type may not have the built-in bypass valve.

Bypass valves are often used with oil coolers for the same reason that they are used with filters. If the oil cooler becomes clogged, oil flows through the valve and back into the lubrication system.

Servicing of Lubricating Valves

Most important in servicing of lubricating valves is to clean them. Where possible, completely disassemble the valve and wash the parts in solvent. Clean the bore in which the valve slides. Use compressed air to dry the parts.

Use a spring tester to check the strength of the spring. A broken or weakened spring will cause the valve to open at a much lower setting.

Inspect the valve poppet for wear and nicks that might cause it to hang up in the bore. It is very important that the valve slides freely in the bore.

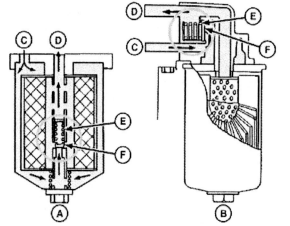

Filter Bypass Valves (Two Locations)

A—Bypass Valve Inside Filter Housing
B—Bypass Valve in Mounting Pad of Filter
C—Oil Inlet
D—Oil Outlet
E—Spring
F—Valve Poppet

Some regulating valves use a bushing in the bore as the seat for the poppet. Inspect the seat for wear and replace it if necessary.

Always check and adjust the engine oil pressure after servicing a lubricating valve.

SS40167,0000116 -19-15JUL09-6/6

Oil Coolers

Lubrication systems of many engines use an oil cooler to help dissipate the heat created by the engine. Coolers may be oil-to-air or oil-to-water types. Oil-to-water coolers can obtain oil temperatures only as low as temperatures in the radiator. In many cases, this temperature is acceptable and water-to-oil coolers use engine coolant to dissipate heat from the oil.

Oil-to-air coolers must be used if lower than radiator oil temperatures are required. These are almost always of the fin-and-tube design and appear much like an automobile radiator.

The oil cooler may be mounted **internally** in the crankcase or **externally** on the outside of the engine block.

When the cooler is mounted in the crankcase, coolant is pumped by the water pump through the cooler to the radiator of the cooling system. The heat of the oil in the crankcase is conducted through the fins of the cooler and is absorbed by the coolant. The heat is then dissipated in the radiator.

When the cooler is mounted externally, both coolant and lubricating oil are pumped through it.

A—Packing Gland
B—Crankcase Oil Cooler
C—Oil Cooler Clamp

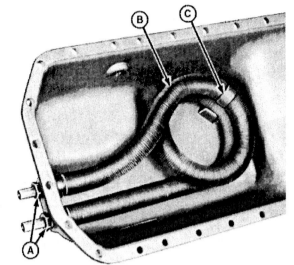

Internal Crankcase Oil Cooler

Oil-to-Water and Oil-to-Air Coolers

SS40167,0000117 -19-15JUL09-1/2

As shown in the illustration, coolant flows through the tubes in the cooler, and oil circulates around the tubes. Heat from the oil is conducted through the tubes to the coolant, which carries it to the radiator for dissipation.

A similar cooler uses a small radiator core instead of tubes within the housing. The oil is pumped through the core, and the coolant is circulated around it.

A bypass valve is used with some oil coolers to ensure circulation of the oil if the cooler should become clogged.

Maintenance of Oil Coolers

Normal maintenance of the cooling system (Chapter 10) will keep the oil cooler passages clean.

When cleaning the lubrication system, remove the cooler and clean the oil passages with a solvent.

When assembling or installing an external cooler, use new gaskets and be sure the cap screws are tight.

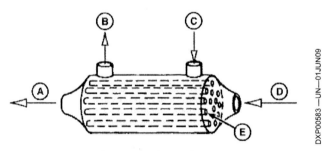

Oil-to-Water Cooler Operation

A—Coolant Out
B—Oil Out
C—Oil In
D—Coolant In
E—Coolant Tubes

SS40167,0000117 -19-15JUL09-2/2

Oil Pressure Indicating Systems

A pressure indicating system is essential for monitoring the lubrication system. These systems may be classified as:

- **Mechanical**
- **Electrical**

Mechanical Indicating Systems

The mechanical indicating system uses a Bourdon tube gauge which is attached by tubing to the pressure source. This is usually the main oil passage in the engine block.

The basic components of the Bourdon tube gauge are a tube made of spring bronze or steel and a pinion and sector mechanism. One end of the tube is permanently attached to the pressure inlet gauge. The other end is attached to the pinion and the sector mechanism.

Oil pressure inside the tube tends to straighten the tube out. This movement acts upon the pinion and sector mechanism, causing the pointer to rotate on the face of the gauge. The pressure can be read directly on the gauge.

Maintenance

1. If a mechanical gauge will not register pressure, loosen the oil line at the engine block (but do not remove it).

2. If oil leaks from the connection while the engine is running, there is pressure in the lubricating system. This means the trouble is in the oil line of the gauge itself.

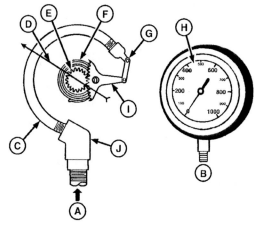

Bourdon Tube Oil Gauge

A—Oil Pressure	F—Hair Spring
B—Complete Gauge	G—Link
C—Bourdon Tube	H—Scale
D—Pointer	I— Sector and Pinion
E—Gear	J— Stationary Socket

3. Check the oil line for restrictions. The pressure must get to the gauge before it can register.

4. If you suspect that the gauge is defective, replace it with one that is known to be accurate. If the new gauge works correctly, this proves that the original was defective.

Continued on next page

SS40167,0000118 -19-15JUL09-1/4

Electrical Indicating Systems

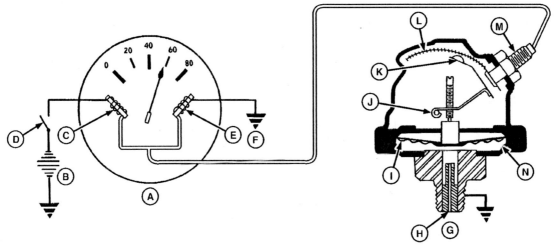

Electromagnetic Coil System for Indicating Oil Pressure

A—Indicating Gauge
B—Battery
C—Coil No. 1
D—Ignition Switch

E—Coil No. 2
F—Ground
G—Sending Unit
H—Oil Pressure Here

I—Spring
J—Lever
K—Wiper
L—Resistor

M—Terminal
N—Diaphragm

An electrical oil pressure indicating system has two parts:

1. Sending Unit (G) — at pressure source
2. Indicating Gauge (A) — on control panel

The sending unit, actuated by the pressure of the lubricating oil, electrically actuates the indicating unit. The two units are connected together by a single wire, and both must be grounded.

There are three kinds of electrical indicating systems:

1. **Electromagnetic coil type**
2. **Heating coil type**
3. **Pressure switch type**

Let's look at the operation of each one.

Electromagnetic Coil Type

The sending unit of the electromagnetic system consists of a resistor, a sliding contact called the wiper, an actuating lever, and a spring-loaded diaphragm enclosed in a housing.

The pressure of the lubricating oil works against the diaphragm, causing the lever to move. The movement of the lever changes the position of the wiper, which varies the resistance of the resistor.

The gauge contains two coils and an armature with a pointer. The magnetic fields of the two coils creates another field, which controls the rotation of the armature. This, in turn, controls the rotation of the pointer on the face of the gauge.

Current flows from the battery to coil No. 1 in the gauge as shown, then through a parallel circuit to coil No. 2 and to the resistor in the sending unit.

As pressure against the diaphragm in the sending unit changes, the resistance of the resistor also changes. The change of the current in coil No. 2 causes its magnetic field to be stronger or weaker than that of coil No. 1.

The armature, of course, is attracted to the stronger of the two fields.

Continued on next page   SS40167,0000118 -19-15JUL09-2/4

Heating Coil System

The sending unit of the heating coil system has a heating coil wound around a bimetal strip. The pressure of the lubricating oil deflects the diaphragm, causing the contact to close. The current flowing through the coil then creates heat.

The heat deflects the bimetal strip until the contact is opened. The bimetal then cools and returns to its original position, again closing the circuit. This cycle of closing and opening, heating and cooling is repeated continuously.

The gauge contains a similar heating coil wound around a bimetal strip. This coil is connected in series with the coil in the sending unit. As the coil of the sending unit heats up, so does the coil of the gauge. The bimetal strips in each unit then deflect at the same time.

The pointer of the gauge is linked to the bimetal strip in the gauge. The deflection of the bimetal strip causes the pointer to rotate on the face of the gauge.

As oil pressure increases, the diaphragm of the sending unit is deflected more. A greater amount of current is then required to heat the bimetal strip enough to open the contact. The increased current in the coil of the gauge,

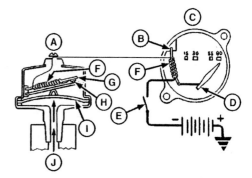

Heating Coil System for Indicating Oil Pressure

A—Sending Unit
B—Bimetal
C—Indicating Gauge
D—Pointer
E—Ignition Switch
F—Heating Coil
G—Bimetal
H—Contacts
I— Diaphragm (Flexed)
J—High Pressure

likewise, causes a greater deflection of the bimetal strip in the gauge. This, in turn, causes the pointer to register the increased oil pressure.

SS40167,0000118 -19-15JUL09-3/4

Pressure Switch System

The pressure switch system is quite different from the electromagnetic coil and heating coil types.

A sending unit is used, but a light bulb is used as the indicator rather than a gauge. This system can only warn of low oil pressure. It cannot tell the actual pressure of the system.

Basically, the sending unit consists of a set of contacts and a diaphragm. When the ignition switch is on and before the engine is started, the contacts are closed and the indicating bulb is lit.

When the engine is started and the oil pressure builds, the diaphragm is pushed up. This separates the contacts and breaks the circuit, causing the bulb to go out.

If the oil pressure drops, the resulting deflection of the diaphragm will close the contacts and again complete the circuit. The lighted bulb warns that the pressure of the lubrication system has dropped below the low pressure setting.

Maintenance

Look for these things when servicing an electrical oil pressure indicating system:

1. Poor grounds

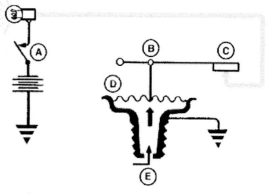

Pressure Switch System for Indicating Oil Pressure

A—Ignition Switch
B—Contact Arm
C—Contacts
D—Diaphragm
E—Oil Pressure

2. Broken or poor connections
3. Defective units

Specific tests vary with the different types of units. See the engine technical manual for details.

SS40167,0000118 -19-15JUL09-4/4

Contamination of Oil

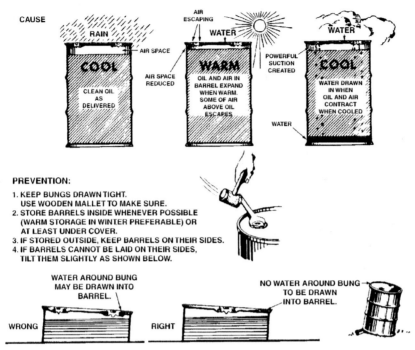

Storage Practices That Prevent Contamination of Oil

Oil contamination will reduce engine life more than any other factor. Some sources of contamination are obvious while others are not.

Let's review some of the sources of contamination and what can be done about them.

1. The most obvious source of contamination is the **storing and handling of the oil itself**. If at all possible, store lubricants in a clean enclosed storage area. Keep all covers and spouts in containers when not in use. These practices not only keep dirt out of the oil, but also reduce condensation of water caused by atmospheric changes.

2. Another obvious source is **dust that is breathed into the engine** with combustion air. It is very important that the air cleaner be cleaned or replaced regularly. At the same time, clean or replace the breather on the oil filler.

Continued on next page

SS40167,0000119 -19-15JUL09-1/3

100709
PN=356

3. A major source of contamination is a **cold engine**. When the engine is cold, its fuel burning efficiency is greatly reduced. Partially burned fuel blows by the piston rings and into the crankcase. Oxidation of this fuel in the oil forms a very harmful varnish, which collects an engine parts. An overchoked or misfiring engine will also create this contamination from unburned fuel.

Water created by a cold engine contaminates the oil, too. Water vapor, a normal product of combustion, tends to condense on cold cylinder walls. This condensation is also blown by the rings into the crankcase. The engine must warm up before this condensation problem is eliminated.

Water not only causes rusting of the steel and iron surfaces, but it can combine with oxidized oil and carbon to form sludge. This sludge can very effectively plug oil screens and passages.

Contamination from a cold engine can be prevented by:

a. Properly warming up the engine before applying load.

b. Making sure the engine is brought up to operating temperature each time it is used.

c. Using the proper thermostat to warm up the engine as quickly as possible.

Varnish Buildup from Contaminated Fuel

Sludge at Bottom of Crankcase

Continued on next page

SS40167,0000119 -19-15JUL09-2/3

100709
PN=357

4. **Antifreeze** can be another sludge-forming source of contamination. To guard against antifreeze contamination:

   a. Tighten the head bolts to specifications during an overhaul.

   b. Use a cooling system sealer when filling the cooling system (if recommended by the engine manufacturer).

   c. Guard against detonation and improper use of starting fluids in diesel engines (both can result in head gasket damage).

Other problems that cause oil contamination are:

5. **Oxidation** is not an obvious source of contamination, but it is a very real one. Oxidation occurs when the hydrocarbons in the oil combine with oxygen in the air to produce organic acids. Besides being highly corrosive, these acids create harmful sludge and varnish deposits.

6. **Carbon particles** are another contaminant created by the normal operation of the engine. The particles are created when oil around the upper cylinder walls is burned during combustion. Excessive deposits can cause the piston rings to stick in their grooves.

7. **Engine wear** also creates a contaminant. Tiny metal particles are constantly being worn off bearings and other parts. These particles tend to oxidize and deteriorate the oil.

All sources of contamination cannot be eliminated. What, then, can be done to protect the engine?

Protecting the Engine Against Oil Contamination

Start by using good quality oil that has additives. Additives are put in the oil for a specific reason based on the service expected for the oil.

Here are some of the important **oil additives**:

ANTI-CORROSION ADDITIVES protect metal surfaces from corrosion attack.

Clean Sludge from Crankcase Suction Screen

OXIDATION INHIBITOR ADDITIVES keep oil from oxidizing even at high temperatures.

ANTI-RUST ADDITIVES prevent rusting of metal parts during storage periods, downtime, or even overnight. They form a protective coating, which repels water droplets and protects the metal. These additives also help to neutralize harmful acids.

DETERGENT ADDITIVES help keep metal surfaces clean and prevent deposits. Particles of carbon and oxidized oil are held suspended in the oil. The suspended contaminants are then removed from the system when the oil is drained. Black oil is evidence that the oil is helping to keep the engine clean by carrying the particles in the oil rather than letting them accumulate as sludge.

However, remember that **additives eventually wear out.**

To prevent this, drain the oil before the additives are completely depleted.

Also, service the filters at regular intervals.

Finally, keep the fuel, cooling, and ignition systems in good condition so the fuel is efficiently burned.

SS40167,0000119 -19-15JUL09-3/3

Oil Consumption

Some oil consumption is natural during normal operation of internal combustion engines.

As we have already learned, lubricating oil provides a seal between the piston rings and cylinders. It is only natural that during the combustion process, some of this sealing oil is burned up.

Excessive oil consumption can be caused by several conditions.

To properly diagnose the problem, follow this procedure:

1. First be sure the correct weight and grade of oil for the type of service and climate is being used. Oil that is too thin may "flood" the piston rings, while oil that is too thick may "starve" the rings.

2. Be sure the engine has run long enough under load to ensure that the rings have had a chance to seat. Some variation in oil consumption can be expected during break-in, but it should be stabilized before 250 hours of operation. If not, assume that a problem exists and correct it at once. (See Chapter 3 for "Engine Break-In.")

3. Check the engine oil pressure and, if necessary, adjust the pressure-regulating valve. High engine oil pressure can cause oil consumption by flooding the rings and valves with oil. Also check the crankcase breather. A plugged breather can increase the crankcase oil pressure.

4. Check for external oil leaks. What appears to be a small leak can add up to a considerable loss of oil. Drops of oil lost externally can add up to quarts of oil between oil change periods. Check the front and rear oil seals, all gaskets, and the filter attaching points.

5. Check for engine blow-by. Watch the fumes expelled through the crankcase vent tube. Fumes should be barely visible with the engine at fast idle with no load. If possible, compare the blow-by with that of identical engines. This can be a guide to determine if the blow-by is excessive.

   Excessive blow-by indicates that piston rings and cylinder liners have worn to the point where the rings cannot seal off the combustion chambers. An overhaul with new rings is then required. New or reconditioned pistons and liners may also be necessary.

6. If blow-by is not excessive, check to see if the oil is being lost through the valve guides. A generous supply of oil is usually maintained in the rocker arm to lubricate the rocker arms, valve stems, and valve guides.

   As shown in the illustration, gravity, inertia, vacuum, and an atomizer effect all combine to force oil down

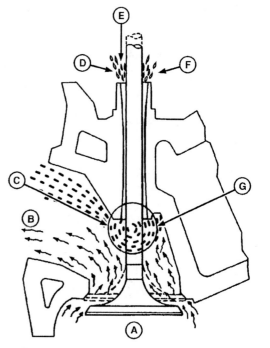

Problems Which Cause Oil Loss Through Valve Guides

A—Combustion Chamber
B—Exhaust Gases
C—Vacuum
D—Gravity

E—Atmospheric Pressure
F—Inertia
G—Atomizer Effect

between the valve stem and guide. Valve guide seals are used to prevent excessive oil from entering the combustion chambers. If these seals become defective, they create a definite oil consumption problem.

To check for this condition, start the engine and warm it up. Let it idle slowly for 10 minutes. Then remove the cylinder head, and inspect the valve ports and undersides of the valve heads. If these are wet with oil, the oil has been drawn through the valve guides.

At the same time, check the piston heads. If they are wet, the oil has been drawn past the piston rings. If excessive blow-by has indicated worn rings, this procedure can be used to confirm it.

7. Worn connecting rod bearings can also contribute to oil consumption. When worn, the bearings throw an excess of oil onto the cylinder walls. The action of the piston forces part of this excess oil into the combustion chambers, where it is burned.

8. Excessive engine speeds are another common cause of oil consumption. Observe the fast idle limits set by the manufacturer.

SS40167,000011A -19-15JUL09-1/1

Crankcase Ventilation

During normal operation, unburned fuel vapor and water vapor are created in the engine. If allowed to condense, these vapors become contaminating liquids that drain into the crankcase. The purpose of the ventilation systems is to circulate fresh air through the engine to carry away these harmful vapors.

As with all ventilating systems, there must be:

- An air inlet
- An air and vapor outlet
- A means of circulating air between the two

The oil filler cap is probably the most common inlet. A small air cleaner built into the cap filters air as it is drawn into the system. Some systems use the main air cleaner as the inlet.

The use of a vent tube is one means of circulating the air through the system. As the machine moves, the air moving past the tube opening creates a lower pressure than that at the breather. Air in the engine naturally flows to this low-pressure area, and the tube becomes an outlet for the system. Some vent tubes use a wire mesh screen to filter oil particles out of the vapor.

Intake Manifold Ventilation System

The intake manifold can serve the double function of circulating the air and being the vent outlet. In this type of ventilation system, a tube connects the crankcase to the intake manifold. The intake vacuum draws the air through the engine, into the manifold, into the cylinders, and out the exhaust system. A ventilation valve is used to regulate the flow of air into the manifold.

At full throttle speed, there is maximum vapor in the crankcase and minimum vacuum in the intake manifold. The valve is held open by the spring, permitting full air flow through the system.

At idle speed, there is minimum vapor in the crankcase and maximum vacuum in the manifold. As the pressure of the air is greater than that of the vacuum, the spring is compressed and the valve is closed. Air flow is restricted by the bleed hole in the valve.

Some ventilating valves do not have a bleed hole. This design eliminates the possibility of a backfire traveling through the ventilating system to the crankcase.

DXP00593 —UN—23DEC08

Crankcase Ventilation Tube

Ventilation System with Circulating Pump

Another type of system uses a vane or impeller pump to circulate the air. The air is taken from the main air cleaner and pumped through the engine. The outlet can be either a vent tube or the intake valve.

Servicing the Ventilation System

Begin with the inlet when servicing the ventilation system. If the oil filler cap is the inlet, periodically clean it with a solvent. If the main air cleaner is the inlet, service it regularly as recommended by the engine manufacturer.

If the system uses a ventilating valve, service it regularly. Some manufacturers recommend cleaning the valve with solvent, while others recommend replacing the valve. If a filter is used with the vent tube, be sure to clean it, too.

SS40167,000011B -19-15JUL09-1/1

Changing Engine Oil

The statement that "oil doesn't wear out" is false. Oil loses many of its good lubricating qualities as it gets dirty and its additives wear out.

Acid formations, sludge, varnish, and engine deposits contaminate oil and make it unfit for continued use. On the other hand, just because crankcase oil is black, doesn't mean it's time for an oil change.

Additives in the oil are supposed to clean and hold deposits in suspension for removal when the oil is drained.

Change the oil sometime **before** the additives wear out which protect the engine against sludge. For the average person, this time is practically impossible to determine. That's why the best policy is to follow manufacturer recommendations on oil and filter changes.

New or rebuilt engines require oil and filter changes after a specified break-in period. Performing this service on time is very important, since foreign material accumulates in the oil at a faster rate during initial operation than later when the engine is broken in.

When changing the oil and filter on any engine, always warm up the engine first. This way the contaminants and foreign materials are mixed with the oil when it is drained.

Replacing some oil filters requires installing new gaskets or sealing rings. Be sure the sealing surfaces on the engine and filter are clean.

After installing the filter and filling the engine with oil, run the engine and check for possible oil leaks.

Keep a record of all oil and filter changes to be sure of regular engine service.

Dispose of Waste Properly

Improperly disposing of waste can threaten the environment and ecology.

Use leak-proof containers when draining the oil. Do not pour the used oil onto the ground, down a drain, or into any water source.

Inquire on the proper way to recycle or dispose of the used engine oil and filters from your local environmental or recycling center.

For more details, refer to the FOS manual on Fuels, Lubricants, and Coolants.

Diesel Engine Oils

Select an oil viscosity based on the air temperature range during the period between oil changes.

Some oil is specially formulated to provide superior protection against high temperature thickening as well

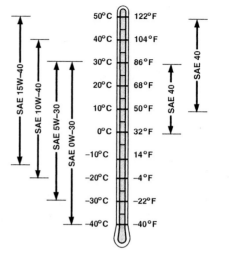

Temperature Range for Diesel Engine Oils

as exceptional cold weather starting performance; these properties may result in longer engine life.

IMPORTANT: **For farm and industrial products, do not extend drain interval beyond the recommendations in the operator's manual.**

For some engine applications, the oil and filter change interval may be extended by 50 hours when using a certain manufacturer's oil and oil filter. Follow the instructions given in the operator's manual.

Consult the engine technical manual or vehicle operator's manual for the recommended oil for the specific engine. In general, oils may be used if they meet one or more of the following:

• **API Service Category CJ-4, CJ-4 PLUS, CI-4, or CI-4 PLUS**
• **ACEA Sequence E7, E6, E5, or E4**

Synthetic lubricants may be used if they meet the engine manufacturer's performance requirements. The temperature limits and service intervals are the same for both synthetic and conventional oils.

Multi-viscosity diesel engine oils are preferred over single viscosity oils. The multi-viscosity oils provide superior cold weather starting performance and protection against high temperature thickening.

Avoid mixing different brands or types of oil. Oil manufacturers blend additives in their oils to meet certain specifications and performance requirements. Mixing different oils can interfere with the proper functioning of these additives and degrade the lubricant performance.

For more information on engine oils, refer to the FOS manual on Fuels, Lubricants, and Coolants.

Continued on next page

SS40167,000011C -19-15JUL09-1/2

Gas Engine Oils

Select the oil viscosity based on the air temperature range during the period between oil changes.

Some oil is specially formulated to provide superior protection against high temperature thickening as well as exceptional cold weather starting performance; these properties may result in longer engine life.

IMPORTANT: For farm and industrial products, do not extend drain interval beyond the recommendations in the operator's manual.

For some engine applications, the oil and filter change interval may be extended by 50 hours when using a certain manufacturer's oil and oil filter. Follow the instructions given in the operator's manual.

Consult the engine technical manual or vehicle operator's manual for the recommended oil for the specific engine. In general, oils may be used if they meet one or more of the following:

• **API Service Classification SJ, SL, SM, or Higher**

Synthetic lubricants may be used if they meet the engine manufacturer's performance requirements. The temperature limits and service intervals are the same for both synthetic and conventional oils.

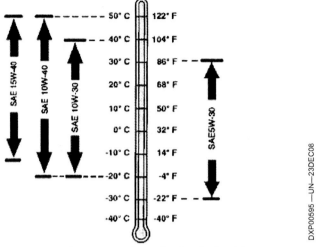

Temperature Range for Gas Engine Oils

The multi-viscosity oils provide superior cold weather starting performance and protection against high temperature thickening.

For more information on engine oils, refer to the FOS manual on Fuels, Lubricants, and Coolants.

SS40167,000011C -19-15JUL09-2/2

Test Yourself

QUESTIONS

1. Name any three of the five jobs that the lubrication system does for the engine.

2. What does a bypass valve do for a full-flow lubrication system?

3. What does a good lubricating oil contain that helps to prevent rust, corrosion, oxidation, and foaming?

4. (True or false?) Oil does not wear out.

SS40167,000011D -19-15JUL09-1/1

Cooling Systems

Cooling System — Introduction

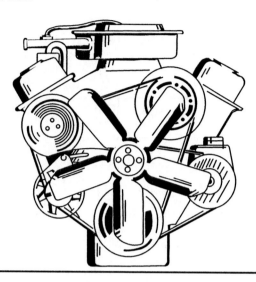

The engine cooling system does two things:

1. **Prevents overheating**
2. **Regulates temperature**

OVERHEATING could burn up the engine parts. Some heat is necessary for combustion, but the engine generates too much heat. So the cooling system carries off the excess heat.

REGULATING TEMPERATURES keeps the engine at the best heat level for each operation. During starting, the engine must be warmed up faster. During peak operations, the engine must be cooled.

Running the engine too hot can cause:

• Preignition
• Detonation
• Knock
• Burned pistons and valves
• Lubrication failure

Running the engine too cold can cause:

• Unnecessary wear
• Poor fuel economy
• Accumulation of water and sludge in the crankcase

DP79986,0000051 -19-16JUL09-1/1

Types of Cooling Systems

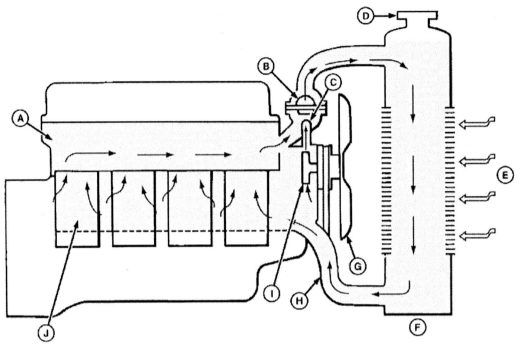

Liquid Cooling System of Engine

A—Coolant Manifold
B—Thermostat
C—By Pass

D—Pressure Cap
E—Air Flow
F—Radiator

G—Fan
H—Hose
I— Water Pump

J— Coolant

Two types of cooling systems are used in modern engines:

- **Air Cooling System — uses air passing around the engine to dissipate heat.**
- **Liquid Cooling System — uses a liquid to cool the engine and air to cool the liquid.**

AIR COOLING is commonly used on smaller engines due to its simplicity and compact design. Metal baffles, ducts, and blowers are used to aid in distributing the air to the heat points of the engine. Air cooling systems do away with many of the components used for liquid cooling, such as the water pump, radiator, liquid coolant, and coolant hoses.

LIQUID COOLING normally uses water or a special solution as a coolant. Antifreeze solutions are added to the water to protect against freezing and boil-over. The coolant circulates in manifold around the cylinders and cylinder head. As heat radiates, it is absorbed into the coolant, which then flows to the radiator. Air flow through the radiator absorbs the heat from the coolant and so dissipates the heat into the air. The coolant then recirculates into the engine to pick up more heat.

Continued on next page   AH98466,00001AD -19-22JUL09-1/3

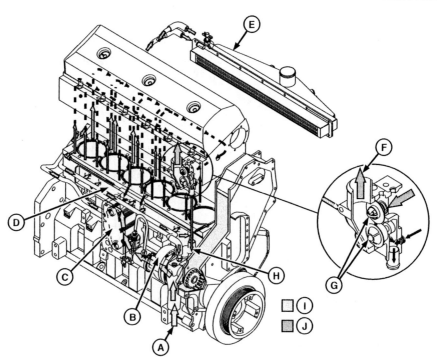

Coolant Flow in a Modern Diesel Engine

A—From Radiator
B—Coolant Pump
C—Oil Cooler

D—Coolant Manifold
E—Aftercooler
F— To Radiator

G—Thermostats
H—Bypass Tube
I— Low-Temperature Coolant

J— High-Temperature Coolant

In this chapter we will cover only the liquid cooling systems used on most modern engines.

Liquid Cooling System Operation

To understand the cooling system operation, let's trace the flow of coolant through a typical modern diesel engine cooling system.

The pressurized cooling system consists of a radiator, water pump, thermostats, thermostat housing, coolant manifold, and coolant passages in the cylinder head and block.

The pump draws coolant from the radiator through the lower radiator hose and discharges it through the coolant manifold. The coolant manifold extends the length of the cylinder block. From the coolant manifold, the coolant flows into the engine oil cooler housing. The coolant flows around the oil cooler and then flows into one or more circuits.

The main circuit flows coolant from the oil cooler into the block and around each cylinder liner. From the liners, the coolant exits the block through a vertical passage into the cylinder head.

Some engine designs vary the size and number of the coolant passages from the block to the cylinder head to optimize the coolant flow through the block and cylinder head. There are more passages and larger passages at the rear of the cylinder head than at the front. This is done to provide ample coolant flow around each liner and to

provide more flow to the rear of the cylinder head than into the front.

The cooling system may have a second circuit that directs coolant to a specific area of the engine. This circuit is called the "directed cooling" circuit. Coolant flows through a small port into a groove at the top of each cylinder liner for top liner cooling. The coolant passes around the groove in the top of the liner, and exits through a passage to either the cylinder head or the thermostat housing.

Once the coolant is in the cylinder head, all flow is toward the front. All coolant then exits the cylinder head and enters the thermostat housing.

When the engine is cold, the thermostats will be closed. Coolant will flow through the bypass tube, into the inlet of the pump to be recirculated. This provides a fast and uniform warm-up for the engine.

When the engine warms to operating temperature, the thermostats will open. Coolant will then flow past the open thermostats and return to the radiator through the upper radiator hose.

Liquid Cooling System Components

A liquid cooling system may consist of the following:

• **Radiator and Pressure Cap**
• **Fan and Fan Belt**
• **Water Pump**
• **Engine Water Jacket**
• **Thermostat**

Continued on next page

AH98466,00001AD -19-22JUL09-2/3

DXP00597 —UN—23DEC08

- **Engine Oil Cooler**
- **Deaeration Tank**
- **Coolant Recovery Tank**
- **Connecting Hoses**
- **Liquid or Coolant**

The RADIATOR is one of the major components of any liquid cooling system. It is here that heat in the coolant is released to the atmosphere. It also provides a reservoir for enough liquid to operate the cooling system efficiently.

The FAN forces cooling air through the radiator to quickly dissipate the heat being carried by the radiator coolant.

The WATER PUMP circulates the coolant through the system. The pump draws coolant from the bottom of the radiator and forces it through the engine water jacket for cooling. The hot coolant flows from the engine into the top of the radiator.

The ENGINE WATER JACKET provides a path for the coolant to flow around the cylinders and through the cylinder head. Some engines have distribution tubes and some have transfer holes that direct extra coolant flow to "hot" areas, such as the exhaust valve seats.

The THERMOSTAT is a heat-operated valve. It controls the flow of coolant to the radiator to maintain the engine

temperature within the range of highest operating efficiency.

The ENGINE OIL COOLER uses the engine coolant to dissipate heat from the engine oil. An option on some engines.

The DEAERATION TANK provides a low velocity area where added coolant is maintained and air is allowed to separate from the coolant. Hoses connect the deareation tank to the top of the cylinder head, the coolant recovery tank, and top and bottom of the radiator.

The COOLANT RECOVERY TANK serves as a reservoir for coolant that could escape the cooling system as a result of coolant expansion caused by overheating.

CONNECTING HOSES are the flexible connections between the engine and other parts of the cooling system.

COOLANT is the liquid that circulates through a cooling system. Coolant carries heat from the engine water jacket into the radiator for transfer to the atmosphere. The coolant then flows back through the engine to absorb more heat.

AH98466,00001AD -19-22JUL09-3/3

Radiators

Typical Radiator

Continued on next page

AH98466,00001AE -19-16JUL09-1/5

DXP00598 —UN—23DEC08

Two types of radiators are used in liquid cooling systems:

- **Cellular-Type Core Radiators — used where air speeds are high and resistance to air flow must be low, as in aircraft or racing cars.**
- **Tubular or Tube-and-Fin Type Core Radiators — used in most other machines.**

Each type of radiator may also be of either the vertical or horizontal (cross-flow) tube design. Both designs work according to the same principles.

In both types of radiators, coolant from the engine enters the radiator by way of the top hose, then passes through a series of small tubes surrounded by fins and air passages.

Illustrations show how air flowing around the fins and tubes removes the heat.

Cooled liquid reaching the bottom of the tank is picked up by the water pump to repeat the cycle.

A—Air
B—Water Passage
C—Tube
D—Fin

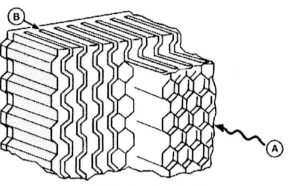

Cellular-Type Core Radiator

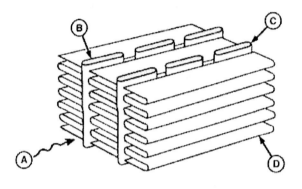

Tube and Fin-Type Radiator

AH98466,00001AE -19-16JUL09-2/5

Testing the Radiator

Because most modern cooling systems are pressurized, all of the components must be tight and in good condition before the system can operate properly.

Overheating and loss of coolant will result unless pressure in the system is maintained.

Test the entire cooling system before servicing it.

⚠ **CAUTION: Prevent possible injury from hot spraying coolant. DO NOT remove radiator cap unless engine is cool. Loosen cap slowly to the stop to release pressure. Then remove the cap.**

Remove the pressure cap and attach a radiator pressure tester to the filler neck according the manufacturer's instructions.

Pressurize the system and observe the pressure gauge reading. If the pressure drops, carefully inspect the

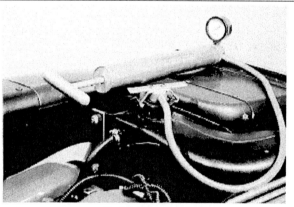

Radiator Pressure Test

radiator, water pump, hoses, drain cocks, and cylinder block for leakage. Mark all leaks to locate them later on.

Coolant may be leaking internally if the pressure drops and external leakage is not evident.

Continued on next page AH98466,00001AE -19-16JUL09-3/5

Overheating problems can result due to restricted air flow or coolant flow through the radiator core. An air flow meter will enable you to diagnose air flow restriction problems.

The air flow meter fan generates a voltage that is read on the AC voltage scale of a digital multimeter. Voltage variations from one section of the radiator core to the next will indicate air flow restrictions.

An electronic thermometer is used to measure temperature variations in the core to determine if blockage exists. Refer to the vehicle technical manual for test procedures and specifications.

Air Flow Meter

AH98466,00001AE -19-16JUL09-4/5

Servicing the Radiator

NOTE: Repairs should be done only by experienced radiator technician.

Inspect the radiator for bent fins. Inspect the tubes for cracks, kinks, dents, and fractured seams.

If radiator leaks and source of leak cannot be visually determined, test it as follows:

1. Install radiator cap. Plug overflow tube and outlet pipe. Attach air hose to inlet connection.

2. Fill radiator with compressed air [not more than 7–10 psi (50–70 kPa) depending on the size of the radiator] and submerge in a tank of water.

3. Look for bubbles, which tell the location of leaks.

Straighten any bent fins. Repair leaks after thoroughly cleaning surfaces, or replace radiator. Inspect rubber cushion washers. Replace as necessary.

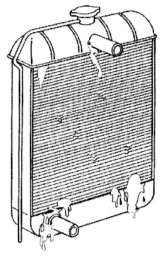

Damaged Radiator

AH98466,00001AE -19-16JUL09-5/5

Radiator Cap

⚠ **CAUTION: Always remove the radiator cap slowly and carefully to avoid a possible fast discharge of hot coolant.**

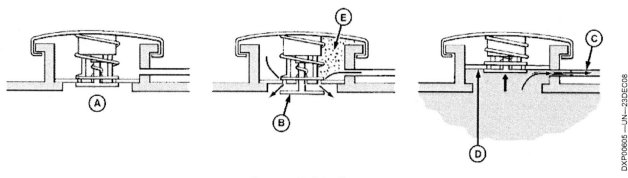

Pressure Radiator Cap

A—Valves Closed **B—Vacuum Valve Open** **C—Overflow Tube** **D—Pressure Valve Open**

The pressure system permits operating the engine at a higher temperature without boiling the coolant or losing it by evaporation.

An increase in pressure of 1 psi (7 kPa) will raise the boiling point of pure water about 3°F (1.67°C).

A special radiator cap is required on most modern cooling systems. The cap has two functions:

- **Allows atmospheric pressure to enter the cooling system.**

- **Prevents coolant escape at normal pressures.**

A pressure valve in the cap permits the escape of coolant or steam when the pressure reaches a certain point.

The vacuum valve in the cap opens to prevent a vacuum in the cooling system.

Hotter operation is desirable for efficient combustion and for evaporating contaminants from the crankcase.

AH98466,00001AF -19-16JUL09-1/2

Testing the Radiator Cap

A radiator and pressure cap tester can be purchased from local tool jobbers. Use instructions with the tester to see if the radiator cap can build and hold pressure.

If the pressure cap is defective, replace it. A radiator cap cannot be repaired.

To prevent damage to the cooling system from either excessive pressure or vacuum, check both valves in the cap periodically for proper opening and closing pressures. Service or replace the cap as required.

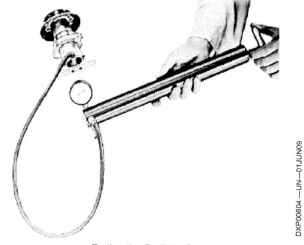

Testing the Radiator Cap

AH98466,00001AF -19-16JUL09-2/2

Fan and Fan Belts

A cooling system fan is usually located between the radiator and the engine. The fan is typically mounted on the front of the engine block and is generally driven by a V-belt from the engine crankshaft. On smaller engines, cooling fans can be driven by an electric motor that is controlled by a thermostat.

A cooling fan can be either a **suction-type** or **blower-type** fan depending on the design of the system.

- **Suction-type fan** pulls air through the radiator and pushes it over the engine. This design permits the use of a smaller fan and a smaller radiator. This type of fan is typically used when machine motion aids air movement through the radiator (such as an automobile).
- **Blower-type fan** pulls air across the engine, then pushes it through the radiator. This design is typically used on slow-moving machines and on machines where harmful materials could be drawn through the radiator.

With either fan design, as the engine runs, the fan moves air across the radiator core to cool the liquid in the radiator. Regardless of the type of fan system used, the ideal location of the fan will be approximately 2-1/2 inches (60 mm) from the radiator core.

Servicing the Fan

The only service on the fan is to make sure each blade on the fan is straight and far enough away from the radiator to ensure the blades do not strike the core.

A bent blade will reduce the efficiency of the cooling system and cause the fan to be out of balance.

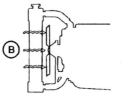
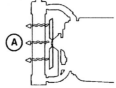

Two Types of Fan

A—Blower Fan   B—Suction Fan

Servicing the Fan Belt

The fan belt should be neither too tight nor too loose.

Too tight a belt puts an extra load on the fan bearings and shortens the life of the bearings as well as the belt.

Too loose a belt allows slippage and lowers the fan speed, causes excessive belt wear, and leads to overheating of the cooling system.

The condition of the belt and its tension should be checked periodically. Adjust fan belt tension as specified by the manufacturer.

AH98466,00001B0 -19-16JUL09-1/1

Water Pump

A water pump is normally a centrifugal-driven type pump. It could be considered the heart of the cooling system.

When the water pump fails to circulate coolant, heat is not extracted from the engine and overheating may occur — possibly resulting in damage to the engine.

Some water pumps can turn at 4,000 rpm, with a flow rate of 125 gpm (470 lpm).

Most pumps are self-lubricated, with sealed bearings.

Water pumps can be either **belt driven** or **gear driven**.

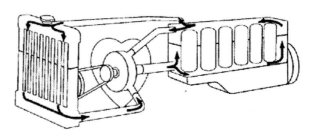

Water Pump

Continued on next page   AH98466,00001B1 -19-16JUL09-1/3

Testing and Servicing Water Pump

A typical belt-driven water pump is made up of a housing, an impeller, and a shaft.

A—Cover   F—Ball Bearing
B—Impeller   G—Fan Pulley
C—Housing   H—Fan Hub
D—Elbow   I— Seal
E—Spring Retainer   J—Gasket

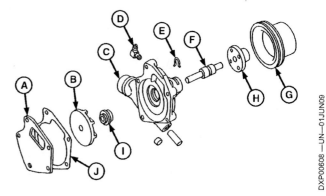

Disassembled Water Pump (Belt Driven)

AH98466,00001B1 -19-16JUL09-2/3

A typical gear-driven water pump is made up of a coolant pump cover, a band clamp, a pump housing, and a drive gear.

When inspecting either type of water pump, look for leaks in the pump housing, broken or bent vanes on the impeller, and damaged seals and bearings. Replace damaged parts.

Always install new seals and gaskets when reassembling a water pump.

A—Pump Cover   C—Pump Housing
B—Band Clamp   D—Drive Gear

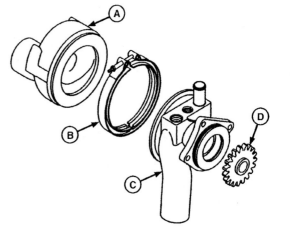

Disassembled Water Pump (Gear Driven)

AH98466,00001B1 -19-16JUL09-3/3

Engine Water Jacket

NOTE: As coolant is heated in the engine block, dissolved minerals separate out in solid form and coat the metal surfaces. The thicker this coating, the less heat is transferred from the engine to the coolant. Prevent rust by adding anti-rust compounds each time coolant is changed.

The engine cylinder block and head both usually contain passages for coolant to flow around the cylinders and valves. Together, they make up the water jacket.

The water jacket holds only a small amount of the total coolant. The advantages of this are:

• Rapid engine warm-up — while the thermostat is closed
• Efficient cooling — when the thermostat opens

AH98466,00001B2 -19-01JUL09-1/1

Coolant Filter

Some engines use a filter in the cooling system.

The coolant filter does two jobs. The outer paper element filters out rust, scale, or dirt particles in the coolant. The inner element releases chemicals into the coolant to soften the water, maintain a proper acid/alkaline condition, prevent corrosion, and suppress cavitation erosion.

The chemicals released into the coolant by the inner element forms a protective film on the cylinder liner surface. The film acts as a barrier against collapsing vapor bubbles and reduces the quantity of bubbles formed.

The coolant filter in the illustration is a spin-on type and should be replaced periodically according to operator's manual recommendations.

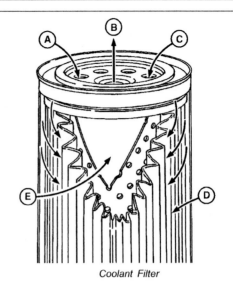

Coolant Filter

A—Coolant In
B—Coolant Out
C—Coolant In
D—Outer Paper Filter
E—Inner Element

AH98466,00001B3 -19-16JUL09-1/1

Coolant Additives — Diesel Engines

IMPORTANT: Do not use a liquid coolant conditioner with a spin-on filter unless the manufacturer's specifications require simultaneous use. Over-concentration may occur and damage the machine cooling system.

Some diesel engines are modified or originally designed to work with coolant additives instead of the coolant filter. This type of conditioner is added directly to the radiator.

Coolant additives are commercially distributed in a powder or pre-mix liquid form. Coolant additives generally give the same protection as the coolant filter. They also contain other anti-rust and pitting inhibitors that cannot be included in the filter alone.

The coolant additives are poured directly into the cooling system as the cooling system is being filled.

They are designed to go to work immediately to help prevent system corrosion, cavitation, and contamination.

The liquid coolant additive may be more reliable than the filter method. Spin-on filters may be worked beyond their capacity. Always see the machine manufacturer's specifications before adding coolant additives to your machine.

AH98466,00001B4 -19-16JUL09-1/1

Thermostats

The thermostat provides automatic control of engine temperature at the correct level. This is necessary in order to get the best performance from an engine.

Some larger engines use dual thermostats for temperature control. The function and operation is the same as for a single thermostat system but allows for more capacity.

Only a small part of the engine's cooling capacity is required under the light loads, even during warm weather.

During warm-up, the thermostat remains closed. The water pump circulates coolant through the engine water jacket only by way of the bypass.

The engine quickly warms up to its operating temperature before the thermostat opens.

When the thermostat opens, hot coolant flows from the engine to the radiator and back.

High-temperature thermostats, which open at 180°F (82°C), improve engine operation, reduce crankcase sludge and corrosive wear of engine parts.

An engine operating in this temperature range is hot enough to:

• Improve combustion
• Burn impurities out of oil in crankcase
• Thin oil to provide good lubrication

Do not use low-boiling-point alcohol or methanol antifreeze with high-temperature thermostats.

When the thermostat is not operating properly, the engine may run too hot or too cold. Overheating may damage the

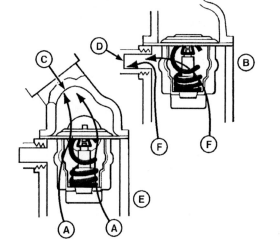

Cooling System Thermostat

A—Thermostat Open
B—Thermostat Closed
C—To Radiator
D—Bypass
E—Coolant Hot
F—Coolant Cold

thermostat so the valve will not function properly. Rust can also interfere with thermostat operation.

Always keep the thermostat in good working condition.

Never operate the engine without a thermostat.

Always use the thermostat design specified for the make and model of engine being used.

AH98466,00001B5 -19-16JUL09-1/3

Thermostat Actuator

The actuator controls the inner workings of the thermostat. It can be made of special wax and granular copper that is sealed in a flexible rubber bollow, or it can be made of bimetallic strips as shown in the Ilustration.

In all cases, heat expands the actuator, which in turn forces the thermostat valve open. Cool temperatures contract the actuator and close the valve. The water pump in the cooling system does not create enough pressure to force the valve open.

G—Bellows Thermostat   H—Bimetallic Thermostat

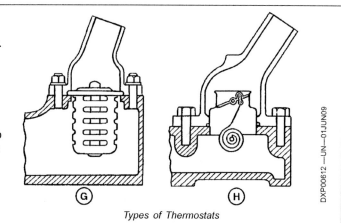

Types of Thermostats

Continued on next page   AH98466,00001B5 -19-16JUL09-2/3

Inspection and Testing of Thermostats

If a thermostat is broken or corroded, discard it.

If it is not of the proper temperature and type, as indicated on an application chart, replace it. The number stamped on it is the approximate temperature that it will reach before starting to open.

- Never use a high-temperature thermostat with an alcohol-base antifreeze.
- Never use a bellows thermostat in high-pressure cooling systems [9 psi (60 kPa) or higher].

Test the thermostat as follows:

1. Suspend the thermostat and a thermometer in a container of water. Do not let them rest against the sides or bottom.

2. Heat and stir the water.

3. The thermostat should begin to open at the temperature stamped on it, plus or minus 10°F (5.5°C). It should be fully open — approximately 1/4 in. (6 mm) at 22°F (12°C) above the specified temperature.

4. Remove thermostat and observe its closing action.

5. If the thermostat is defective, discard it.

Installing Thermostats

When installing a thermostat in the engine water jacket, position the thermostat with the expansion element toward the engine.

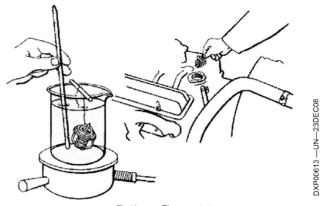

Testing a Thermostat

Some thermostats are marked with arrows that point to the radiator or to the engine block or are marked "top" or "T." "Front" is indicated on some models. **The frame must not block the water flow.**

To prevent leakage, clean the gasket surfaces on the thermostat. Use a new gasket; normally it need not be cemented.

When the outlet casting and gasket have been properly located, tighten the nuts evenly and securely.

AH98466,00001B5 -19-16JUL09-3/3

Engine Oil Cooler

Coolant flows through coolant tubes to the cylinder block. As oil flows through the cooler, heat is transferred to the coolant tubes and coolant.

Some engines are not equipped with engine oil coolers.

AH98466,00001B6 -19-01JUL09-1/1

Cooling System Hoses

Flexible hoses are used in connecting cooling system components because they stand up under vibration better than rigid pipes do. However, hoses have weak points, too.

Radiator hoses can be damaged by air, heat, and water in two ways:

- **Hardening or cracking** — destroys flexibility, causes leakage and allows small particles of rubber to clog the radiator.
- **Softening and swelling** — produces lining failure and hose rupture.

Replace hoses often enough to be sure they are always pliable and able to pass coolant without leakage.

Examine hoses at least twice a year for possible replacement or tightening.

Damaged Hoses

Check any reinforcing springs inside the hoses for corrosion.

AH98466,00001B7 -19-16JUL09-1/3

Hoses can deteriorate on the inside and still appear all right on the outside.

When hoses are removed, check them for wear. Hoses may harden and crack, allowing the system to leak, or they may soften, then collapse. A softened hose may also collapse during high-speed operation or restrict circulation enough to cause overheating.

Some hoses are manufactured with reinforcing springs inside the hose to help prevent the hose from collapsing.

The quality of the hose will affect its service life more than the cooling liquid. **Use only the best hoses available.**

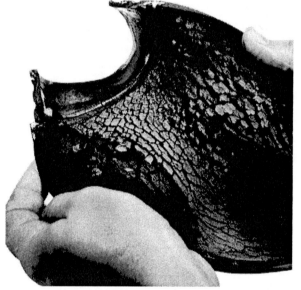

Interior of Damaged Hose

Continued on next page   AH98466,00001B7 -19-16JUL09-2/3

Installing Hoses

Clean the pipe connections and apply a thin layer of non-hardening sealant when installing hoses.

Locate the hose clamps properly over the connections as shown to provide a secure fastening. A pressurized cooling system will blow off an improperly installed hose.

Other Rubber Components

Heat and water can also damage other rubber sealing parts, such as the rubber seals in water pumps and O-rings that seal the lower end of wet cylinder sleeves.

Damage or improper installation of these parts results in serious coolant leakage.

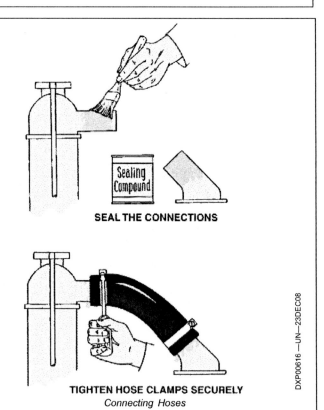

SEAL THE CONNECTIONS

TIGHTEN HOSE CLAMPS SECURELY
Connecting Hoses

AH98466,00001B7 -19-16JUL09-3/3

Cooling Liquid

Water is a good coolant because:

- It is plentiful and readily available
- It absorbs heat well
- It circulates freely at all temperatures between its freezing and boiling points — 32°F to 212°F (0°C to 100°C)

But water has the following disadvantages:

- It freezes readily when cold
- It boils and evaporates when hot
- It may corrode metal parts
- It may cause deposits in water jackets

Modern pressurized systems overcome the low boiling temperatures of water, and additives can offset many of its harmful properties.

Antifreeze is used to reduce the high freezing point of water.

Preparing System for Use

Most engine manufacturers recommend that solutions of quality water and antifreeze be used year-round. Solutions of water and antifreeze not only protect against freezing in cold weather operation, they also provide boil-over protection in warm weather operation.

Water quality is important to the performance of the cooling system. Distilled, deionized, or demineralized water is recommended for mixing with the antifreeze.

Antifreeze contains chemical additives that help provide a stable, non-corrosive environment for seals, hoses, and metal engine parts. Some manufacturers recommend adding supplemental coolant additives to protect heavy-duty diesel engines from corrosion, cylinder liner erosion and pitting, and other damage to the engine and cooling system. Follow the manufacturer's instructions carefully when adding supplemental coolant additives.

Fill the cooling system with a solution of quality water and antifreeze. The proportion of water to be used depends upon the lowest freeze protection temperature desired. See the equipment operator's manual for complete details.

After adding new coolant solution, run the engine until it reaches operating temperature. This mixes the coolant solution uniformly and circulates it through the entire system.

After running the engine, check coolant level and entire cooling system for leaks.

Antifreeze Selection

IMPORTANT: Do not use automotive-type antifreeze in a diesel engine — damage to engine could result.

Use only one of these types of antifreeze:

- **Ethylene Glycol Base**
- **Propylene Glycol Base**

Alcohol-base antifreeze provides protection against freezing in cold weather operation, but it does not provide boil-over protection in warm weather operation. Modern engines usually operate at a temperature above the boiling point of alcohol, causing a loss of coolant during normal operation. Because of this, permanent type (ethylene glycol) and (propylene glycol) antifreeze is recommended by most engine manufacturers.

The term "permanent antifreeze" means only that the solution will not boil away at normal engine operating temperatures and does not mean that it is good for use for more than one season.

For more information about coolant and antifreeze, refer to the FOS Manual Fuels, Lubricants, and Coolants.

AH98466,00001B8 -19-16JUL09-1/1

Servicing the Cooling System

⚠ **CAUTION: Explosive release of fluids from a pressurized cooling system can cause serious burns. Shut off the engine. Only remove the filler cap when cool enough to touch with bare hands. Slowly loosen the cap to its first stop to relieve pressure before removing cap completely.**

Correct cooling system servicing is vital for a smooth-running, long-lasting engine.

Overheating is the big danger. It can be caused by:

• Clogging of the cooling system
• Lack of coolant
• Defective water pump or thermostat
• Restricted air flow through the radiator

Check the coolant level and temperature frequently. Service the entire cooling system at least twice a year.

Cleaning the Cooling System

Efficient operation of the cooling system requires an occasional cleaning, particularly at seasonal changes when antifreeze solution is added or removed.

Several methods of cleaning the system are available, the proper one depends upon the amount of corrosion present in the system.

Rust-Proofing the System

Rust is formed by corrosion of iron parts:
IRON + WATER + OXYGEN = RUST

Make it a rule to check the cooling liquid in the spring and in the fall.

FOR BEST COOLING SYSTEM PERFORMANCE, DRAIN THE SYSTEM AT LEAST ONCE PER YEAR.

Use a cooling system cleaner if the liquid drains out rusty. Otherwise, use a radiator flush or plain water. Corrosion inhibitors do not clean out rust already formed.

In the spring, if the antifreeze is drained, renew the rust proofing by adding a reputable anti-rust compound to a filling of fresh water or by installing a new filling of antifreeze.

Regardless of the summertime practice, corrosion protection can best be provided in winter by inhibitors built into the antifreeze coolant.

IMPORTANT: Adding rust inhibitors or fresh antifreeze to used solutions will not restore full strength corrosion protection. Some mixtures may even do harm.

Some manufacturers recommend the year-round use of a highly inhibited antifreeze coolant. However, remember that the service life of the inhibitor may be drastically reduced as the machine gets older and the coolant may have to be replaced more often.

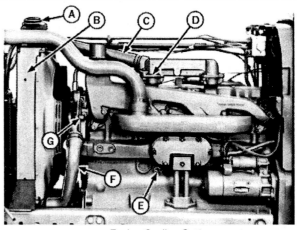

Engine Cooling System

A—Radiator Cap	E—Engine Block Drain Cock
B—Radiator	F—Lower Radiator Hose
C—Upper Radiator Hose	G—Water Pump
D—Thermostat	

Some of the specific factors involved are:

• Accumulation of contaminants, including rust and corrosion products.
• Dilution with water required to replace losses, and possible absorption of air and exhaust gases.

For these reasons, be sure to:

1. Check the cooling system and coolant periodically.

2. Maintain a 50% solution concentration year-round for adequate corrosion protection

3. Install new antifreeze at least once per year.

Rust Clogging

Rust clogging is a fairly common cause of cooling system trouble. It can be avoided entirely by periodic rust proofing and cleaning when necessary.

The most common clogging materials are:

• **Rust**
• **Scale**
• **Grease**
• **Lime**

Rust accounts for 90% of the clogging. It forms on the walls of the engine water jacket and other metal parts.

Grease and oil enter the system through:

• Cylinder head joint
• Water pump
• Leaking oil cooler

Coolant circulation loosens rust particles that settle in the water jacket and build up in layers inside the radiator water tubes.

Continued on next page   AH98466,00001B9 -19-20JUL09-1/10

As the rust layer becomes thicker, it cuts down heat transfer from the radiator until the engine overheats and boiling starts in the water jacket.

This boiling stirs up more rust in the block and forces it into the radiator — eventually clogging it.

AH98466,00001B9 -19-20JUL09-2/10

Mineral Deposits

Mineral deposits on the coolant side of the engine cylinders can cause hot spots due to the insulation quality of the deposits.

Overheating, engine knock, and eventually engine damage can result from the buildup of rust and mineral scale on coolant side of the combustion chamber.

To avoid excessive formation of rust or mineral deposits:

1. Use only distilled, deionized, or demineralized water in the cooling system.

2. Mix quality antifreeze with the water in the proportion recommended by the engine manufacturer.

3. Maintain full-strength corrosion protection at all times.

Cooling System Cleaners

Well-maintained cooling systems seldom, if ever, require corrective cleaning.

However, if periodic rust-proofing or other mechanical preventive maintenance is neglected, deposits will build up and cleaners must be used to restore cooling capacity.

Use a cleaner to remove:

• Hard rust deposits

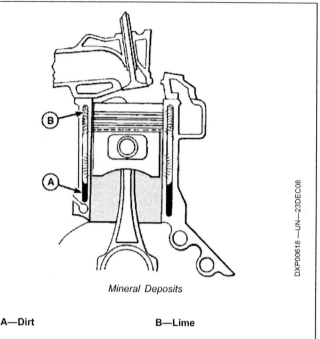

Mineral Deposits

A—Dirt   B—Lime

• Scale
• Grease

Continued on next page   AH98466,00001B9 -19-20JUL09-3/10

Iron rust and water scale build up together in the cooling system. Clogging material is usually made up of grease, rust, and water scale, but is 90% iron rust.

Remove both hard rust scale and grease with a double-action cleaner. It is harmless to cooling system metals and hose connections when used according to directions.

IMPORTANT: If rust and grease are not completely neutralized and flushed out, they can destroy the corrosion inhibitors in later fills of antifreeze and anti-rust solutions.

NOTE: Some radiators are made of aluminum or other metals, which may be affected by certain compounds used for immersion cleaning. Follow the manufacturer's instructions for cleaning these radiators.

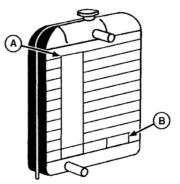

Rust in Clogging Material

A—90% Rust   **B—10% Grease and Water Scale**

Flushing the Cooling System

Incomplete flushing, such as hosing out the radiator, closes the thermostat and prevents thorough flushing of the water jacket. For complete flushing, take the following steps:

1. Fill the system completely with fresh water.

2. Run the engine long enough to open the thermostat (or remove the thermostat).

3. Open all drain points to drain the system completely.

Clean out the overflow pipe and remove insects and dirt from radiator air passages, radiator grille and screens.

Also check the thermostat, radiator pressure cap, and the cap seat for dirt or corrosion.

For more information, refer to the FOS Manual Fuels, Lubricants, and Coolants.

Continued on next page   AH98466,00001B9 -19-20JUL09-4/10

Dispose of Waste Properly

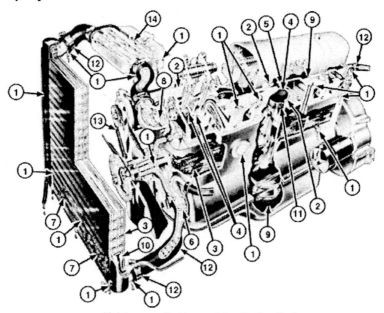

Maintenance Problems of the Cooling System

1—External Leakage
2—Internal Leakage
3—Rust Deposits
4—Heat Cracks

5—Exhaust Gas
6—Air Suction
7—Clogged Air Passages
8—Stuck Thermostat

9—Sludge Formation in Oil
10—Transmission Oil Cooler
11—Heat Damage
12—Hose Failure

13—Worn Fan Belt
14—Pressure Cap Leakage

Improperly disposing of waste can threaten the environment and ecology.

Use leak-proof containers when draining the coolant. Do not use food or beverage containers that may mislead someone into drinking from one of the containers.

Do not pour the used coolant onto the ground, down a drain, or into any water source. The coolant can be toxic to animals and fish.

Inquire on the proper way to recycle or dispose of the used engine coolant from your local environmental or recycling center.

For more details, refer to the FOS Manual Fuels, Lubricants, and Coolants.

Cooling System Leakage

Leakage is the most common problem in a cooling system. During the winter it can involve the loss of valuable antifreeze.

But leakage actually does increase during the winter due to metal shrinkage and cooling system pressure.

Air pressure testers can be helpful in locating external leaks, but they cannot be depended upon to locate small combustion leaks.

Only practical experience enables a service technician to tell whether the location and size of a leak can be corrected with a sealing solution.

All leakage exposed to combustion pressure must be repaired mechanically.

Follow instructions when using sealing solutions. Some sealing solutions react chemically with antifreeze and rust inhibitors, and seriously affect coolant performance.

Radiator Leakage

Most radiator leakage is due to mechanical failure of soldered joints. This is caused by:

• Engine vibration
• Frame vibration
• Pressure in the cooling system

Examine radiators carefully for leaks and before and after cleaning. Cleaning may uncover leakage points already existing but plugged with rust.

White, rusty or colored leakage stains indicate previous radiator leakage. These spots may not be damp if water or alcohol is used since such coolants evaporate rapidly, but ethylene glycol antifreeze shows up because it does not evaporate.

Always stop any radiator leakage before installing antifreeze coolant.

Leakage Outside Engine Water Jacket

Inspect the engine cylinder block before and after it gets hot while the engine is running.

Leakage of the engine block is aggravated by:

Continued on next page

AH98466,00001B9 -19-20JUL09-5/10

- Pump pressure
- Pressurized cooling systems
- Temperature changes of the metal

Remember: Small leaks may appear only as rust, corrosion, or stains, due to evaporation.

Watch for these troubles:

1. **Core-Hole Plugs.** Remove old plug. Clean plug seat, and coat with sealing compound. Drive new plug into place with proper tool.

2. **Gaskets.** Tighten joint or install new gaskets. Use sealing compound when required.

3. **Stud Bolts and Cap Screws.** Apply sealing compound to threads.

Continued on next page

AH98466,00001B9 -19-20JUL09-6/10

Water Jacket Leakage into Engine

Coolant leaks into the engine through:

- A loose cylinder head or sleeve joint
- A cracked or porous casting
- The push rod compartment

Water or antifreeze mixed with engine oil will form sludge, which causes:

1. Lubrication failure
2. Sticking piston rings and pins
3. Sticking valves and valve lifters
4. Extensive engine damage

The amount of damage depends upon the amount and duration of leakage, service procedures, and seasonal operating conditions.

High-temperature thermostats let the engine heat remove moisture (formed from blow-by) from the crankcase.

Head gasket leakage is more likely to occur in winter than summer because of more metal contraction and expansion in winter.

Give special attention to the cylinder head joint gasket, no matter what type or material.

An improperly installed gasket can cause:

- Coolant and oil leakage
- Overheating

It may be necessary to pressurize the cooling system and tear down part of the upper part of the engine to find a coolant leak in the push-rod compartment.

Replacing Head Gaskets

To replace a cylinder head gasket:

1. Use only a new gasket designed for this particular engine.

2. Make sure the head and block surfaces are clean, level and smooth.

3. See Chapter 3 for details.

IMPORTANT: Leading engine manufacturers strongly recommend the use of torque wrenches to tighten cylinder heads.

Always follow the engine manufacturer's instructions on the cylinder head bolt tension and order of tightening.

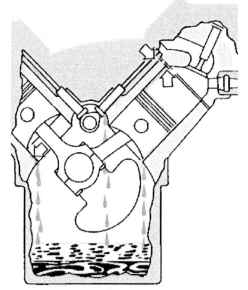

Water Jacket Leakage into Engine

Overheating Damage

In modern high-performance engines, the intense heat of combustion can cause engine components such as valves, pistons, and rings to operate near their critical temperature limits — even when the cooling system is operating normally.

Overheating severely affects the lubrication of the engine. High metal temperature can destroy the lubricating film, accelerate oil breakdown, and cause formation of varnish.

These in turn may cause:

- Excessive wear
- Scoring
- Valve burning
- Seizure of moving parts

Continued operation above the normal temperature range may result in:

- Lubrication failure
- Heat distortion
- Engine knocking

Continued on next page   AH98466,00001B9 -19-20JUL09-7/10

Cylinder heads and engine blocks are often warped and cracked by terrific strains set up in the metal by overheating, especially when followed by rapid cooling.

Hot spots in an overheated engine can cause engine knocking. If allowed to continue, knocking results in:

1. "Blown" head gaskets
2. Damaged pistons
3. Ring and bearing failure

See Chapter 3 for details.

Exhaust Gas Leakage

A cracked head or a loose cylinder head joint allows hot exhaust gas to be blown into the cooling system under combustion pressures; even though the joint may be tight enough to keep liquid from leaking into the cylinder.

The cylinder head gasket itself may be burned and corroded by escaping exhaust gases.

Exhaust gases dissolved in coolant destroy the inhibitors and form acids, which cause corrosion, rust, and clogging.

Excess pressure may also force coolant out of the overflow pipe.

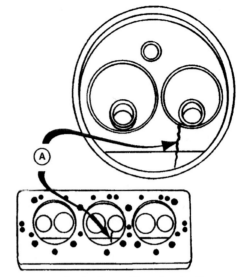

Cracks Caused from Overheating

A—Heat Crack

Continued on next page   AH98466,00001B9 -19-20JUL09-8/10

Testing for Exhaust Gas Leakage (Blow-By)

NOTE: Make this test quickly, before boiling starts, since steam bubbles give misleading results.

1. Remove the radiator cap while the engine is cold.

2. Warm up the engine and keep it under load.

3. Look for excessive bubbles in the coolant.

4. Either bubbling or an oil film on the coolant is a sign of blow-by in the engine cylinders.

Corrosion in the Cooling System

Water solutions will corrode the cooling system metals when they are not protected by inhibitors added to water for summer operation.

Corrosion coating on metal surfaces reduces heat transfer at the walls of the engine water jacket and in the tubes of the radiator even before clogging occurs.

Galvanic action is a form of corrosion that can occur when two different metals (in contact) are suspended in a liquid which will carry a current.

It is similar to what happens in a battery when electric current flows from one area to another, removing metal from one electrode and depositing it on another in the process.

Galvanic corrosion or direct attack by the coolant will affect the following metals in the cooling system: stainless steel, brass, solder, aluminum, and copper.

Aeration in Cooling System

Aeration, or mixing of air with water, speeds up the formation of rust and increases corrosion of cooling system metals.

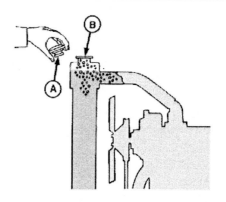

Testing for Exhaust Gas Leakage (Blow-By)

A—Pressure Cap Removed   **B—Gas Bubbles Mean Blow-By**

Aeration may also cause:

- Foaming
- Overheating
- Overflow loss of coolant

Air may be drawn into the coolant because of:

- A leak in the system
- Turbulence in the top radiator tank
- Too-low coolant level

Check the cooling system for exhaust gas leakage and air suction when you find these conditions:

- Rusty coolant
- Severe rust clogging
- Corrosion
- Overflow losses

Continued on next page
AH98466,00001B9 -19-20JUL09-9/10

Testing for Air in the Cooling System

If air leaks in the cooling system are suspected, the following checks can be made.

1. Adjust coolant to correct level.

2. Replace pressure cap with a plain, air-tight cap.

3. Attach rubber tube to lower end of overflow pipe. Be sure radiator cap and tube are air-tight.

4. With transmission in neutral gear, run engine at high speed until temperature gauge stops rising and remains stationary.

5. Without changing engine speed or temperature, put end of rubber tube in bottle of water.

6. Watch for a continuous stream of bubbles in the waste bottle, showing that air is being drawn into the cooling system.

A—Air Bubbles   B—Air

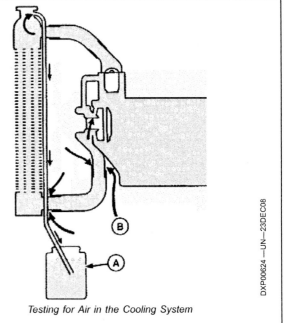

Testing for Air in the Cooling System

AH98466,00001B9 -19-20JUL09-10/10

Test Yourself

Questions

1. What are the eight parts of the liquid cooling system?

2. (True or false?) Permanent antifreeze should be changed every year.

3. What is the basic difference between a bellows-type thermostat and a bimetallic-type?

4. What part of the cooling system could be referred to as the "heart" of the system?

5. (True or false?) Thermostats are not required when operating the engine in hot weather.

AH98466,00001BA -19-16JUL09-1/1

Governing Systems

Governing Systems — Introduction

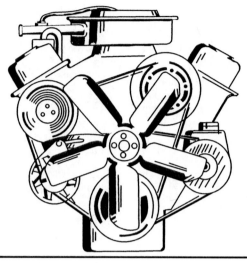

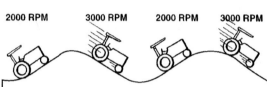

WITHOUT GOVERNOR (FIXED THROTTLE)

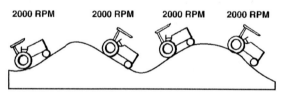

WITH GOVERNOR (KEEPS SAME SPEED)

What the Governor Does for the Engine

The governor is a device that automatically controls the speed of an engine under varying loads.

Governors can do any of three jobs:

- **Maintain a selected speed**
- **Limit the slow and fast speeds**
- **Shut down the engine when it overspeeds**

Note in the illustration how the governor keeps the engine at a constant speed when the tractor is going both uphill and downhill.

Most modern governors are the centrifugal spring-loaded type and adjust fuel intake according to demand on the engine. Others use a vacuum to control a throttle valve.

Electronic governors are used on some modern engines. These governors precisely control the fuel delivery to provide consistent engine performance and reduced exhaust emissions.

SS40167,000011E -19-23JUN09-1/1

Features of Governors

Let's discuss some of the main features of governing:

Stability — to maintain a desired engine speed without fluctuations. Indicated by the number of corrective movements the governor makes and the time required to correct engine speed for a given load change.

Lack of stability will result in hunting, or oscillations due to over-correction. Too much stability will result in a dead-beat governor, or one that does not correct sufficiently.

Another term for this ability of the governor to keep a steady speed is isochronous governing.

Sensitivity — the percent of speed change required to produce a corrective movement of the fuel control mechanism.

Promptness of Response — the time in seconds required for movement of the fuel control from no load to full-load position.

Speed Drift — a gradual deviation of the main governed speed above or below the desired governed speed.

Work Capacity — the work that the governor can perform to overcome the resistance in the fuel controls, expressed in inch-pounds or foot-pounds.

Load-Limit Changer — an adjustable device for limiting maximum fuel flow to the engine, thus limiting power output to any desired maximum value.

Speed Changer — a device for adjusting the speed governing system to change engine speed, or power-output relationship, with other engine generator units when in parallel operation. Normally this can be adjusted while the engine is in operation.

Speed-Regulation Changer — a device by which the steady speed regulation (speed droop) can be adjusted while the engine is in operation.

Speed Droop — All mechanical governors and some hydraulic governors have permanent speed droop and are non-isochronous because the engine's steady speed is slightly different at different speeds.

Speed droop is the variation in governor rotating speed from no load to full load (sometimes referred to as steady-state speed regulation).

Speed droop refers to a change in the steady speed when the load on an engine is reduced from rated output to zero output, without adjustment of the governor. It is usually expressed as a percent of rated speed.

For example: An engine equipped with a governor that has been set for a no-load speed of 2,000 rpm will not run this fast under load. The difference between the governor setting (no-load speed) and full-load speed is called speed droop. This difference may be as much as 10 percent less than the no-load speed.

In this example, no-load speed will be 2,000 rpm, but all engine speeds under partial loads will be somewhat less than this and full-load speed will drop down to 1,800 rpm.

Speed droop is useful in preventing overshooting or hunting by the governor when the engine loads change.

Let's discuss the governing of engines in two parts — first spark-ignition, then diesel.

SS40167,000011F -19-16JUL09-1/1

Governing of Spark-Ignition Engines

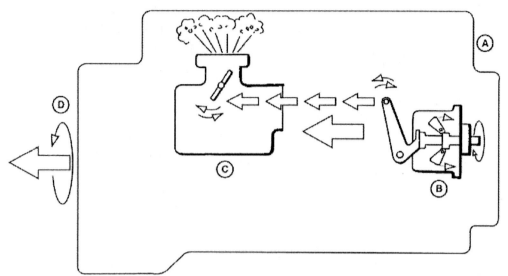

Governing System (Gasoline Engine with Centrifugal Governor)

A—Engine Speed Varies with Load B—Governor Reacts C—Fuel-Air Mixture Is Changed D—Result: Engine Speed Is Controlled under Variable Loads

Governing of spark-ignition engines is more critical than for diesel engines. This is partly due to the design of the air and fuel control system, but chiefly to:

• **The small variation in fuel-air ratio which is permissible.**
• **The metering lag of nearly one cycle for complete response to governor action.**

The governor quickly feels a reduction in speed, and increases fuel to the engine about one cycle before air is supplied. This results in erratic operation.

If the engine over-speeds, a quick reduction in fuel causes the engine to misfire. Therefore, the governor response must be slow enough to ignore quick speed variations caused by occasional misfiring, yet be sensitive to the sustained speed changes resulting from variations of load.

Most gasoline engines use a throttle governor.

The throttle governor permits the engine to fire steadily regardless of the load but varies the charge of fuel-air and so the intensity of the combustion. This speeds up or slows down the engine to allow for different loads.

The governor actuates a throttle valve in the intake passage between the carburetor and the engine manifold. The position of the throttle valve at any given time controls the mixture to the engine (see Chapter 3 for details).

The two most common types of throttle governor systems are:

• **Automatic or vacuum governors**
• **Centrifugal governors**

Let's discuss each type.

Continued on next page   SS40167,0000120 -19-16JUL09-1/5

Automatic or Vacuum Governors

Many farm and industrial machines are equipped with a governor for maintaining a set maximum operating speed.

An automatic or vacuum governor is used for this purpose. Located between the carburetor and the intake manifold, it has no mechanical connection to any other parts of the engine.

This type of governor consists of a housing and a throttle-butterfly valve connected to a cam and lever mechanism that is spring-controlled.

The unit operates by closing the throttle-butterfly valve, causing a decrease in the fuel flow, as engine speed and suction increase.

When engine speed and suction decrease, the spring opens the throttle-butterfly valve and the fuel flow increases.

A simple adjustment of spring tension maintains the desired speed range.

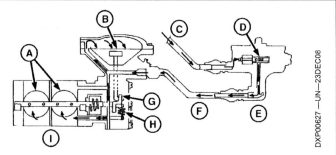

Vacuum Governor

A—Throttle-Butterfly Valve
B—Diaphragm
C—Air In
D—Valve
E—Adjusting Device
F—Air
G—Lever
H—Spring
I—Governor

Efficient operation of this governor depends largely upon precise machining, freedom of movement of the parts, and proper installation and adjustment of the engine.

SS40167,0000120 -19-16JUL09-2/5

Centrifugal Governors

The centrifugal governor has two weights, called flyweights, suspended on a weight carrier, which is mounted on the governor shaft. The governor is driven directly from the engine.

As the governor shaft rotates, the weights are thrown outward by centrifugal force, pushing a thrust sleeve against the governor fork, which in turn actuates the governor lever and, through governor-to-carburetor linkage, controls the amount of fuel and air supplied to the engine.

This control of the fuel-air mixture keeps the engine at the desired speed under varying loads.

The governor is linked to the hand throttle and will maintain any engine speed, unless the engine is overloaded.

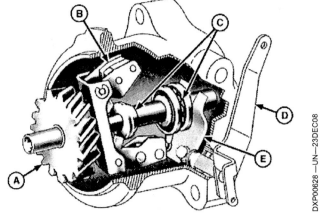

Cutaway of Centrifugal Governor

A—Drive Gear and Shaft
B—Flyweights
C—Sleeve and Thrust Bearing
D—Lever
E—Fork

Continued on next page

SS40167,0000120 -19-16JUL09-3/5

The centrifugal force of the weights decreases as the speed falls and the governor spring pushes them inward until a new balance point is reached. For any given speed, the flyweights will always take a definite position a certain distance from the axis of rotation.

"Hunting" of Centrifugal Governors

A problem with centrifugal governors is hunting.

Frequently when an engine is first started or after it is warmed up and is working under load, its speed will become uneven or irregular. The engine speeds up quickly, the governor suddenly responds, the speed drops quickly. The governor again suddenly responds, and the speed drops quickly, the governor responds again, and the action is repeated. This is known as hunting.

Hunting is usually caused by the wrong carburetor adjustment and can be corrected by making the mixture either slightly leaner or slightly richer. The governor itself may also cause hunting by being too stiff or by striking or binding at some point so that it fails to act freely.

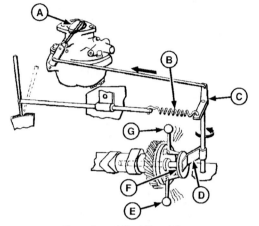

Operation of Centrifugal Governor

A—Throttle (Butterfly) Valve
B—Governor Control Spring
C—Governor Lever
D—Governor Arm
E—Governor Flyweights
F—Governor Drive Plate
G—Flyweight Levers

SS40167,0000120 -19-16JUL09-4/5

Governor Linkage and Adjustment

Remember these two points when connecting or adjusting the governor to the engine:

1. **Amount of work** — Governors are capable of doing a specified amount of work, say 8 foot-pounds, when moving through full control travel.

2. **Speed droop** — Governors may have an adjustable speed droop up to 10 % for full control travel.

If the linkage is set to give full control of fuel during one quarter of governor lever travel, the governor can do only one quarter of its maximum work. So the maximum speed droop obtainable will be one quarter of rated maximum.

For smoother operation of the engine, change the lengths of the governor and fuel control rods so that 70 to 80 % of full governor lever travel is used between no load and full load. This means that the governor can perform 70 to 80 % of its work rating.

Maintenance of Governor

The two chief requirements for trouble-free operation of a governor are:

- **Vibration-free drive**
- **Cleanliness of moving parts**

The governor is a sensitive device which demands a non-fluctuating speed drive with no vibration. This is the reason why many governors are driven through a spring or rubber coupling to smooth out any roughness in the engine drive.

Governor and Linkage (Centrifugal Type Shown)

A—Governor Control Lever
B—Throttle Rod
C—Governor Attaching Points
D—Linkage

The smoothest governor drive is obtained through gearing from the crankshaft. The free end of the camshaft will give the roughest drive, the greatest variation in speed, and the most vibration.

SS40167,0000120 -19-16JUL09-5/5

Governing of Diesel Engines

The principles of governing a diesel engine are the same as for gasoline engines, but there are several different ways in which governors perform their jobs:

- **Mechanical** — a speed measuring device, which directly actuates the linkage to the throttle.
- **Hydraulic** — the speeder rod actuates a small valve controlling fluid under pressure; this fluid operates a piston or servomotor which actuates the fuel control.
- **Variable speed** — maintains any selected engine speed from idling to maximum speed.
- **Over-speed** — shuts down the engine in case it runs too fast. It is only a safety device.
- **Torque converter regulator** — controls the maximum speed of the output shaft of a torque converter attached to an engine, independently of engine speed.
- **Pressure regulator** — used on any engine driving a pump, to maintain a constant inlet or outlet pressure on the pump.
- **Load control** — adjusts the amount of load applied to the engine to suit the speed at which it is set to run. This actually improves fuel economy and reduces engine strain when the engine is not under load.
- **Electric load sensing** — detects variations in the electrical load and quickly adjusts fuel quantity to match. This governor is most generally used in electrical power plants.
- **Electric speed** — controls speed by measuring the frequency in cycles-per-second of the electrical current produced by a generator driven by the engine.
- **Electronic** — controls the fuel delivery to the engine based on throttle position, engine speed, and fuel temperature. This system uses an electronic actuator solenoid to move the fuel pump rack.

Mechanical Governors (Diesel)

Mechanical governors provide full fuel for starting when the speed control lever is in the idle or run position. Immediately after starting, the governor moves the injector rack to the position required for idling.

Mechanical governors are simple centrifugal types, with few moving parts. They work best when engine speed is not critical.

Disadvantages are:

1. Dead Band — the change in speed the engine must make before the governor makes a corrective movement of the throttle. Caused mainly by friction.

2. Speed Droop— the change in governor speed required to cause the throttle rod to move from full-open to full-closed or full-closed to full-open.

Types of Mechanical Governors

The two types of mechanical governors are:

- Speed Limiting— used when a maximum and minimum speed plus manually-controlled intermediate speed is required.
- Speed Varying— used for a near-constant engine speed under varying loads, with speeds adjustable by the operator.

Hydraulic Governors (Diesel)

Hydraulic governors have more parts and are generally more expensive than mechanical governors. They can also be made isochronous because they are more sensitive and have greater power to move the fuel control mechanism than mechanical governors.

The power which moves the engine throttle does not come from the speed measuring device of hydraulic governors. Instead, it comes from a hydraulic power piston or servomotor. This is a piston that is acted upon by oil under pressure to operate the throttle.

The speed measuring device is attached to a small pilot valve as shown. This valve slides up and down in a bushing containing ports that control the oil flow to the piston.

The force needed to slide the port valve is very small. Therefore, a small ball head is able to control a large amount of power in the piston.

The engine should be equipped with a separate over-speed governor to prevent a runaway should the governor fail.

Continued on next page

SS40167,0000122 -19-23JUL09-1/6

100709
PN=392

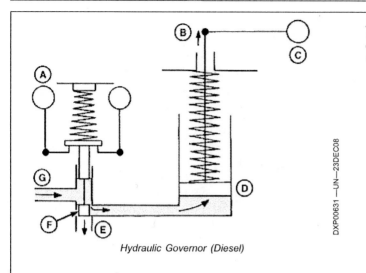

A—Ball Head
B—Increased Fuel
C—Throttle
D—Power Piston

E—Drain
F—Pilot Valve
G—Oil Supply

Hydraulic Governor (Diesel)

SS40167,0000122 -19-23JUL09-2/6

"Hunting" of Hydraulic Governors

A basic fault of hydraulic governors is that they have a tendency to hunt. The cause is the unavoidable time delay between the time the governor acts and the engine responds. The engine cannot instantly come back to the speed called for by the governor.

For example: If the engine speed is below the governor control speed, the pilot valve moves the power piston or servomotor to increase the fuel flow.

By the time the speed has increased to the control setting so that the port valve is centered and the power piston or servomotor dropped, the fuel has already been increased too much, but the engine continues to speed up.

This over-speed opens the port valve the other way to decrease the fuel, but by the time engine speed falls to the right level, the fuel control has again traveled too far. The engine over-speeds, and the whole cycle is repeated continuously.

To overcome hunting, speed droop must be provided.

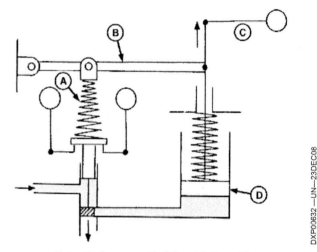

Hydraulic Governor with Adjustable Speed Droop

A—Speeder Spring
B—Speed-Droop Lever

C—Increased Fuel
D—Piston Signals More Fuel

Continued on next page

SS40167,0000122 -19-23JUL09-3/6

Variable-Speed Governors

A variable speed governor can control the engine speed at different settings between minimum and maximum. It resembles a constant-speed governor in its ability to hold the engine at steady speed. The difference is that the operator can adjust the governor to maintain any desired steady speed.

Speed adjustment is obtained in most variable speed governors by changing the spring force that resists the centrifugal force of the flyweights.

Isochronous Governors

Governors which hold steady speeds are classed as constant speed governors. The only truly constant speed governors are isochronous governors.

An isochronous governor is able to maintain a constant speed without changing it. It does this by employing speed droop to give stability while fuel intake is being corrected.

Over-Speed Governors

Over-speed governors are safety devices which protect engines from damage due to over-speeding.

When an engine is equipped with a regular speed governor, the over-speed governor will function only if the regular governor fails to operate. The over-speed governor merely slows the engine down while allowing it to keep on running at a safe speed.

Over-speed governors use a preloaded spring to overbalance the centrifugal force of the flyweights.

When the maximum speed is passed, centrifugal force overcomes the spring and activates the fuel controls which slow down the engine.

Torque Converter Governors

A torque converter governor is used on a torque converter in conjunction with the engine governor.

This governor takes its speed reading from the output shaft of the torque converter.

Engine speed is governed by the engine governor as long as the torque converter speed is below its maximum level.

But if the torque converter overspeeds, its governor reduces the engine fuel and keeps converter speed from rising further.

Pressure Regulating Governors

Pressure regulating governors are used on pipe line pumps to ensure a constant pumping pressure.

An ordinary variable speed governor can be converted into a pressure regulating governor by adding a device which varies the speed setting of the governor in accordance with the pumping pressure.

The pumping pressure acts on a piston or diaphragm, balanced by a spring connected to a speed adjuster on

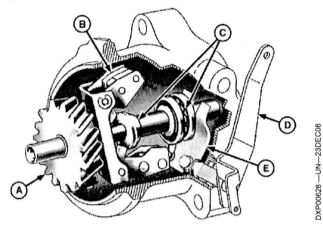

Cutaway of Centrifugal Governor

A—Drive Gear and Shaft   D—Lever
B—Flyweights   E—Fork
C—Sleeve and Thrust Bearing

the governor. If the pressure drops below the set level, the resulting movement of the diaphragm moves the speed adjuster to increase the speed.

When the speed is increased, the engine runs faster and pumping pressure is increased. When pumping pressure rises, engine speed is reduced, causing the pressure to return to normal.

Load-Limiting Governors

Load-limiting governors prevent the loss of engine speed due to heavy loading by limiting the load to the capacity of the engine. They cannot always be used because it is not always practical to limit the load.

A prime application of load limiting is on diesel electric locomotives. Here the load-limit device is attached to the governor, which, when the governor terminal shaft reaches the maximum safe fuel position, operates an electric control, reducing the output of the electric generator.

The electric control is a rheostat, which is used to reduce generator excitation. This reduces the load on the engine enough to permit it to run at full speed and develop full power.

Load-limiting governors are required when a series of various capacity engines are working together. The governor limits the fuel consumed by each engine to utilize its full power capability without the danger of overloading any of the others.

Load Control Governors

Load control governors maintain the engine speed required and adjust the load limiting device so that the amount of load applied to the engine is the amount predetermined as suitable for each speed.

Continued on next page   SS40167,0000122 -19-23JUL09-4/6

When partial power is required, the engine not only runs at a reduced speed, but also uses less fuel per stroke. This greatly reduces engine strain.

SS40167,0000122 -19-23JUL09-5/6

Electric Load-Sensing Governors

Electric load-sensing governors find their greatest use in electric power generating plants where an absolutely stable frequency must be maintained regardless of wide, rapid changes in load.

Load-sensing governors have an advantage over speed governors because they can divide the load among several generating units. Using this method, governors in all engines can be set for the same speed, and the load-sensing elements can be adjusted so the engines will always share the load, per their capacities, through the entire load range.

In operation, electric load-sensing governors detect changes in the load and adjust the fuel supply to maintain the same engine speed under the new load. They are able to respond faster to load changes than speed governors because the load change itself directly operates the governor mechanism

An electric load-sensing device consists of an electric network connected to an engine-driven generator. The network measures the generator load and controls a solenoid within the governor which operates the fuel control mechanism.

Load-sensing governors always include speed-sensing and load-sensing elements, since the speed would gradually float away from normal (due to the fact that normal is constantly changing) if the engine fuel were not controlled.

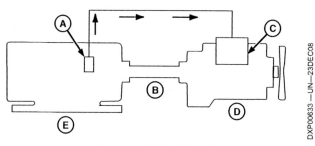

Governor with Load Sensing

A—Load-Sensing Device
B—Coupling
C—Governor
D—Engine
E—Generator

Electric Speed Governors

Electric speed governors respond to engine speed by measuring the frequency of the electric current produced by the generator. When the measuring network notes that the frequency has departed from normal, it delivers an electric signal to the solenoid which operates the fuel control mechanism. The frequency measuring network replaces the flyweights and speeder spring used on mechanical governors.

SS40167,0000122 -19-23JUL09-6/6

Electronic Governors (Diesel)

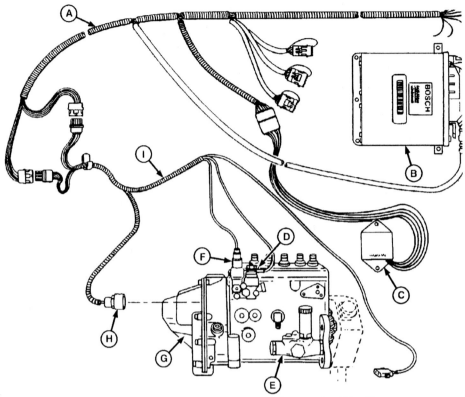

Electronically Controlled Diesel Fuel Injection System with Wiring Harness

A—Vehicle Wiring Harness
B—Engine Control Unit (ECU)
C—TVP Module
D—Fuel Shutoff Solenoid
E—Fuel Injection Pump
F—Fuel Temperature Sensor
G—Electronic Governor Actuator
H—Actuator Connector
I—Engine Wiring Harness

The electronic governor serves as an engine governor by regulating the fuel delivery of the injection pump. A typical governor system consists of an engine control unit (ECU), auxiliary speed sensor, a transient voltage protection module (TVP), and an injection pump/actuator assembly.

The injection pump/actuator includes the injection pump, actuator solenoid, rack position sensor, primary speed sensor, fuel shutoff solenoid, and fuel temperature sensor.

Refer to Chapter 7 in this manual for more information about electronically controlled fuel injection systems.

Engine Control Unit (ECU)

The ECU is a self-contained module that is the brains of the electronic governor system. It contains electronic circuitry and computer software that together perform governor and diagnostic functions. The ECU is remotely located from the engine and connects to the engine application through the wiring harness.

The ECU receives input from the engine sensors, throttle position sensor, and engine speed sensor. Based on this information, it directs an output signal to the actuator solenoid which moves the injection pump control rack to control the fuel delivery. The ECU also controls the fuel limiting for torque curves and the governing speed control.

Refer to Chapters 6 and 7 in this manual and to the FOS manual Electronic Ignition and Fuel Injection for more information about the functions performed by the Engine Control Unit (ECU).

Continued on next page   SS40167,0000123 -19-17JUL09-1/3

Fuel Injection Pump

The electronically controlled fuel injection system uses the same basic pumping mechanism used in mechanically governed injection pumps.

The mechanical governor is replaced with the electronically controlled actuator assembly. The actuator assembly includes the actuator solenoid to move the control rack, rack position sensor, primary speed sensor, and a toothed speed wheel. The actuator responds to signals from the engine control unit (ECU) to control the position of the fuel injection pump rack and, therefore, fuel delivery.

The throttle lever mechanism used on mechanically governed injection pumps is removed and its function is implemented by a throttle position sensor. The ECU converts the input signals from the sensor into a percentage of full-throttle command and controls engine speed and fuel quantity accordingly.

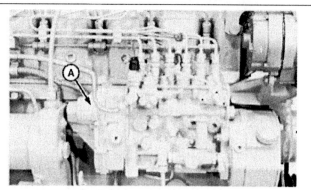

Fuel Injection Pump with Electronic Governor

A—Electronic Governor
     Actuator

Continued on next page   SS40167,0000123 -19-17JUL09-2/3

Actuator Solenoid

The actuator solenoid is contained within the actuator housing mounted on the rear of the fuel injection pump. It responds to signals received from the engine control unit (ECU) to control the position of the fuel injection pump rack.

The injection pump control rack is spring-loaded to the fuel shutoff position of zero rack. When the engine key switch is turned to the START position, the ECU powers the actuator solenoid, causing the rack to move to the starting fuel position.

Once the engine has started, the ECU determines the optimum fuel delivery based on various inputs (primarily throttle position and engine speed). The ECU adjusts the current level to the actuator solenoid, causing a corresponding movement of the rack to provide the correct fuel delivery. The ECU has the ability to control the current to the solenoid in order to position the control rack anywhere between zero rack (fuel shutoff) and full rack (maximum fuel delivery).

The actuator also provides feedback to the ECU on engine speed and rack position. The ECU monitors this feedback and adjusts the current level to the actuator solenoid until the rack position signal from the actuator matches the commanded signal.

When no fuel is desired, the ECU turns off current to the actuator solenoid.

Rack Position Sensor

The rack position sensor supplies rack position information to the ECU so that a specific rack position can be controlled.

The sensor includes an electronic module mounted in the actuator housing that provides a voltage to the engine control unit (ECU) indicating the position of the rack. The ECU monitors the signal from the sensor and adjusts the current to the actuator solenoid until the rack position signal matches the commanded signal.

If a problem occurs where the ECU cannot control the rack position, it will shut off current to the fuel shutoff solenoid in addition to the actuator solenoid to shut down the engine.

Primary Speed Sensor

The primary speed sensor is also located within the actuator housing. It is a magnetic pickup that generates voltage pulses to the engine control unit (ECU) as the teeth on the speed wheel attached to the injection pump camshaft pass by the tip of the sensor. The ECU monitors the voltage pulses to determine the engine speed. The

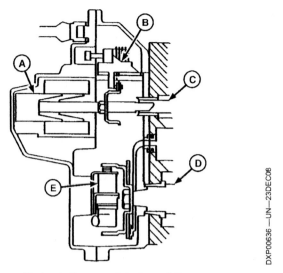

Electronic Governor Actuator and Sensors

A—Actuator Solenoid
B—Rack Position Sensor
C—Rack

D—Pump Camshaft
E—Speed Sensor

ECU continually adjusts current to the actuator solenoid to maintain the desired engine speed.

If this sensor were to fail completely, the ECU would use the signal from the auxiliary speed sensor to govern the engine.

Electronic Governor Benefits

Some of the benefits of the electronic governor system are:

- Fewer mechanical parts in governor to encounter wear
- Different governor modes are possible with specific operating conditions
- Precise fuel control based upon engine speed and fuel temperature
- Quick start-up with less smoke and faster warm-up
- Improved fuel economy and reduced smoke
- Consistent power and performance
- Engine performance is fine tuned to meet a specific load requirement
- Engine control unit (ECU) is programmable to provide precise torque curve characteristics designed for each application
- Capable of communicating with other machine equipment such as transmission controllers
- Built-in self-diagnostics allows quick and accurate repairs

SS40167,0000123 -19-17JUL09-3/3

Malfunctions of Governors

Governors are relatively trouble-free, but need adjusting, servicing, etc.

Listed below are a few of the symptoms and causes of governor problems.

Erratic Engine Operation, Hunting, or Misfiring	
Possible Cause	Possible Solution
1. Governor idle spring missing or adjusted incorrectly.	Replace or adjust spring.
2. Governor control spring worn or broken.	Remove and replace.
3. Governor not operating — parts worn, sticking, binding, or incorrectly assembled.	Disassemble and inspect parts.
4. Wrong governor spring.	Remove and replace with proper spring.
5. Governor high-idle adjustment incorrect.	Adjust governor to diesel pump specifications.
6. Governor adjusting screw not properly adjusted.	Adjust governor screw.
Engine Idles Erratically	
1. Governor idle spring missing or adjusted incorrectly.	Replace or adjust spring.
2. Governor not operating — parts worn, sticking, binding, or incorrectly assembled.	Disassemble and inspect parts.
3. Wrong governor spring.	Remove and replace with proper spring.
Engine Not Receiving Fuel	
1. Governor not operating — parts worn, sticking, binding, or incorrectly assembled.	Disassemble and inspect parts.
Engine Does Not Develop Full Power Or Speed	
1. Governor not operating — parts worn, sticking, inspect parts. binding, or incorrectly assembled.	Disassemble and inspect parts.
2. Wrong governor spring.	Remove and replace with proper spring.
3. Governor adjusting screw not properly adjusted.	Adjust governor screw.

SS40167,0000125 -19-17JUL09-1/1

Summary

Here are some key facts to help you understand governors:

- A governor is a speed-sensitive device that controls the speed of an engine under varying loads.
- Governors can do any of three jobs:
  - Maintain a selected speed.
  - Limit the slow and fast speeds.
  - Shut down the engine before it overspeeds.
- Isochronous governors maintain a constant engine speed independent of load requirements.
- Governor speed droop is a change in the steady state of speed.
- Speed drift is a gradual deviation from desired governor speed.
- A speed changer is a device for adjusting the speed governing system to change engine speed.
- A governor's two functions are: to measure speed and to operate the throttle.
- The two most common types of throttle governors are automatic or vacuum, and centrifugal.
- The different types of diesel engine governors are: mechanical, hydraulic, variable speed, over-speed, torque converter, pressure regulator, load control, electric load-sensing, electric speed, and electronic.
- Hydraulic governors are more expensive and have more parts than mechanical governors but are more sensitive to speed changes.
- Variable speed governors maintain any selected engine speed from idle to maximum speed.
- Over-speed governors are safety devices for controlling maximum engine speed.
- A torque converter governor controls the maximum speed of the output shaft of a torque converter attached to an engine, independently of the engine speed.
- A load control converter adjusts the amount of load applied to the engine to match the speed at which it is set to run.
- A pressure regulator governor is used on any engine driving a pump for maintaining a constant pump pressure.
- Electronic governors provide the most precise governing action for diesel engines. They can be programmed to allow the injection pump system to achieve many different horsepower and performance ratings.

SS40167,0000126 -19-23JUL09-1/1

Test Yourself

Questions

1. What are the three jobs that governors can do?

2. What is speed droop?

3. Name the two most common types of throttle governor systems.

4. What is an isochronous governor?

5. (True or false?) The electronically controlled diesel fuel injection system uses the same basic pumping mechanism used in mechanically governed injection pumps.

6. Name the two primary functions of the engine control unit (ECU) used with the electronic governor system.

SS40167,0000127 -19-17JUL09-1/1

**Engine Test Equipment and Service
Tools — Introduction**

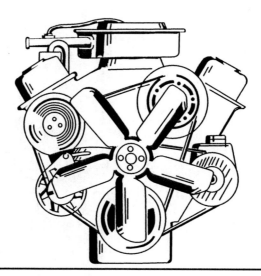

12

DXP01073 —UN—08JUN09

Continued on next page

SS40167,0000128 -19-17JUL09-1/2

100709

Many tools are available today for servicing engines, but they all have one purpose — to help the service technician do the job better and faster.

Two general types of engine tools are used:

- **Testing Tools — to diagnosis the engine**
- **Servicing Tools — to fix the engine**

Testing tools help to find out what must be done, while servicing tools help to do the actual service.

In this manual we will cover many of the basic testing and servicing tools needed to properly diagnosis and repair most engines.

In addition to the testing and servicing tools listed in this chapter, various measuring tools and hand tools also are available to help you do the job better and faster. The FOS manual, Shop Tools, describes these tools and gives detailed instructions for their proper use.

Good Test Equipment Does the Job Better and Faster

Servicing Tools Help to Fix the Engine After It Is Tested

SS40167,0000128 -19-17JUL09-2/2

Testing Tools

Before servicing the engine, test it to locate the problem. Here are some prime tools you will need.

Cylinder Compression Tester

Use a cylinder compression tester to check the compression of each cylinder. Poor compression indicates faulty piston rings, valves, or head gasket. See Chapter 14 for details.

Cylinder Compression Tester

SS40167,0000129 -19-17JUL09-1/36

Cylinder Bore Gauge

The cylinder bore gauge checks the roundness and taper of a cylinder bore or cylinder liner bore. You can also use it to find the exact amount of oversize needed for honing or reboring cylinders.

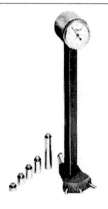

Cylinder Bore Gauge

Continued on next page

SS40167,0000129 -19-17JUL09-2/36

Micrometers

Micrometers are precision tools which measure in thousandths or tenths of thousandths of an inch, or hundredths or thousandths of a millimeter. Use these measuring tools to determine if engine parts are worn beyond their serviceable limits.

NOTE: The FOS manual, Shop Tools, gives detailed instructions for using micrometers and other precision measuring tools.

Use an outside micrometer to measure the size of parts, such as diameter and thickness. The standard micrometer shown in the illustration is numbered on the thimble for direct reading. Digital micrometers (not shown) display a digital readout of the measurement. They are quicker and easier to read, but are more expensive.

Use an inside micrometer to measure bores or openings. Uses include: measuring cylinder and main bearing bores for taper and out-of-roundness.

Telescoping gauges allow easy measurement in small bores. These gauges have no scale markings. Measure the setting with an outside micrometer.

Use a depth micrometer to measure the depth of openings or the difference in the levels of two surfaces. The depth micrometer is numbered on the thimble for direct reading.

Outside Micrometer

Inside Micrometer

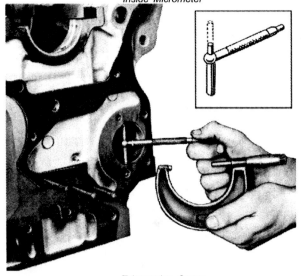

Telescoping Gauge

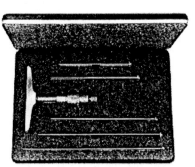

Depth Micrometer

DXP00641 —UN—01JUN09

DXP00642 —UN—23DEC08

DXP00643 —UN—23DEC08

DXP00644 —UN—23DEC08

Continued on next page

SS40167,0000129 -19-17JUL09-3/36

Piston Pin Bore Gauge

A piston pin bore gauge is used to measure piston pin bores for precision pin fits. The gauge shown in the illustration can also be used to measure bores in the journal end of a connecting rod.

Piston Pin Bore Gauge

SS40167,0000129 -19-17JUL09-4/36

Ring Groove Wear Gauge

Use a wear gauge to make a fast check for wear in the piston ring grooves. Gauges are available for both rectangular and keystone grooves.

NOTE: A new piston ring and a feeler gauge can also be used to measure rectangular ring groove wear.

1—Wear Gauge

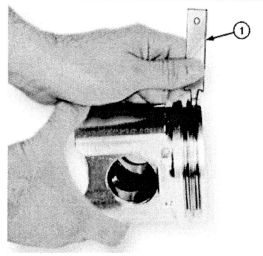

Ring Groove Wear Gauge

Continued on next page

SS40167,0000129 -19-17JUL09-5/36

Piston and Liner Height Gauge

Use the piston and liner height gauge to accurately measure:

- Height of the piston in the bore
- Height or protrusion of the cylinder liner above the top surface of the cylinder block

Piston and Liner Height Gauge

SS40167,0000129 -19-17JUL09-6/36

Plastigage®

Using Plastigage is a fast and accurate way to check engine bearing clearance. Determine clearance by crushing the Plastigage between the bearing and the crankshaft, and then comparing the width of the crushed Plastigage with a graduation scale provided with the package.

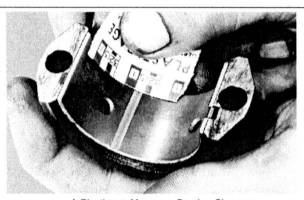

A Plastigage Measures Bearing Clearance

Plastigage is a trademark of AE Clevite, Inc.

SS40167,0000129 -19-17JUL09-7/36

Valve Spring Tester

Use the valve spring tester to check the tension or strength of valve springs in service.

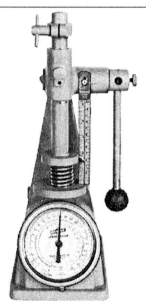

Valve Spring Tester

Continued on next page

SS40167,0000129 -19-17JUL09-8/36

Valve Inspection Center

Use the valve inspection center to determine if valves are out-of-round, bent, or warped.

A—Valve Inspection Center

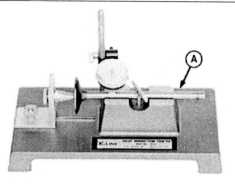

Valve Inspection Center

Eccentrimeter

Use an eccentrimeter to check valve seat runout. The results will tell you if the valve seat is concentric with the valve guide.

A—Eccentrimeter

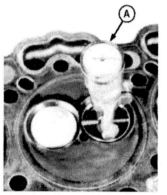

Eccentrimeter

Injection Nozzle Tester (Diesel)

The bench-style injection nozzle tester shown in the illustration is mounted to a workbench. With the proper adapters, this low-cost nozzle tester will test all major brands of fuel injection nozzles.

Bench-Style Injection Nozzle Tester

Continued on next page

The injection nozzle tester shown is totally portable for field use, or it can be bench mounted.

An injection nozzle tester is essential to properly test and calibrate a diesel fuel injection nozzle.

An injection nozzle tester will check the condition of:

• Needle valve and seat within the nozzle
• Spray pattern of the nozzle tip
• Cracking or pop-off pressure of the nozzle
• Leak-off through the nozzle
• Nozzle valve lift

Portable-Style Injection Nozzle Tester

SS40167,0000129 -19-17JUL09-12/36

Injection Pump Tester (Diesel)

An injection pump tester is essential to properly test and calibrate a diesel fuel injection pump. The tester shown in the illustration is used to test leakage, vacuum, pressure, and delivery. It is also used to make idle and torque control adjustments.

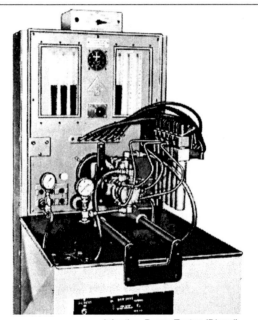

Injection Pump Tester (Diesel)

Continued on next page   SS40167,0000129 -19-17JUL09-13/36

Dynamometer

A dynamometer is used to apply a load to an engine for testing purposes. Three common types of dynamometers are found in repair shops:

- Chassis Dynamometers
- Engine Dynamometers
- Power Take-Off (PTO) Dynamometers

The chassis dynamometer is driven by the vehicle's drive wheels. This type dynamometer can simulate road conditions to test the engines used in automotive applications. It gives a true reading of usable horsepower — the horsepower available to the drive wheels to push the vehicle down road.

The engine dynamometer connects directly to the flywheel of an engine. It measures the flywheel or brake horsepower and the maximum horsepower the engine can produce.

The power take-off (PTO) dynamometer shown in the illustration measures the horsepower at the power take-off shaft of a tractor. A power take-off usually has some gear reduction between the engine and the PTO output shaft. This reduction increases the torque value but reduces the speed. When measuring PTO horsepower, the speed

PTO Dynamometer

is held at a constant value (usually rated speed). The horsepower can be read directly on a gauge measuring torque, but having its scale calibrated in horsepower.

Measure engine horsepower, torque, and fuel consumption with a dynamometer. The results will tell you if the engine efficiency can be restored by a tune-up or whether reconditioning is required.

SS40167,0000129 -19-17JUL09-14/36

Water Manometer

Use a water manometer to check the air intake system of an engine. High depression readings indicate a restriction. A vacuum gauge calibrated in inches of water can also be used to make this check.

Water Manometer

Continued on next page

SS40167,0000129 -19-17JUL09-15/36

Tachometer

Most tachometers measure rotary speed in revolutions per minute (rpm).

The hand-held digital tachometer shown in the illustration measures either rotary speed or surface speed. To measure rotary speed, the tachometer uses a photo (light) probe and a strip of reflective tape attached to a shaft. As the shaft rotates, the tachometer measures the time between each pass of the tape. The speed of the shaft can then be read on the digital display of the meter.

This tachometer is also equipped with a mechanical speed wheel that is placed directly on the surface of a moving object to measure surface speed.

Hand-Held Digital Tachometer

SS40167,0000129 -19-17JUL09-16/36

Electronic tachometers measure speed in a variety of ways. The tachometer shown in the illustration has an inductive pickup sensor that is clamped to the diesel fuel injection pump line. The sensor reads the fuel pulsations in the injection line and sends a signal to the digital meter.

For spark-ignition engines, the tachometer has an inductive pickup sensor that is clamped to the high-tension ignition wire. The sensor picks up the electrical pulses in the wire. The meter times the pulses and then displays the rpm.

This type tachometer can be used for two- or four-stroke cycle engines as well as diesel or spark-ignition engine applications.

Portable Electronic Tachometer

SS40167,0000129 -19-17JUL09-17/36

Exhaust Gas Analyzer

An exhaust gas analyzer is used to check carburetor or fuel injector performance, combustion efficiency, and air-fuel ratio. The model shown in the illustration measures:

- Hydrocarbons (HC)
- Carbon Monoxide (CO)
- Carbon Dioxide (CO_2)
- Oxygen (O_2)
- Oxides of Nitrogen (NO_x)

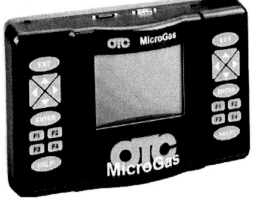

Exhaust Gas Analyzer Set

Continued on next page

SS40167,0000129 -19-17JUL09-18/36

100709
PN=410

Ignition Coil and Condenser Meter

This combination tester checks the condition of the coil and condenser in the conventional breaker point-type ignition system.

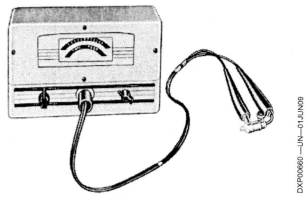

Ignition Coil and Condenser Meter

SS40167,0000129 -19-17JUL09-19/36

Distributor Tester (Spark Ignition)

The all-purpose distributor tester shown in the illustration will test:

- Centrifugal advance
- Vacuum advance
- Cam dwell angle
- Variations in cam lobes
- Breaker point contact resistance
- Breaker point synchronization
- Breaker point operation
- Breaker point spring tension
- Vacuum chamber condition
- Distributor shorts
- Excessive wear
- Distributor high-speed performance

These testers are also known as synchrographs. Bench models of the tester are also available.

Distributor Tester

SS40167,0000129 -19-17JUL09-20/36

Engine Analyzer (Spark Ignition)

The spark-ignition engine analyzer, shown in the illustration, can be used to test the entire ignition system:

- Point resistance
- Cam angle
- Basic timing
- Generator voltage
- Starter motor voltage
- Mechanical condition of distributor

Spark-Ignition Engine Analyzer

Continued on next page   SS40167,0000129 -19-17JUL09-21/36

Spark Plug Voltage Tester

Use this tester to check the condition of the spark plugs, plug wires, ignition system, and other factors affecting the spark plug voltage. The tester measures the following:

- Spark plug firing voltage
- Spark plug burn time
- Spark plug burn voltage

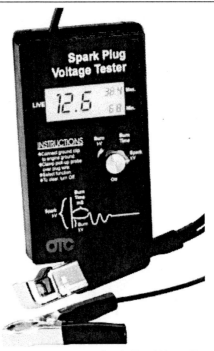

Spark Plug Voltage Tester

SS40167,0000129 -19-17JUL09-22/36

Electronic Ignition Spark Tester

Use the electronic ignition spark tester as a quick check for a no-start problem. Connect the plug wire to the tester, clamp the tester to an engine ground, crank the engine, and watch for a spark. A properly working electronic ignition system will be able to produce a spark that will jump across the tester's preset 0.250 inch (0.64 mm) spark gap.

This tester may not work on conventional breaker point-type ignitions because they may not generate enough spark to jump the wide gap.

Electronic Ignition Spark Tester

SS40167,0000129 -19-17JUL09-23/36

Spark Plug Gap Gauge

Use a spark plug gap gauge to check and set the gap between spark plug electrodes. A plug can't do a good job unless it is set to exact engine specifications.

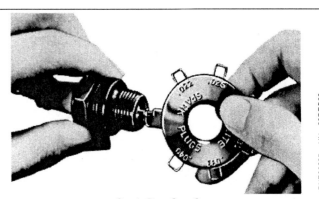

Spark Plug Gap Gauge

Continued on next page

SS40167,0000129 -19-17JUL09-24/36

Power Timing Light

Use the power timing light to check the ignition timing while the engine is running. The timing light shown in the illustration will check both conventional and electronic ignition systems. It has an adjustable advance knob with a scale reading directly in degrees and allows checking:

- Basic ignition timing
- Vacuum advance
- Centrifugal advance
- Computer advance

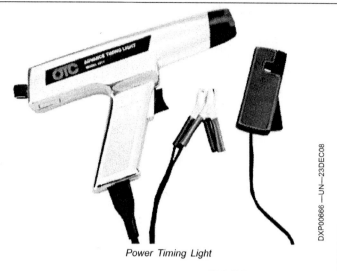

Power Timing Light

SS40167,0000129 -19-17JUL09-25/36

Electronic Stethoscope

Use the electronic stethoscope to determine the cause of irregular sounds in the engine.

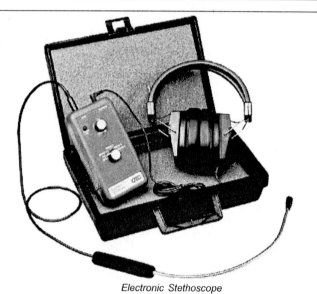

Electronic Stethoscope

SS40167,0000129 -19-17JUL09-26/36

Battery Hydrometer

DXP00668 —UN—23DEC08

Use the hydrometer to check the specific gravity of the battery. All good hydrometers have a built-in thermometer.

Battery Hydrometer

Continued on next page   SS40167,0000129 -19-17JUL09-27/36

Digital Multimeter

The digital multimeter is used to test electrical components for voltage, resistance, or current flow. It is especially good for measuring low voltage or high-resistance circuits.

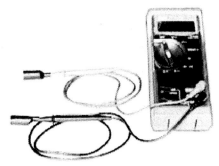

Digital Multimeter

SS40167,0000129 -19-17JUL09-28/36

Air Flow Meter

The air flow meter is a special fan that is used with a digital multimeter to measure air flow through radiator cores. The fan motor generates a voltage that is read on the digital multimeter. Voltage variations from one section of the radiator core to the next will indicate air flow restrictions.

Air Flow Meter

SS40167,0000129 -19-17JUL09-29/36

Electronic Governor Tester

The electronic governor tester is a diagnostic reader that is able to read and display the diagnostic codes output by the electronic engine control module (ECU). It is also capable of recalling or clearing stored diagnostic codes.

NOTE: Other electrical tools are listed in the FOS manual, Electrical Systems.

Electronic Governor Tester

Continued on next page   SS40167,0000129 -19-17JUL09-30/36

Magnetic Inspection Kit

Use the magnetic inspection kit to detect surface cracks in ferrous metal by the magnetic particle method. Cylinder heads, cylinder blocks, tools, welds, forgings, and castings can be checked for cracks and other defects using the magnetic detector.

Magnetic Inspection Kit

DXP00672 —UN—23DEC08

SS40167,0000129 -19-17JUL09-31/36

Oil Pressure Gauge

As most engine lubricating systems now use an oil pressure indicating light, an oil pressure gauge is required to check the system pressure. A typical gauge with fittings is shown in the illustration.

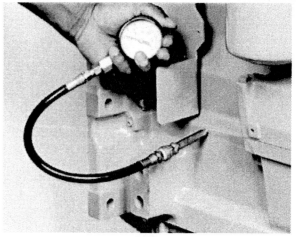

Oil Pressure Gauge

DXP00673 —UN—01JUN09

SS40167,0000129 -19-17JUL09-32/36

Thermostat Tester

A thermostat tester is used to check the temperature at which a thermostat valve starts to open. This is an important check because the proper engine operating temperature is maintained by the thermostat (A).

A—Thermostat

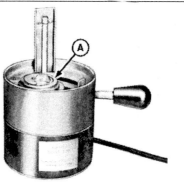

Thermostat Tester

DXP00674 —UN—23DEC08

Continued on next page

SS40167,0000129 -19-17JUL09-33/36

Radiator and Cap Tester

This tester is used to check the radiator and radiator cap for leaks and for correct opening pressure. An efficient pressure cooling system must be free from leaks.

Radiator and Cap Tester

SS40167,0000129 -19-17JUL09-34/36

Spectrometer

Maintaining the correct concentration of antifreeze can save thousands of dollars on engine repairs and maintenance. Unlike other models, this meter is very easy to use. The focusing eyepiece provides a sharp, easy-to-read scale. More importantly, our model uses a semi-transparent blue background, which means you don't have to point the instrument in a certain direction to get a reading: ambient room lighting will provide plenty of illumination. Gives quick, accurate determination of the freezing point of propylene and ethylene glycol-based antifreeze solutions.

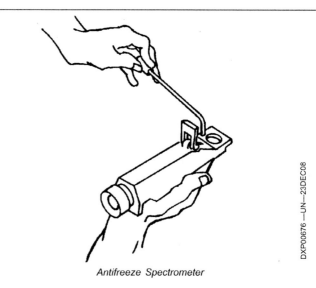

Antifreeze Spectrometer

SS40167,0000129 -19-17JUL09-35/36

Vacuum Gauge

A vacuum gauge, calibrated in inches of mercury, will help determine the condition of a spark-ignition engine. Various malfunctions can be identified by how they cause the gauge to react. The gauge shown in the illustration has an auxiliary scale to check fuel pump pressure.

Vacuum Gauge

SS40167,0000129 -19-17JUL09-36/36

Servicing Tools

After the engine is tested and diagnosed, service it using the proper tools. Here is a list of the major ones.

Repair Stand

Use the repair stand to safely support the engine during service. The stand is equipped with a gearbox that allows full rotation of the engine with the turn of a handle. Adapters are available for mounting many different engines.

Repair Stand

SS40167,000012A -19-17JUL09-1/19

Valve Refacer

A valve refacer is used to grind an exact angle on the face of a valve. The accuracy of this machine is critical as the angle of the valve face must match the angle of the valve seat.

Valve Refacer

Continued on next page　　SS40167,000012A -19-17JUL09-2/19

Valve Seat Grinder

The valve seat grinder is used to recondition the valve seat. Both rough and finish grinding stones are required.

Valve seat cutters are also available to recondition the valve seat. The tungsten carbide cutter blades retain their shape and cutting angle, eliminating the need to redress grinding stones and avoiding the dust produced by valve seat grinders.

Valve Seat Grinder

SS40167,000012A -19-17JUL09-3/19

Valve Spring Compressor

This tool is used to compress the spring when removing or installing a valve. Various types of spring compressors are available.

Valve Spring Compressor

SS40167,000012A -19-17JUL09-4/19

Valve Seat Puller

Use the valve seat puller to remove valve seat inserts.

A—Valve Seat Puller

Valve Seat Puller

Continued on next page   SS40167,000012A -19-17JUL09-5/19

Valve Guide Knurler

Use the valve guide knurler to resize the valve guides. The tool reduces the inside diameter of a valve guide by displacing the metal of the guide inner surface. The guide is then reamed to accept a standard valve. The closer clearances stop oil consumption caused by valve guide wear and allow improved valve seating.

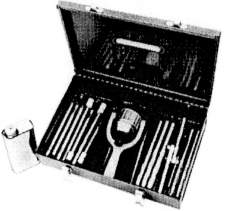

Valve Guide Knurler

SS40167,000012A -19-17JUL09-6/19

Cylinder Ridge Reamer

Use a cylinder ridge reamer to remove the ridge found at the end of the ring travel at the top of a cylinder or liner. Pistons can then be easily and safely removed.

Cylinder Ridge Reamer

Continued on next page

SS40167,000012A -19-17JUL09-7/19

Cylinder Liner Puller-Installer

Removal and installation of cylinder liners or sleeves is easier and quicker with a puller-installer.

Both hydraulic and manual models are available. A great variety of engines can be serviced by using different adapter plates.

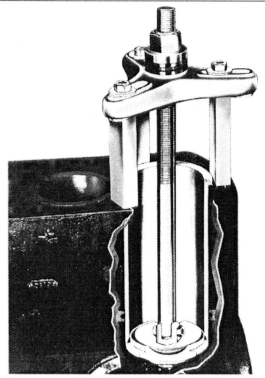

Cylinder Liner Puller-Installer

Continued on next page   SS40167,000012A -19-17JUL09-8/19

Cylinder Reboring Bar

A reboring bar is used to machine the cylinders of large engines that have integral cylinders.

This boring operation is necessary if oversized pistons are to be used.

A typical reboring machine is shown in the illustration.

Cylinder Reboring Bar

SS40167,000012A -19-17JUL09-9/19

Cylinder Deglazer

DXP00687 —UN—01JUN09

Use a cylinder deglazer to deglaze and finish the cylinder or cylinder liner bore. Stone, pad, and brush-type deglazers are available for use with an electric drill.

The brush-type deglazer shown in the illustration is flexible, selfcentering, and self-aligning. Use with honing oil or SAE 10W-30 oil and a 3/8- or 1/2-inch drill motor at 300–500 rpm with 30–40 strokes a minute. This produces a 45-degree crosshatch finish, ensuring rapid ring seating and good oil retention.

Cylinder Deglazer

Continued on next page   SS40167,000012A -19-17JUL09-10/19

PN=421

Piston Ring Groove Cleaning Tool

Clean piston ring grooves using a piston cleaning tool. Soaking pistons in a recommended piston cleaner or hot water with a liquid detergent soap is another good way to clean pistons.

IMPORTANT: Avoid damage to pistons. Do not use a wire wheel to clean piston ring grooves; use a stiff bristle brush.

Piston Ring Groove Cleaning Tool

SS40167,000012A -19-17JUL09-11/19

Piston Ring Groove Cutting Tool

Worn piston ring grooves on some pistons can be machined with a ring groove cutting tool. New standard-width rings can then be used with flat steel spacers. Refer to the engine technical manual to determine if this service is recommended.

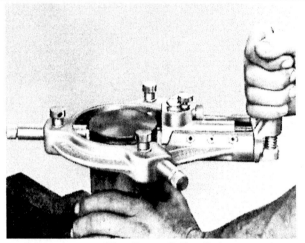

Ring Groove Cutting Tool

SS40167,000012A -19-17JUL09-12/19

Piston Knurling Tool

A piston knurling tool is used to resize pistons. This tool increases the diameter of a piston by displacing the metal of the skirt thrust faces.

Piston Knurling Tool

Continued on next page

SS40167,000012A -19-17JUL09-13/19

Piston Ring Expander

Use a piston ring expander to remove and install piston rings without damaging them. The expander should limit the ring opening so that the rings are not stretched or broken.

Piston Ring Expander

SS40167,000012A -19-17JUL09-14/19

Piston Ring Compressor

Use a piston ring compressor to compress the rings when installing pistons into the cylinders. This prevents breaking the rings.

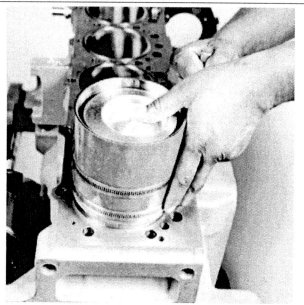

Piston Ring Compressor

Continued on next page   SS40167,000012A -19-17JUL09-15/19

Piston and Rod Aligning Tool

A piston and rod aligning tool is used to check the piston and connecting rod alignment. Bending bars and clamps are provided to correct any bend, twist, or offset in the rod.

Piston and Rod Aligning Tool

SS40167,000012A -19-17JUL09-16/19

Carburetor Service Tools (Spark-Ignition Engines)

Special tools are available to service most gasoline and LP-gas carburetors. Tools are usually designed for one particular carburetor, and some carburetors require more tools than others.

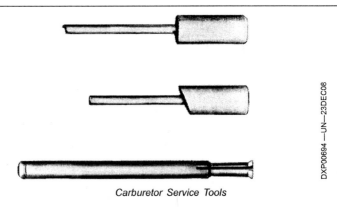

Carburetor Service Tools

SS40167,000012A -19-17JUL09-17/19

Injection Nozzle Removal Tools (Diesel)

Removal and installation kits are available for some injection nozzles. A typical kit includes hose clamp pliers, nozzle puller, bore cleaning tool, and a guide.

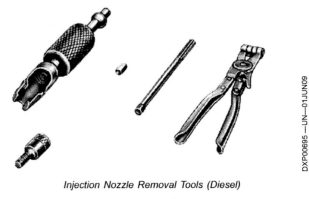

Injection Nozzle Removal Tools (Diesel)

Continued on next page   SS40167,000012A -19-17JUL09-18/19

Injection Nozzle Cleaning Kit (Diesel)

Cleaning kits are usually designed to service just one particular make of nozzle. These kits contain such items as cleaning wires, brushes, drills, and lapping compounds.

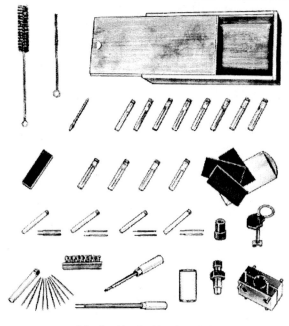

Injection Nozzle Cleaning Kit (Diesel)

SS40167,000012A -19-17JUL09-19/19

Summary

Most of the tools we have listed can be used for any engine. However, there is another category of tools that we have not covered — the special tools designed for each engine.

Be sure that your shop has all the special tools needed to service the models of engines you normally work on. These tools are listed in the technical manual for each engine.

The tools we have shown in this chapter are basic to almost all engines. Some are simple tools, others complex testers. Yet each one, regardless of size or cost, is designed to make your job easier and faster.

Special Tools for Each Engine Are Also Needed to Completely Equip a Service Shop

SS40167,000012B -19-17JUL09-1/1

Test Yourself

Questions

1. What are the two major types of engine tools?

2. Which type of tool is normally used first?

3. What other category of tools is not covered in this chapter?

SS40167,000012C -19-17JUL09-1/1

Diagnosing and Testing Engines — Introduction

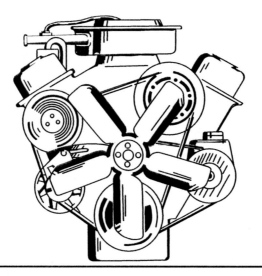

13

Hit-or-Miss Parts Exchanger or Knowledgeable Modern-Day Service Technician — which would you rather be?

The hit-or-miss parts exchanger dives into an engine and starts replacing parts helter-skelter until the problem is found — maybe — after wasting a lot of the customer's time and money.

The knowledgeable service technician starts out by getting all the facts and examines them until the trouble is pinpointed. The technician then verifies the diagnosis by testing it and only then starts replacing parts.

The hit-or-miss parts exchanger is fast becoming a person of the past. With the complex systems of today, diagnosis and testing by the knowledgeable modern-day service technician is the only way.

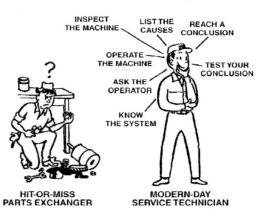

Which Would You Rather Be?

SS40167,000012D -19-17JUL09-1/1

Seven Basic Steps of Diagnosis

A good program of diagnosis and testing has seven basic steps:

1. **Know the System**
2. **Ask the Operator**
3. **Inspect the Engine**
4. **Operate the Engine**
5. **List the Possible Causes**
6. **Reach a Conclusion**
7. **Test Your Conclusion**

1. Know the System

In other words, do your homework. Study the engine technical manuals. Know how the engine works, how it can fail, and the three basic needs — fuel-air mixture, compression, and ignition.

Keep up with the service bulletins. Read them and then file in the proper place. The solution to your problem may

Know the System

be in this month's bulletin. You can be prepared for any problem by knowing the engine.

SS40167,000012E -19-17JUL09-1/7

2. Ask the Operator

A good reporter gets the full story from a witness — the operator.

The operator can tell how the engine acted when it started to fail, what was unusual about it.

What work was the engine doing when the trouble occurred? Was the trouble erratic or constant?

What did the operator do after the trouble? Did the operator attempt to repair the problem?

Ask how the engine is used and when it was serviced. Many problems can be traced to poor maintenance or abuse of the engine.

Be tactful, but get the full story from the operator.

Ask the Operator

Continued on next page   SS40167,000012E -19-17JUL09-2/7

3. Inspect the Engine

Inspect the engine for all items listed in the illustration. Use your eyes, ears, and nose to spot trouble.

- Look for water leaks at the radiator, water pump, hoses, and around the cylinder head gasket.
- Check for oil leaks at the oil pan, drain plugs, and gaskets. Also inspect around crankshaft bores, oil seals, and tappet cover. Look inside the flywheel housing for signs of oil.
- Look for fuel leaks at the tank, lines, filters, and pumps. Also check for restrictions or evidence of water in the fuel.
- Inspect for ignition problems such as loose or bare wires, cracked distributor cap, shorted wires at plugs.
- Check for electrical problems such as battery corroded, sulfated, or low on water; battery cables loose or corroded; low charge in battery.
- Inspect for clutch problems that might affect the engine. Free travel should be adequate.
- Look for other trouble signs that could lead to future problems if not corrected early.

INSPECT FOR:

✓ **WATER LEAKS**

✓ **OIL LEAKS**

✓ **FUEL LEAKS**

✓ **IGNITION PROBLEMS**

✓ **ELECTRICAL PROBLEMS**

✓ **CLUTCH PROBLEMS**

✓ **OTHER TROUBLE SIGNS**

Check List for Inspecting the Engine

In general, look for anything unusual. Keep a list of all the trouble signs.

SS40167,000012E -19-17JUL09-3/7

4. Operate the Engine

If the engine can be run, start it and warm it up. Then run it through its paces. Don't completely trust the operator's story — check it yourself.

Test the engine on a dynamometer. This is the only way to get a full picture of the engine's condition. For details, see Testing the Engine.

If a dynamometer is not available, look, smell, and listen for engine problems:

- Are the gauges reading normal?
- Hear any funny sounds? Where? At what speeds?
- Smell anything? Any signs of unusual exhaust smoke? Are the breathers smoking?
- How do the engine controls work?
- How is the power under load?

Operate the Engine

- Does the engine idle okay?

Use your common sense to find out how the engine is operating.

Continued on next page   SS40167,000012E -19-17JUL09-4/7

5. List the Possible Causes

Now you are ready to make a list of the possible causes of the engine's troubles.

What were the signs you discovered while inspecting and operating the engine?

- Did the engine lack power?
- Any smoke from the crankcase vent?
- Did the engine run too hot or too cold?
- How was the oil pressure?

Which of the signs tell you the most likely cause, which is second, etc.?

Are there any other possibilities? (One failure often leads to another.)

List the Possible Causes

SS40167,000012E -19-17JUL09-5/7

6. Reach a Conclusion

Look over your list of possible causes and decide which are most likely and which are easiest to verify.

Use the Troubleshooting Charts at the end of this chapter as a guide.

Reach your decision on the leading causes and plan to check them first — after making the easy checks.

Reach a Conclusion

Continued on next page   SS40167,000012E -19-17JUL09-6/7

7. Test Your Conclusion

Before you start repairing the system, test your conclusions to see if they are correct.

Many of the items on your list can be verified without further testing.

Maybe you can isolate the problem to one system of the engine — lubrication, cooling, etc.

But the location within the system may be harder to find. This is where testing tools can help.

The next part of this chapter will tell how to test the engine and locate failures.

But first let's repeat the seven rules for good troubleshooting:

1. **Know the System**
2. **Ask the Operator**
3. **Inspect the Engine**
4. **Operate the Engine**

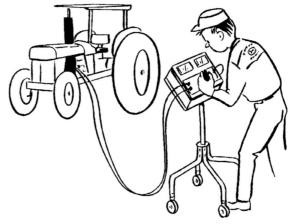

Test Your Conclusion

5. **List the Possible Causes**
6. **Reach a Conclusion**
7. **Test Your Conclusion**

SS40167,000012E -19-17JUL09-7/7

Testing the Engine

The use of engine test equipment is most effective once the failure has been isolated to one system or area of the engine.

Chapter 12 lists all the testing tools needed for normal engine trouble shooting. Other chapters of this manual show how to use most of this equipment.

Dynamometer Test

A complete dynamometer test is the best way to troubleshoot the engine.

Remember, however, the engine must be reasonably well-tuned to pass the dynamometer tests.

If the engine was tested on a dynamometer, the following tests should have been made and the results listed:

- Engine horsepower
- Exhaust smoke analysis
- Fuel consumption
- Crankcase blowby
- Air cleaner restriction
- Oil pressure
- Battery condition
- Alternator or generator output
- Clutch operation

Refer to Chapter 12 for the testing equipment used in the above tests. Chapter 14 gives a general guide to the tests.

Dynamometer Test

Compare the test results with the correct readings given in the engine technical manual.

Continued on next page   SS40167,0000131 -19-20JUL09-1/2

Engine Horsepower

Horsepower is the fundamental measure of engine efficiency.

Three basic things are needed for the engine to produce horsepower:

1. **Fuel-air mixture**
2. **Compression**
3. **Ignition**

A good supply of AIR is essential to engine combustion. The air cleaner must be kept clean and the air intake system kept free of restrictions.

There must be a good supply of FUEL. The tank, lines, filters, and pumps must be open and clean.

COMPRESSION must be adequate. Low compression can be from bad valves, leaking head gaskets, or blowby at the pistons. In a diesel engine, compression is extra vital because the heat of compression rather than a spark is what ignites the fuel.

IGNITION must be adequate and properly timed. In spark-ignition engines, the battery and ignition circuit are vital in producing a strong and timely spark for ignition. In diesel engines, the injection pump must be timed properly.

Unless all three, fuel-air, compression, and ignition, are doing their job in the right sequence, engine horsepower will be low.

The basic dynamometer test will tell you whether the engine is low on horsepower. Further tests will isolate the cause to one of the three major areas.

Mechanical Problems

Beyond the problems of engine efficiency, there are many mechanical things that can cause an engine to fail.

Here are a few of the basic problems and their likely results:

Worn Bearings — can increase oil consumption, create noise, cause eventual failure of bearings and damage to crankshaft.

Worn Valve Guides — can increase oil consumption, eventually cause valves to fail.

Worn Pistons and Rings — also increase oil consumption, cause loss of compression and power.

Worn Camshaft Lobes — loss of power because valves open too late and close too early.

Summary

The tests we have given you are only basic guidelines.

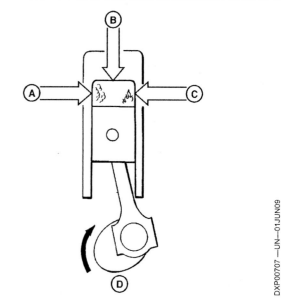

Three Basic Things are Needed to Produce Engine Horsepower

A—Fuel-Air Mixture　　　　C—Ignition
B—Compression　　　　　　D—Horsepower

Once you start testing actual engines, use the engine technical manual for detailed tests and test results.

And remember that the best testing tools have no value unless the technician at the controls knows how to interpret the results.

GOOD DIAGNOSIS

WHAT THE CUSTOMER WANTS FROM TODAY'S TECHNICIAN

- That only NECESSARY work be done
- That cost be REASONABLE
- That work be completed QUICKLY
- That problems be fixed RIGHT — THE FIRST TIME

THE ONLY WAY TO SATISFY THIS CUSTOMER:
- Quick, complete, and accurate DIAGNOSIS

AND THE WAY TO GOOD DIAGNOSIS:
- Discover the problem, with the customer's help
- Write a clear and accurate repair order
- Test the diagnosis in a logical sequence

SS40167,0000131 -19-20JUL09-2/2

Troubleshooting Charts

Use the charts on the following pages to help in listing all the possible causes of trouble when you diagnose and test an engine.

Once you have located the cause, refer to the chart again for the possible remedy.

Some of the problems are discussed in detail following each chart.

Other chapters of this manual give more details on diagnosing a particular component.

Also use the technical manual for each engine for specific tests and specifications.

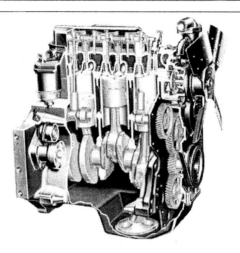

1. Engine Hard to Start or Will Not Start	Solution
No Fuel or Improper Fuel	Fill tank. If wrong fuel, drain and fill with proper fuel.
Water or Dirt in Fuel or Dirty Filters	Check out fuel supply. Replace or clean filters.
Air in Fuel System (Diesel)	Bleed air from the system.
Low Cranking Speed	Charge or replace battery, or service starter as necessary.
Poor Timing	Check distributor or injection pump timing.
Defective Coil or Condenser (Spark-Ignition)	Replace coil or condenser.
Pitted or Burned Distributor Points (Spark-Ignition)	Clean or replace points.
Cracked or Eroded Distributor Rotor (Spark-Ignition)	Replace rotor.
Distributor Wires Loose or Installed in Wrong Order (Spark-Ignition)	Push wires into sockets. Install in correct firing order.
Defective or Wrong Spark Plugs (Spark-Ignition)	Replace plugs.
Poor Nozzle Operation (Diesel)	Clean, repair, adjust, or replace.
Liquid Fuel in Lines (LP-Gas)	Always turn on the vapor valve when starting the engine.
Too-Heavy Oil in Air Cleaner (LP-Gas)	Use correct weight of oil.
Fuel Tank Manual Shutoff Valve Is Closed (Natural Gas)	Turn valve to open position.
Fuel Lockoff Device Is Closed (Natural Gas)	Inspect, repair, or replace.
Fuel Lockoff Device Is Leaking (Natural Gas)	Inspect, repair, or replace.

Fuel Problems

In spark-ignition engines, the volatility of the fuel has much to do with the starting ability of the engine.

NOTE: Volatility is what causes the fuel to vaporize for combustion.

Liquid gasoline, LP-gas, or natural gas must be in vapor form to mix with the air when it is taken into the engine.

If the flash point of the fuel is too low, not enough vapor is formed to mix with the air for easy ignition.

NOTE: Flash point is the temperature at which vapor from the fuel will ignite.

In diesel engines, since ignition is from the heat of compression, the flash point of the fuel affects the ability of the engine to start.

Generally speaking, the lower the grade of diesel fuel, the higher the flash point. This means that as the grade of fuel is lowered, the heat created in the engine must be hotter. This can contribute to hard starting.

Low Cranking Speed

Low cranking speed is a result of 1) low charge in the battery, 2) a defective battery or terminal connection, or 3) binding in the starter or engine.

In spark-ignition engines, when any of these conditions exist, so much of the battery voltage is used by the starter in cranking the engine that the remaining voltage is not sufficient to create an ignition spark.

If, while cranking the engine, the voltage of a 12-volt battery falls below 9 volts, the battery is either defective or needs recharging.

In diesel engines, when the battery is low or defective, the starter cannot turn the engine fast enough to create enough heat within the cylinders to provide ignition.

Also, worn starter bearings can allow the armature to drag on the field poles. This results in low cranking speed and more voltage drop at the battery.

Any binding within the engine will produce the same effect as starter drag.

It is also possible the battery cables are too small or the terminals are corroded. In this case, the resistance of the cable is too high to carry the amount of current required to crank the engine at the proper speed.

Defective Coil or Condenser (Spark-Ignition)

When either the coil or condenser is defective, not enough spark will be created to start the engine.

Continued on next page   SS40167,0000132 -19-05AUG09-1/7

Pitted or Burned Distributor Points (Spark-Ignition)

When the distributor points become pitted or burned, their ability to conduct current is diminished. This results in a poor spark from the ignition coil.

Poor Nozzle Operation (Diesel)

The nozzles must be in good working order. If they are partially clogged or if the cracking pressure is low, much of the fuel is injected as a stream. This will cause hard starting.

Heavy Oil in Air Cleaner (LP-Gas)

Oil with a weight heavier than what is specified in the air cleaner restricts the flow of air through the air cleaner, resulting in a choking action at the carburetor. This, in turn, creates a air-fuel mixture which is too rich to burn.

Fuel Lockoff Device Faulty (Natural Gas)

The fuel lockoff device consists of a solenoid-actuated valve that is located upstream of the injectors. The lockoff device is electronically controlled by the engine control unit (ECU). The lockoff valve will be closed anytime the ignition is off or is on but hasn't been in the crank position for a predetermined length of time. The normally closed valve opens when the engine control unit (ECU) energizes the lockoff solenoid. If the fuel lockoff device fails to open, fuel flow will be stopped upstream of the injectors and the engine will not start.

A fuel lockoff valve that leaks when the ignition is off could cause the intake manifold to fill with a noncombustible (excessively rich) mixture. Prolonged cranking would then be required to clear this mixture from the manifold.

2. Engine Starts But Will not Run	Solution
Fuel Problems — Air Restriction or Clogged Filter(s)	Check fuel supply, bleed system (diesel), check for line restrictions, or clean or replace filters and screens.
Carburetor Idle Fuel Needs Adjustment	Adjust correctly.
Defective Coil or Condenser (Spark-Ignition)	Replace.
Defective Ignition Resistor or Key Switch (Spark-Ignition)	Replace resistor or key switch.

Fuel Problems

When the flow of fuel to the engine is restricted, not enough fuel can get to the engine to keep it running. Check fuel supply, repair or replace faulty fuel lines, clean or replace fuel filters and screens.

Ignition Coil and Condenser (Spark-Ignition)

In some instances a coil or condenser will provide a spark to start the engine. Then, after the engine starts, due to a loose connection or internal short, it will not keep on running. Also a coil or condenser may function when cold and then fail when heated.

Defective Ignition Resistor Key Switch (Spark-Ignition)

When the key switch is turned to crank the engine, full battery voltage is applied to the ignition coil for easier starting. When the switch is released, the ignition current flows through a resistor, limiting the coil voltage to about six volts. So, if the resistor is burned out or has a loose connection, the engine will not continue to run when the switch is released.

3. Engine Misses	Solution
Water or Dirt in Fuel	Drain and refill with clean fuel.
Gasoline in Diesel Fuel	Drain and refill with proper fuel.
Dirty Spark Plugs or Faulty Cables (Spark-Ignition)	Clean or replace plugs. Replace cables.
Cracked Distributor Cap (Spark-Ignition)	Replace cap.
Carburetor Not Adjusted Correctly (Spark-Ignition)	Adjust carburetor.
Air in Fuel System (Diesel)	Bleed the system.
Poor Nozzle Operation (Diesel)	Clean the injection nozzle and check nozzle spray pattern.
Faulty Injection Pump (Diesel)	Check and calibrate the fuel injection pump.
Nozzles Not Seated Properly in the Cylinder Head (Diesel)	Reposition nozzles and tighten retaining screws to specified torque.
Faulty Engine Control Unit (ECU) or Ignition Control Unit (ICU) Wiring (Natural Gas)	Check wiring for loose connections or damaged wires.
Clogged Fuel Injectors (Natural Gas)	Clean, repair, or replace injectors.

Gasoline in Diesel Fuel

Gasoline in diesel fuel can cause the engine to miss. The reason is this: The diesel fuel flowing through the nozzle provides lubrication for the nozzle valve. Gasoline does not have this lubricating quality and the valve sticks, causing misfiring. In many cases the nozzle valve will be scored and need replacement.

Air in Fuel System (Diesel)

Air in the diesel fuel system can cause the engine to miss. Fuel cannot be compressed while air can be. Consequently, if there is a bubble of air in the system, the air will compress and no fuel will be injected.

Nozzles Not Seated Properly (Diesel)

It is very important that the nozzles be correctly installed. They should seat squarely on the seating surface and the retaining clamp cap screws or nuts be tightened to the specified torque.

If the nozzle clamp is too loose, the nozzle will leak. This causes a buildup of carbon around the nozzle, making it very difficult to remove. In addition, the escape of burning gases will cause the nozzle to overheat, and the nozzle valve to stick, causing the cylinder to misfire.

Faulty ECU/ICU Wiring (Natural Gas)

Continued on next page

SS40167,0000132 -19-05AUG09-2/7

PN=433

The engine control unit (ECU) constantly monitors electrical signals from various sensors to determine the engine operating conditions. It makes decisions on the best air flow, fuel flow, and ignition timing based on the information from the sensors.

The ECU controls the "on" time (pulse width) of the fuel injectors to deliver the correct amount of fuel to the engine. The ECU determines the desired timing advance based on engine speed and piston position, then outputs a trigger signal to the ignition control unit (ICU) when an ignition coil should be fired.

Damaged wires or loose connections can interrupt the transfer of electrical signals between the engine sensors and the ECU, resulting in erratic control of fuel delivery and ignition timing.

Clogged Fuel Injectors (Natural Gas)

Clogged fuel injectors could cause a miss or rough-running engine.

If clogged injectors are suspected, follow the recommendations of the engine manufacturer to clean and test the injectors. Refer to the engine technical manual for details.

Replace the injectors if cleaning fails to correct the problem.

4. Engine Detonates (Gasoline)	Solution
Wrong Type of Fuel	Use fuel of correct octane.

Detonatation of Gasoline Engine

Detonation in an engine is usually recognized by sharp pinging in the engine cylinders and erratic firing. The engine appears to be working against itself.

In gasoline engines, detonation is usually caused by the use of fuel with too-low octane rating.

NOTE: The octane rating of a fuel is its ability to avoid detonation or self-ignition. The higher the octane rating, the more the mixture can be compressed without detonation when it burns.

Detonation is an explosion with sudden violence. When the compressed fuel mixture in the cylinder is ignited, instead of burning evenly across the combustion chamber, the entire charge literally explodes. This is partially due to the heat of compression. That is to say, when the mixture starts to burn and the pressure starts to rise, the entire charge ignites. This action is very harmful to the engine.

5. Engine Preignition (Spark-Ignition)	Solution
Distributor Timed Too Early	Time distributor.
Distributor Advance Mechanism Stuck	Free up and test.
Carbon Particles in Cylinder	Remove carbon buildup.
Faulty Spark Plugs or Spark Plug Heat Range Too High	Install proper plugs.

Preignition of Gasoline Engines

Preignition is the ignition of the fuel mixture before the proper time. Its effect on the engine is similar to that of detonation.

Preignition can be caused by incandescent particles of carbon in the cylinder that will ignite the mixture as the heat rises due to compression.

If the distributor advance mechanism sticks in the advanced position at certain engine speeds, the spark will occur too early.

Preignition can be caused by spark plugs that are eroded at the electrodes or by plugs with too high a heat range. In either case, the electrodes become red hot and the action is the same as that caused by incandescent carbon particles.

6. Engine Backfires (Spark-Ignition)	Solution
Spark Plug Cables Installed Wrong	Install in correct firing order.
Carburetor Mixture Too Lean	Adjust carburetor for correct mixture.

Backfiring of Spark-Ignition Engines

When the spark plug wires are incorrectly installed, the spark will not occur in the cylinder at the right time. This can cause backfiring.

Engine backfiring can also be caused by a too-lean carburetor mixture. A lean fuel mixture is a slow-burning mixture and may still be burning when the intake valve opens and admits a new charge.

7. Engine Knocks	Solution
Improper Distributor or Injection Pump Timing	Check and time correctly.
Worn Engine Bearings or Bushings	Replace.
Excessive Crankshaft Endplay	Adjust to specification.
Loose Bearing Caps	Tighten caps.
Foreign Matter in the Cylinder	Remove.

Knocking of Engines

When an engine knocks, there is one of two causes:

1. Mechanical Interference of Parts. The parts are striking each other due to loose parts, worn bearings, or a loose bolt or other object loose in the cylinder. The remedy is to repair the engine at once.

2. Improper Adjustment of Timing, Fueling, or Poor Fuels. If the ignition or injection timing is bad, the engine gets out of rhythm and knocks. Using the wrong fuels can also cause knocking, especially with low-octane gasoline. The remedy is either using proper fuels or readjusting the timing or fuel settings.

Continued on next page SS40167,0000132 -19-05AUG09-3/7

8. Engine Overheats	Solution
Defective Radiator Cap	Install new gasket or replace the cap.
Radiator Fins Bent or Plugged	Straighten fins or clean out dirt.
Defective Thermostat	Replace.
Loss of Coolant	Check for leaks and correct.
Loose Fan Belt	Adjust.
Cooling System Limed Up	Use good scale remover to clean, then flush.
Overloaded Engine	Adjust load.
Faulty Engine Timing	Time distributor or injection pump correctly.
Distributor Advance Mechanism Stuck (Spark-ignition)	Free up and lubricate.
Engine Low on Oil	Add oil to the proper level or change oil.
Wrong Type of Fuel	Use correct fuel.

Overheating of Engine

Radiator Cap:

The radiator cap must be in good condition so that it does not leak. The pressure cap allows pressure to build up in the cooling system, and this causes the boiling point of the coolant to rise.

Using the pressure cap thus prevents the coolant from boiling away in hot weather or during heavy loads.

Radiator:

Foreign matter between the fins of the radiator core or bent fins can cause overheating.

Thermostat:

If the thermostat develops a leak or sticks closed, it stops the circulation of coolant through the radiator core. This will cause overheating.

Engine Overloaded:

Overloading an engine often causes the peak firing pressure in the cylinders to be too high, creating more heat. The overload causes the engine speed to slow and the piston does not move in the cylinder as fast as it does under normal load. This can also contribute to overheating.

Faulty Engine Timing:

When the ignition is timed too late, a normal engine load becomes an overload. This is because the engine does not develop as much power. Even if the load is light, the engine may still run hot, due to the late firing.

Distributor Advance Mechanism Stuck (Spark-Lgnition):

If the distributor advance mechanism should stick in the retarded position, it has the same effect on the operation of the engine as late timing.

Engine Low on Oil:

One of the functions of the oil in the engine is to help carry off the heat created in the cylinders. If the oil level is too low, if the oil pressure is too low, or even if the oil is too heavy, not enough oil will be circulated to carry off the heat of combustion. This can lead to overheating.

Wrong Type of Fuel:

Using the wrong type of fuel in an engine can cause overheating. As the fuel detonates or does not burn efficiently, loss of engine power also results.

9. Engine Runs Too Cold (Diesel)	Solution
Defective Gauge or Sending Unit	Replace gauge or sending unit.
Defective Thermostat	Replace.
Defective Radiator Cap	Install new gasket or replace the cap.
Fan Speed Too High	Replace belts and/or pulleys.

Engine Runs Too Cold (Diesel)

Allowing an engine to run with low coolant temperature could result in:

• Carbon buildup throughout engine
• Unburned fuel washes cylinder wall
• Gummy deposit left on pistons, rings, and valves
• Piston rings fail to seat properly during break-in period
• Premature bearing failure
• Sludge accumulation throughout engine exhaust system

Continued on next page

SS40167,0000132 -19-05AUG09-4/7

10. Lack of Power	Solution
Air Cleaner Dirty or Otherwise Obstructed	Clean and check main element for restriction.
Restricted Air Flow in Intake System	Clean intake hoses and tubes.
Restriction in Fuel Lines or Filters	Disconnect line at carburetor or injection pump and check it. Remove any restriction. Also check filters.
Wrong Type of Fuel	Use correct type.
Frost at Fuel-Lock Strainer (LP-Gas)	Clean strainer.
Governor Binds (Spark-Ignition)	Free up and lubricate. Adjust if necessary.
Valve Failure	Recondition.
Incorrect Valve Tappet Clearance	Adjust.
Low Engine Speed	Adjust.
Crankcase Oil Too Heavy	Use correct weight of oil.
Low Compression	Check for leaky valves or worn or stuck piston rings. Replace or recondition.
Improper Hitching or Belting of Machine	Adjust or correct as required.
Incorrect Timing	Check and time correctly.
Faulty Carburetor, Float Level Too Low, Plugged Orifices, or Wrong Fuel Adjustment	Clean and repair.
Wrong Spark Plugs (Spark-Ignition)	Replace with proper plugs.
Distributor Points Burned (Spark-Ignition)	Replace and adjust.
Incorrect Camshaft Timing	Retime.
Low Operating Temperature	Check thermostat.
Faulty Injection Pump Delivery (Diesel)	Service pump.
Turbocharger (If Used) Not Functioning Correctly	Check turbocharger for physical damage. Repair or replace.
Regulated Gas Supply Pressure Too Low (Natural Gas)	Adjust, repair, or replace pressure regulator.
Faulty Throttle Pedal Position Sensor or Sensor Wiring (Natural Gas)	Check wiring for loose connections or damaged wires.
Faulty Engine Control Unit (ECU) or Wiring	Check wiring for loose connections or damaged wires. Replace ECU if necessary.

Lack of Power

Air Cleaner:

If the air cleaner is dirty or the air passages obstructed, the engine will not receive enough air to develop its horsepower.

Fuel Lines:

When the flow of fuel through the fuel lines is restricted, the engine cannot get enough fuel to develop its power.

Frost at Fuel-Lock Strainer (LP-Gas):

Frost at the fuel-lock strainer restricts the flow of gas, resulting in a power loss.

Governor Binds (Spark-Ignition):

When there is binding in the governor, the action will be slow. Engine speed is pulled down before the governor can react to increase the fuel supply to carry the load.

Wrong Type of Fuel:

See Troubleshooting Chart No. 4 in this chapter for details.

Low Compression:

Leaky valves or blowby at the pistons results in low compression. As a result, the burning of fuel in the cylinders cannot produce enough pressure to develop full power.

Valve Clearance:

Valve adjustment can affect the compression pressure. If the adjustment is too tight, the valve cannot entirely close, allowing leakage.

If the valve clearance is too great, the valve opens late and closes early. This results in it not being open long enough for complete intake of air or air-fuel mixture or the complete expulsion of exhaust.

Too-Heavy Crankcase Oil:

Many times, when an engine is consuming too much oil, the operator will change to a heavier weight. This can affect the power output of the engine due to the increased drag on the moving parts.

Improper Hitching or Belting:

Improper hitching or adjustment of equipment behind a tractor or other machine can result in an apparent loss of engine power. Actually the improper hitching really results in increasing the load on the machine.

Raising or lowering the hitch point can change this load.

For example, by correctly hitching a tool to the drawbar, a tractor may operate in a higher transmission gear range.

Improper belting can also reduce the ability of the engine to pull its load.

If the drive pulley is too large, the engine can easily handle the load, but the machine speed will be too slow. If the drive pulley is too small, the machine speed will be too fast and more engine power will be required to work the pulley.

Turbocharger not Functioning Correctly:

The main function of the turbocharger is, by compressing the air, to force more air into the engine cylinders. This allows the engine to efficiently burn more fuel, thereby producing more horsepower. Failure of the turbocharger will cause a reduction in the amount of power that the engine can produce.

Continued on next page SS40167,0000132 -19-05AUG09-5/7

On some engines, the amount of boost pressure that the turbocharger is allowed to generate is controlled by a wastegate mechanism. When the wastegate is closed, the turbocharger generates maximum boost. When the wastegate is opened, the engine exhaust gases are allowed to bypass the turbocharger turbine, resulting in a reduction in turbocharger boost pressure. If the wastegate becomes stuck in the open position, the engine will not receive enough air to develop its horsepower.

Regulated Gas Supply Pressure Too Low or Too High (Natural Gas):

A pressure regulator reduces the natural gas tank pressure to the working pressure required by the specific engine. If the pressure is too low or too high, it may cause the air-fuel mixture to be too lean or too rich for proper combustion. Loss of engine power will result.

Faulty Throttle Foot Pedal Position Sensor or Sensor Wiring (Natural Gas):

The throttle on some natural gas fuel systems is a drive-by-wire, electronically-controlled throttle/actuator. The engine control unit (ECU) monitors electrical signals from the foot pedal position sensor to determine the operator's desired engine load. The ECU then provides a signal to command the throttle opening based on the foot pedal position input and other sensor inputs.

Power output of a spark-ignited natural gas engine is controlled by varying the air flow to the engine. More air flowresults in more power. A faulty foot pedal position sensor or wiring may cause the ECU to command an incorrect throttle opening, resulting in loss of engine power.

Faulty Engine Control Unit (ECU) or Wiring:

The engine control unit (ECU) is the brains of the electronically-controlled fuel system used on some engines. The ECU monitors electrical signals from various sensors, then determines the best airflow, fuel flow, and ignition or fuel injection timing based on information from the various sensors.

A faulty ECU, loose wiring connections, or faulty sensors can result in poor engine performance. The ECU is designed to be trouble-free, and the cause of the problem will usually be found elsewhere, such as a faulty sensor or damaged wires.

Refer to Chapters 6, 7, and 11 in this manual and to the FOS Manual Electronic and Electrical Systems for additional information.

11. Engine Uses Too Much Oil	Solution
Crankcase Oil Too Light	Use correct weight of oil.
Worn Pistons and Rings	Recondition.
Worn Valve Guides or Stem Oil Seals	Replace.
Loose Connecting Rod Bearings	Replace bearings.
External Oil Leaks	Eliminate leaks.
Internal Oil Leaks	Locate and correct.
Oil Pressure Too High	Adjust to specifications.
Engine Speed Too High	Adjust to specifications.
Crankcase Ventilating Pump Not Working	Repair or replace.
Restricted Air Intake or Breather	Remove the restriction.
Restricted Oil Return Passage from Valve Cover	Remove the restriction.

Engine Uses Too Much Oil

Pistons and Rings:

Worn pistons and rings can cause an engine to use too much oil. Reconditioning is normally required.

However, many other conditions can cause excessive oil consumption, as given below.

Valve Guides:

Worn valve guides can be a cause of oil consumption. Every time the piston goes down in the cylinder it creates a partial vacuum. To equalize this vacuum, air tries to enter the cylinder. If the valve guides are worn, this leaves an opening for the air to enter. As it enters, it carries oil from around the valve stem with it.

Valve Stem Oil Seals:

Some engines are equipped with valve stem oil seals, usually located on the intake valves. These seals prevent loss of oil at this point. If worn, these seals can make the engine use oil.

Loose Connecting Rod Bearings:

When the bearings are worn, more oil is pumped through them. As a result, more oil than needed is thrown into the cylinder from the crankshaft. The extra oil escapes past the piston and rings.

Much the same action occurs when the engine oil pressure is too high.

Leakage:

External leaks can usually be seen and corrected. These include leaks at the crankshaft rear bearing, crankshaft front oil seal, valve cover gasket, and cylinder head gasket.

Internal leaks are not so easily detected. Generally, the engine must be disassembled while carefully checking points at which leakage could occur.

Continued on next page   SS40167,0000132 -19-05AUG09-6/7

When an internal leak is causing oil consumption, it will usually be found that the leak is located so that the escaping oil strikes the crankshaft. This results in too much oil being thrown into the cylinders.

Crankcase Ventilating Pump (If Equipped):

Most ventilating pumps draw in air, mix it with oil, and force the mixture into the engine crankcase, pushing out vapors.

When the pump fails, the oil is drawn into the air intake and then on into the engine.

To find if the ventilating pump is at fault, remove the air intake pipe. If the pump is defective, the inside of the pipe will be oily.

Another method of checking the pump is to remove the pipe between the pump and the air intake pipe. Be sure to cap both openings.

Then fill the engine crankcase with a measured amount of oil and operate the engine under load for four or five hours. Drain the oil and measure it. If too much oil is lost, the engine is using oil and the vent pump is not the cause.

Restricted Air Intake System:

When the air flow into the engine is restricted, there is more vacuum or suction in the cylinders. This results in increased suction around the valve stems and pistons and rings, and can cause excessive oil consumption.

Restricted Oil Return Passage from Valve Cover:

When this passage is restricted, the oil collects in the valve tappet area. Much of this oil enters the engine around the valve stems.

12. Oil Pressure Too High	Solution
Stuck Relief Valve	Free up valve.

Oil Pressure Too High

This condition rarely occurs unless it has been deliberately set too high.

However, if the relief valve sticks closed, the oil pressure will be too high.

Other possible causes could be a defective pressure gauge or sending unit.

13. Oil Pressure Too Low	Solution
Too-Light Oil	Use correct weight of oil.
Engine Low on Oil	Fill to proper level.
Worn Bearings	Replace engine bearings.
Poor Relief Valve Seating	Replace valve, valve seat or both.
Worn Oil Pump	Repair.
Loose Connections or Leaking Seals at Oil Filter, Pump, or Oil Cooler	Tighten connections or replace seals.

Oil Pressure Too Low

Worn bearings, poor relief valve action, loose connections, or leaking seals can all contribute to low oil pressure.

Any of these items can allow the escape of so much oil that the pump cannot keep up the pressure.

Worn Oil Pump:

When the oil pump becomes worn, the oil tends to recirculate within the pump, preventing the normal pressure buildup.

Too-Light Oil:

Oil is actually composed of many tiny globules. The heavier the oil, the larger the globules.

Therefore, a lighter oil with smaller globules can flow more freely through the bearings and other openings.

While lighter oil provides better lubrication, using too light an oil can contribute to high oil consumption.

However, do not attempt to eliminate excessive oil consumption by using a heavier oil than recommended.

SS40167,0000132 -19-05AUG09-7/7

The Future: Self-Diagnosis of Engine Systems

The future has arrived on most new vehicles with sophisticated computer controlled engine systems. Most vehicles now come with electronic systems that monitor and diagnose engine problems (see Chapters 6 and 7).

If a fault occurs, the control system can record and store the problem where is can be retrieved later on as a

diagnostic code. If the operator cannot determine the problem from the machine operator's manual, the service dealer or associated customer support can be contacted to help diagnose the fault code over the telephone or by way of the Internet.

SS40167,0000133 -19-17JUL09-1/1

Test Yourself

Questions

1. List the seven basic steps for good troubleshooting.

2. During which of the seven steps should you begin replacing parts?

3. What test is the most comprehensive way to tell the condition of the engine?

4. What three basic things are needed for the engine to run?

SS40167,0000134 -19-02JUL09-1/1

Engine Tune-Up

Engine Tune-Up — Introduction

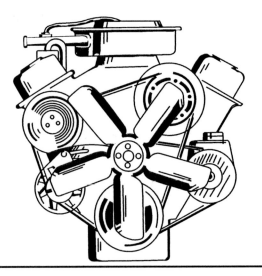

Making Checks and Minor Adjustments

Visual Inspection

What Is Tune-Up?

Tune-up is the process of making checks and minor adjustments to improve the operation of the engine.

Tune-up is also preventive maintenance. Troubles can be caught early and prevented by checking out the engine before it actually fails.

When Should an Engine Be Tuned?

The intervals for tune-up may vary from 500 to 1000 hours or seasonally, depending upon the operating conditions.

But regularity is the key to maintaining the engine so that major problems are prevented.

A badly worn engine cannot be tuned up. This is why the engine must first be tested to see if:

1. A tune-up will restore it, or
2. Major overhaul is needed.

Let's go through the visual inspection first and then we'll follow with the tune-up.

DP79986,0000050 -19-17JUL09-1/1

Visual Inspection

By inspecting the engine before you tune it, you can learn a lot about its general condition.

Check out the following items:

1. Oil And Water Leakage

Inspect the engine for any oil or water leaks. If the engine has been using too much oil, this often means an external oil leak. If the engine overheats, look for leaks in the cooling system.

2. Electrical System

Inspect the battery for corrosion, cracked case, or leaks at the cell covers.

Remove the cell caps and examine the tops of the battery plates. if they are covered with a chalky deposit, this means one of three things:

• Electrolyte level has been too low.
• Battery charge has been too low, causing sulfation.
• Battery was charged at too high a rate, boiling out water.

Any of these conditions can reduce the life of the battery. If they have gone too far, the battery must be replaced.

Check the battery cables and connections for damage and looseness.

Make sure the battery cables are the right size. Many complaints of poor starting can be traced to cables that are too small.

To check for this, operate the starter with the engine cold. If the battery cable becomes hot, the cable is most likely too small.

Inspect the wiring harnesses. If they are too oil-soaked, frayed, or corroded, replace them.

On spark-ignition engines, check the distributor or coils for a cracked cap, excessive grease, or other damage.

Check the operation of the alternator or generator indicator light. It should light when the starter switch is turned on.

Failure to light can be due to a burned-out bulb, an incomplete circuit, or the alternator or generator is not producing current. (Lack of current to the battery will show up a discharged battery.)

If the oil pressure indicator light does not go out when the engine is running, check for low or no oil pressure, or a short circuit.

Stop the engine at once and find the cause.

Lack of engine oil pressure can severely damage expensive parts inside the engine due to lack of lubrication.

3. Cooling System

Wait until the engine has been off for several hours and the crankcase oil is cold. Then loosen the crankcase drain plug and carefully turn it out to see if any water seeps out. If water is present, locate the cause of the cooling system leak.

Inspect the cooling system for leaks, deteriorated hoses, bent or clogged radiator fins, slipping fan belt, or any other condition that could result in improper cooling.

4. Air Intake System

Inspect the air intake system for possible leaks or restrictions. If the proper amount of clean air does not reach the engine, performance and durability will be affected.

5. Fuel System

Check the fuel system for leaks and for bent or dented lines, that might cause a restriction.

Check the fuel transfer pump sediment bowl. On diesel engines, inspect the fuel filters for dirt, water, or other foreign matter.

6. Pressure Washing

After checking for leaks, pressure wash the engine. This not only helps to recondition the engine, it makes tune-up easier and problems easier to spot.

JB06590,0000485 -19-17JUL09-1/1

Visual Inspection Checklist

✓	Visual Inspection Checklist	Comments
	Oil and Water Leakage	
	Electrical System	
	Battery	
	Cables	
	Wiring	
	Indicator Lights	
	Cooling System	
	Water in Crankcase	
	Hoses	
	Slipping Fan Belt	
	Clogged Radiator Fins	
	Air Intake System	
	Air Leaks	
	Restrictions	
	Fuel System	
	Leaks	
	Restrictions	
	Clogged Filter	
	Pressure Washing	

JB06590,0000486 -19-01JUL09-1/1

Dynamometer Test

If possible, test the engine on a dynamometer before it is tuned. This test gives you the horsepower output and fuel consumption of the engine as it is. This will help you to determine if a tune-up can restore the engine or whether an overhaul is needed.

Good performance by the engine depends on these basic things:

- **An adequate supply of clean air and fuel**
- **Good compression**
- **Proper valve and ignition timing for good combustion**

Failure or low performance on any of these factors makes it impossible to fully restore the engine.

Therefore, if the dynamometer test shows that any of these three factors is bad, the engine will have to be reconditioned.

Make the dynamometer test as follows:

1. Observe all safety rules for dynamometer testing outlined in Chapter 2 of this manual as well as those of the dynamometer manufacturer.

2. Check the engine oil and coolant levels before beginning the test. Make sure there is an adequate supply of fuel to the engine during the test.

3. Connect the engine to the dynamometer using the manufacturer's instructions (see illustration).

IMPORTANT: Perform the test in a well-ventilated area, outdoors if possible. If it is necessary to run an engine in an enclosed area, remove

PTO Dynamometer Test

the exhaust fumes from the area with an exhaust pipe extension.

4. Run the engine until the coolant and crankcase oil temperatures are at normal operating conditions. (This will take several minutes — but is very important to a good test.)

5. Operate the engine at full speed and check the fast idle (no load) speed against the specifications given in the engine technical manual. If fast idle speed deviates from the specifications, adjust or repair the speed control linkage until fast idle speed is within specifications. Engine horsepower will be affected if fast idle speed deviates significantly.

6. Gradually increase the load on the engine until its speed is reduced to the rated load speed as given in the engine technical manual.

Continued on next page   JB06590,0000487 -19-22JUL09-1/3

7. Most dynamometers have a direct-reading analog gauge or digital display that is calibrated to indicate the horsepower. Follow the instructions provided by the dynamometer manufacture to read the horsepower. Compare the measured horsepower with that given in the engine technical manual.

Do not expect engines to always equal these specifications. However, if the engine rates much lower than normal, this is a signal that service is needed.

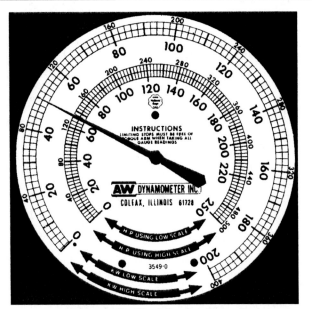

Dynamometer Direct-Reading Analog Gauge

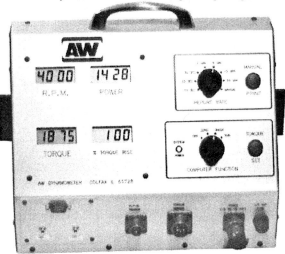

Dynamometer Direct-Reading Digital Display

Continued on next page

JB06590,0000487 -19-22JUL09-2/3

8. The dynamometer can also measure the engine's torque output and torque rise or lugging ability. The following procedure is a general guide for measuring torque.

Set the dynamometer to record torque according to the instructions provided by the dynamometer manufacturer.

Operate the engine at full speed. Gradually increase the load on the engine until its speed is reduced to rated load speed as given in the engine technical manual. Record the torque reading indicated on the dynamometer gauge. Depending on the dynamometer, it may be necessary to convert the horsepower reading of the dynamometer gauge to a torque value. Continue to increase the load until peak torque rpm is achieved.

NOTE: Peak torque can be observed on the dynamometer gauge when the torque reading stops increasing or if the torque reading starts to fall below maximum torque observed. Most engine manufacturers give a specific rpm at which to check peak torque.

To determine the engine's torque rise, subtract the rated torque reading from the peak torque reading to determine the increase in torque. Example: Peak torque of 525 lb-ft (710 N·m) minus 425 lb-ft (575 N·m) rated torque equals 100 lb-ft (135 N·m) torque increase.

Divide the increase in torque by the recorded rated torque reading to calculate the percentage of torque rise. Example: 100 lb-ft (135 N·m) divided by 425 lb-ft (575 N·m) equals 24% torque rise. A high torque rise percentage indicates that the engine has good lugging ability.

While the engine is operating under load, note the outlet of the crankcase ventilating system. If too much vapor appears, also remove the crankcase oil filler cap.

If an excessive amount of vapor or smoke appears here as well as at the vent, there is blowby in the

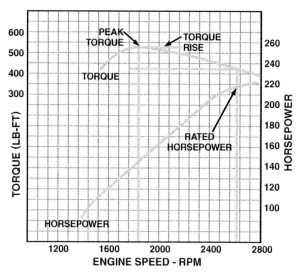

Rated Horsepower and Torque Chart

engine cylinders and they must be reconditioned before the engine will perform at its best.

NOTE: Instruments are available to measure the flow of air and gases through the crankcase ventilating system.

The normal rate of engine vapor flow is specified in the engine technical manual.

Any increase in vapor flow over the specified amount indicates crankcase blowby.

If the blowby is excessive, recondition the engine for good operation.

Even though the engine develops its rated horsepower using a normal amount of fuel, a tune-up may still improve its efficiency. Consider both hours of operation and the conditions under which the engine has been operated. It is far more economical in the long run to tune the engine before a lack of performance makes it mandatory.

JB06590,0000487 -19-22JUL09-3/3

Engine Tune-Up

Most manufacturers suggest a regular period of operation between tune-ups.

For proper engine tune-up, use the sequence listed in the tune-up chart.

A detailed description of each operation follows the tune-up chart.

JB06590,0000488 -19-17JUL09-1/1

Tune-Up Chart

NOTE: The numbers in the Item No. column refer to numbered groups in the detailed instructions.

For best results and time savings, follow the sequence given in the tune-up chart.

Item No.	Operation	Gasoline	Diesel	LP-Gas, Nat. Gas
1	**Air Intake and Exhaust System**	•	•	•
	Clean out pre-cleaner (if used)	•	•	•
	Remove and clean air cleaner	•	•	•
	Swab out inlet pipe in air cleaner body	•	•	•
	Inspect exhaust system and muffler	•	•	•
	Check crankcase ventilating system for restrictions	•	•	•
2	**Basic Engine**			
	Recheck air intake for restrictions	•	•	•
	Check radiator for air bubbles or oil indicating compression or oil leaks	•	•	•
	Cylinder head gasket leakage	•	•	•
	Retighten cylinder head cap screws if recommended	•	•	•
	Check compression pressure in each cylinder	•	•	•
3	**Ignition System (Spark-Ignition Engines)**			
	Spark Plugs			
	Clean, test, and adjust spark plugs	•		•
	Ignition Coil			
	Test ignition coil	•		•
	Check for proper coil connections	•		•
	Distributor (If Used)			
	Cap and rotor	•		•
	Condenser	•		•
	Breaker points	•		•
	Point gap (cam dwell)	•		•
	Breaker point spring tension	•		•
	Lubrication	•		•
	Distributor timing	•		•
4	**Fuel Systems**			
	Check fuel lines for leaks or restrictions	•	•	•
	Clean fuel pump sediment bowl	•	•	
	Test fuel pump pressure	•	•	
	Check and clean LP-gas fuel-lock strainer (observe caution)			•
	Replace fuel filter (natural gas)			•
	Drain carburetor sump			•
	Check radiator for LP-gas leaking from converter into cooling system			•
	Drain gasoline carburetor, clean inlet strainer	•		
	Check carburetor choke disk operation	•		•
	Check speed control linkage	•	•	•
	Service diesel fuel filters		•	
	Check diesel injection pump		•	
	Check and clean diesel injection nozzles		•	
	Bleed diesel fuel system		•	
	Check diesel injection pump timing		•	
5	**Lubrication System**			
	Check operation of pressure gauge or light	•	•	•
	Service oil filter	•	•	•
	Check condition of crankcase oil	•	•	•
	Check engine oil pressure	•	•	•

Continued on next page   JB06590,0000489 -19-20JUL09-1/2

Item No.	Operation	Gasoline	Diesel	LP-Gas, Nat. Gas
6	**Cooling System**			
	Check water pump for leaks and excessive shaft endplay	•	•	•
	Inspect radiator hoses	•	•	•
	Clean and flush cooling system	•	•	•
	Test thermostat and pressure cap	•	•	•
	Test radiator for leaks	•	•	•
	Check condition of fan belt	•	•	•
7	**Electrical System**			
	Battery			
	Clean battery, cables, terminals, and battery box	•	•	•
	Tighten battery cables and battery hold-down	•	•	•
	Apply a terminal protectant to battery posts and cable clamps	•	•	•
	Check specific gravity of electrolyte and add water to proper level	•	•	•
	Make light load test of battery condition	•	•	•
	Generator or Alternator			
	Check belt tension	•	•	•
	Test alternator or generator output	•	•	•
	Starting Circuit			
	Check safety starter switch	•	•	•
	Check current draw of starting motor	•	•	•
8	**Clutch Free Travel**			
	Check free travel at clutch pedal or lever	•	•	•
9	**Dynamometer Test**			
	Use a dynamometer to check engine performance	•	•	•
	Use an exhaust analyzer while engine is on test to check for accurate carburetor adjustment	•		•

JB06590,0000489 -19-20JUL09-2/2

Detailed Tune-Up Procedures

The following pages have detailed instructions for the checks and adjustments given in the tune-up chart.

The tune-up chart is divided into groups and each group is numbered. This number matches the number given in the detailed procedures which follow.

Where specifications are required, always refer to the engine technical manual for details.

1. Air Intake and Exhaust System

Pre-Cleaner

Check the pre-cleaner collector bowl for collected foreign matter. Clean the bowl whenever 3/4-inch (19 mm) of dirt collects in the bowl.

Check the pre-cleaner screen and clean as required.

A—Collector Bowl
B—Pre-Cleaner

C—Screen

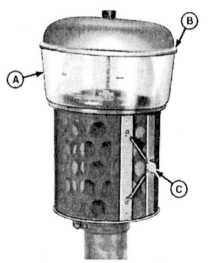

Pre-Cleaner and Screen

Continued on next page

JB06590,000048A -19-05AUG09-1/32

Air Cleaner (Dry Type)

Remove and check the dry air cleaner element. If the element is ruptured or the seal is damaged, replace it.

Most dry elements can be cleaned a few times before they must be replaced. Two cleaning methods are used: One is for dusty elements. The other is for oily or sooty elements. Refer to Chapter 8 in this manual for recommended cleaning methods. See the machine operator's manual for additional details.

IMPORTANT: **Never wash dry elements in gasoline, fuel oil, or solvents. Do not oil the element unless instructed.**

Check the operation of the restriction indicator and dust unloading valve (if used). Flex the valve to clean out debris. Also wipe out the inside of the air cleaner housing. If the cleaner has a dust cup, clean it.

Install the cleaner element (usually fins first as shown in illustration) and tighten it securely. This will prevent air leaks past the end of the element. Then install the cover on the air cleaner housing.

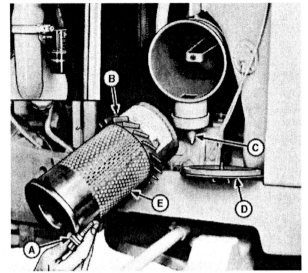

Air Cleaner (Dry Type)

A—Wing Nut
B—Fins
C—Dust Unloading Valve
D—Cover
E—Element

Continued on next page

JB06590,000048A -19-05AUG09-2/32

Air Cleaner (Oil Bath Type)

Remove the oil cup from the air cleaner. Pour the oil from the cup and clean the cup.

Clean the tray (if used) or oil trap. Swab out the air intake tube.

If the main cleaner element is extremely dirty, remove it and clean by dashing it up and down in a pail of solvent. Do not use pressure washing equipment.

Install the tray or oil trap and other parts removed. To ensure clean air, check to see that all connections are tight and leakproof.

Fill the oil cup to proper level with the correct amount of new oil of the same weight as is used in the engine. Do not overfill the cup. Install the oil cup and tighten it securely.

IMPORTANT: Never check the air cleaner with the engine running.

Exhaust System

Check the exhaust system for leaks. Leaks at the engine can be dangerous to the operator when the engine is used with an enclosed machine.

Check the exhaust pipe and muffler for restrictions which could prevent the free flow of exhaust gases.

Crankcase Ventilating System

Check for restrictions which could prevent free air flow through the crankcase ventilating system. Lack of ventilation causes sludge to form in the engine crankcase. This can cause clogging of oil passages, filters, and screens, resulting in expensive damage to the engine.

2. Basic Engine

Air Intake System

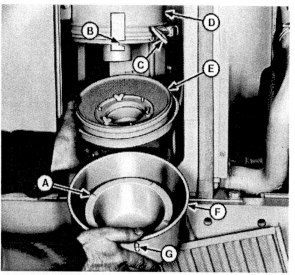

Air Cleaner (Oil Bath Type)

A—Oil Level	E—Tray (if used)
B—Slot	F—Oil Cup
C—Clamp	G—Pin
D—Air Cleaner Body	

A further check of the air intake system should now be made. Do this with a water manometer or vacuum gauge calibrated in inches (mm) of water.

NOTE: If the diesel engine, on the dynamometer test, produced a reasonable amount of horsepower and the air cleaner appears to have been regularly serviced, this operation can be omitted. However, on diesel engines that have air cleaners equipped with restriction indicators, check the operation of the indicator against the gauge.

Continued on next page JB06590,000048A -19-05AUG09-3/32

Intake Vacuum Test (Diesel Engines)

Test as follows:

1. Warm up the engine.

2. On engines with restriction indicators, remove the indicator, install a pipe tee fitting, and reinstall the indicator. Connect the vacuum gauge to the tee fitting.

3. On naturally aspirated engines without restriction indicators, connect the gauge to the intake manifold (see illustration).

4. Set the diesel engine speed at fast idle and note the reading on the gauge. Too high a reading means that there is a restriction in the air intake system. Check the engine technical manual for correct specifications.

5. On engines with restriction indicators, check the operation of the indicator. Use a board or metal plate to very slowly cover the air intake opening. Note the action of the indicator in relation to the reading on the gauge. If the indicator does not operate properly, replace it.

Manometer or Gauge Connection Port

A—Intake Manifold   **B—Gauge Connection Port**

Continued on next page   JB06590,000048A -19-05AUG09-4/32

Manifold Depression Test (Spark-Ignition Engines)

Much can be learned about the internal condition of a spark ignition engine by checking the intake manifold depression between the carburetor and the engine. Use a vacuum gauge calibrated in inches of mercury.

1. Connect the vacuum gauge (C) to the intake manifold.

2. Warm up the engine and operate it at idle speed.

3. Note the reading on the vacuum gauge. Check the engine technical manual for exact specifications.

4. Interpret the gauge reading as follows:

   - If the reading is steady and low, loss of power in all cylinders is indicated. Possible causes are late ignition, incorrect valve timing, or loss of compression at valves or piston rings. A leaky carburetor gasket will also cause a low reading.
   - If the needle fluctuates steadily, a partial or complete loss of power in one or more cylinders is indicated. This can be due to an ignition defect, loss of compression due to stuck piston rings, or a leaky cylinder head gasket.
   - Intermittent needle fluctuation indicates occasional loss of power due to an ignition defect or a sticking valve.
   - Slow needle fluctuation is usually caused by improper carburetor idle mixture adjustment.
   - A gradual drop in the gauge reading at idle engine speed indicates back pressure in the exhaust system due to a restriction.

Since the gauge readings can indicate more than one fault, be very careful in analyzing abnormal readings.

Cylinder Head Gasket Leaks

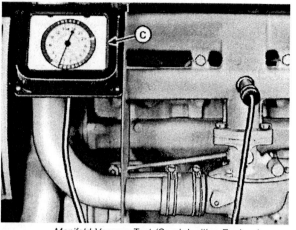

Manifold Vacuum Test (Spark-Ignition Engines)

C—Vacuum Gauge

To check for cylinder head gasket leaks, first remove the radiator cap while the engine is cold. Start the engine and run at rated speed until it warms to normal operating temperature. Check for bubbles or oil at the surface of the coolant.

If bubbles appear, cylinder pressure is leaking past the cylinder head gasket (or LP-gas or natural gas is leaking from the converter or vaporizer into the engine coolant). Replace the head gasket or repair the converter or vaporizer.

If oil is present, it can be leaking past the cylinder head gasket or leaking from an engine oil cooler. Determine the cause and correct it.

JB06590,000048A -19-05AUG09-5/32

Cylinder Head Cap Screws

Some engine manufacturers may recommend that the cylinder head cap screws be retightened to specified torque at the time of an engine tune-up. However, retightening the cylinder head cap screws is not recommended for most modern engines. Consult the engine technical manual for the specific engine.

Operate the engine until it is at normal operating temperature. While the engine is still hot, loosen each cylinder head cap screw and retighten to specified torque in the proper sequence given in the engine specifications.

One sequence for tightening cylinder head cap screws is to work from the center outward on alternating sides of the head in a circle as shown in the illustration.

A—Start at the Center

B—Tighten toward Each End, Alternating from Side to Side in a Circle as Shown

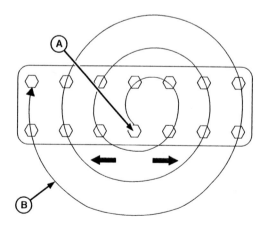

Cylinder Head Cap Screw Tightening Sequence

Continued on next page   JB06590,000048A -19-05AUG09-6/32

The illustration shows a tightening sequence for newer engines. However, use the exact sequence given in the engine specifications. Refer to Chapter 3 of this manual for additional information.

NOTE: A film of oil on the threads and under the head of each cap screw will ensure better tightening.

Importance of Using a Torque Wrench

Many inexperienced service technicians do not realize the importance of using a torque wrench to tighten cap screws or nuts. This is particularly true on cylinder head cap screws or nuts.

Unevenly tightened cylinder head cap screws can distort the valve seats (cause leakage) or the cylinder head or block surfaces (which could result in cylinder head gasket failure).

DXP00722 —UN—01JUN09

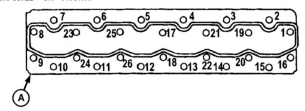

Sequence for Tightening Cylinder Head Cap Screws (Newer Engines)

A—Front of Engine

Tightening the cap screws or studs beyond their specified torque may cause them to break or fail to maintain their torque setting.

Continued on next page   JB06590,000048A -19-05AUG09-7/32

Valve Tappet Clearance

Check the clearance on the engine valves. This is the distance between the end of the tappet lever and the end of the valve stem when the valve is closed (piston is at top dead center of compression stroke).

One sequence for adjusting the valve tappet clearance is as follows:

NOTE: The No.1 cylinder on an in-line engine is normally located at the front (end opposite the flywheel) of the engine. The No.1 cylinder on a V-block engine is also normally at the front of the engine, but could be located either on the right or left cylinder bank. Usually, numbers corresponding to the cylinders are cast into the intake manifold to identify the cylinder arrangement on V-block engines. See the engine technical manual.

Turn the crankshaft in its normal running direction (counterclockwise as viewed from rear) until the No. 1 front piston is at top dead center (TDC) on compression stroke.

Many engines have a timing mark on the front pulley or a timing pin on the flywheel to set the engine at TDC. The engine is at No. 1 TDC-compression if the rocker arms for No. 1 cylinder are loose (valves closed).

A positive way to find when a piston is at TDC-compression is to remove the spark plug or injection nozzle and hold your finger over the opening. On the compression stroke, air will be forced out against your finger until the piston reaches the TDC position.

Adjust the clearance on the two valves for the No. 1 cylinder to specifications. Use a feeler gauge as shown in the illustration to check the setting.

NOTE: Be sure to determine which are intake and which are exhaust valves because the clearances are usually different for the two. To identify the valves, observe the position of the valves in relation to the intake and exhaust ports in the cylinder head.

Rotate the engine crankshaft in normal running direction until the next piston in the firing order reaches TDC of its compression stroke. Check and adjust the valve clearances for this cylinder. Repeat this procedure for the remainder of the cylinders.

NOTE: The firing order will vary, depending on the design of the engine and the number of cylinders. Consult the engine technical manual for specifications.

THE IMPORTANCE OF ACCURATE VALVE CLEARANCE:

- The importance of accurate valve adjustment cannot be too highly stressed.
- If there is not enough valve clearance, the valve may not entirely close, resulting in loss of compression or burned exhaust valves.

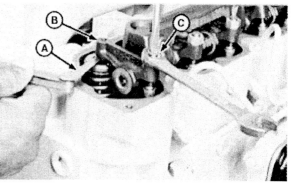

Adjusting Valve Tappet Clearance

A—Feeler Gauge   C—Adjusting Screw
B—Rocker Arm

- If the valve clearance is too great, the valve opens late and closes early. Thus the valve does not remain open long enough to allow the full charge of air or fuel-air mixture to enter the cylinder. This results in loss of engine efficiency.

Compression Test

Weak compression results in loss of horsepower and poor engine performance. This is even more true in diesel engines, which depend upon the heat of compression to ignite the fuel in the cylinders.

Check the engine compression as follows:

IMPORTANT: Compression pressures are affected by the cranking speed of the engine. Before beginning the test, ensure that the batteries are fully charged.

1. Start the engine and run until it warms up to normal operating temperature.

2. Remove the spark plugs or injection nozzles.

3. On spark-ignition engines, be sure that both the carburetor and choke valves are in the wide-open position. On conventional ignition systems, disconnect the ignition coil-to-distributor wire at the distributor. Ground the wire by placing it in contact with some part of the engine. On electronic ignition systems, disable the ignition by disconnecting the electrical connector between the engine control unit (ECU) and the ignition control unit (ICU). Refer to the engine technical manual for specific instructions.

4. On natural gas engines, close the fuel shutoff valve and disconnect the electrical connector to the low-pressure lock-off solenoid.

5. On diesel engines, the air intake system has no throttle valve, so merely be sure that the engine speed control is in the STOP position.

Continued on next page

JB06590,000048A -19-05AUG09-8/32

6. Install the compression tester into the spark plug hole or injection nozzle bore (see illustration).

NOTE: The cranking speed specified in the engine technical manual (usually 275–-325 rpm) must be maintained during the test to get a valid pressure reading.

7. Crank the engine with the starter until the compression gauge registers no additional increase in pressure. It is a good practice to count the number of compression strokes (indicated by movement of the gauge needle) and check each cylinder with the same number of strokes. The engine must be at full cranking speed to get a good reading.

8. Check the pressure reading against the engine technical manual.

9. If the compression is very low, apply oil to the ring area of the piston through the spark plug or injector openings. Do not use too much oil to avoid getting oil on the valves. Then check the compression again.

10. Judge the compression readings as follows.

JUDGING THE ENGINE COMPRESSION READINGS:

All cylinder pressures should be approximately alike. There should be less than 50 psi (340 kPa) difference between cylinder pressures.

If compression is too low, it is possible that the valves are worn or sticking, the cylinder head gasket is leaking, or the piston and its rings are worn or damaged.

If the pressure readings increased significantly after engine oil was applied to the piston and rings, worn or stuck piston rings are indicated.

If compression is higher, carbon buildup could be indicated.

Altitude affects compression pressures. Normally there is about a 4% loss for every 1000 ft (300 m) of altitude above sea level. Specified pressures are usually given for an altitude of 600–1000 ft (180–300 m) above sea level.

Compression pressures are affected by the cranking speed of the engine. Therefore, be sure the batteries

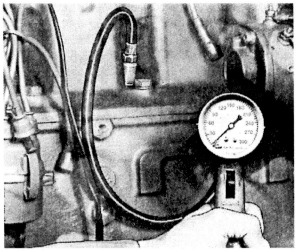

Checking Engine Compression

are in good condition and fully charged when making a compression test.

How Compression Affects Engine Performance

The expansion of gases in the cylinder caused by the burning of a mixture of fuel and air creates horsepower in an engine.

For example, in a spark-ignition engine, peak firing pressures are roughly four times higher than compression pressures. In an engine with 160 psi (1100 kPa) compression pressure, peak firing pressure will be approximately 640 psi (4400 kPa).

If the pressure is only 150 psi (1000 kPa) in each cylinder when checked, there is approximately a 9% loss in horsepower. This, of course, is a loss of engine efficiency and results in higher operating costs.

In diesel engines, peak firing pressures do not rise as high, comparatively, as in spark-ignition engines. They are about double that of the compression pressure.

This is because diesel engines only take in air. The fuel is then sprayed into the hot air in the cylinder.

There is less pressure rise but the peak pressure is maintained over a longer period of time. This gives the diesel engine its greater lugging ability.

Continued on next page   JB06590,000048A -19-05AUG09-9/32

These principles are shown in the illustration. Note that while firing pressure rises very rapidly in the spark-ignition engine, it also falls off very rapidly. In diesel engines, the pressure rises, is maintained over a longer period of time, and then falls off.

Causes of Poor Compression

- Worn cylinders
- Worn piston rings
- Leaking gaskets
- Worn pistons
- Leaking valves that can be caused by:
  - seats burned
  - valves burned
  - valve springs weak or broken
  - valve stem warped or sticking in valve guide
- Loose spark plug
- Loose cylinder head bolts
- Warped cylinder head
- Worn camshaft lobes
- Not enough valve tappet clearance

For remedies, see Chapter 3 of this manual.

3. Ignition System (Spark-Ignition Engines Only)

An oscilloscope is an instrument used to electronically check the performance of the ignition system. It is fast, accurate, and often pinpoints troubles that might be overlooked.

Details on servicing the ignition system are given in the FOS manual, Electronic and Electrical Systems. Here we will only outline the ignition services as they relate to engine tune-up.

Ignition Coil

Use a reliable meter to check the ignition coil.

If the performance is doubtful, replace the coil. Refer to the technical manual for tests and specifications.

Check the coil for proper connections.

Use this rule of thumb:

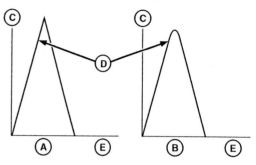

Pressure Curves for Engine Compression

A—Spark-Ignition Engine
B—Diesel Engine
C—Pressure
D—Ignition Occurs Here
E—Time

- On negative-ground system, the negative (–) primary terminal is connected to the distributor.
- On positive-ground systems, the positive (+) primary terminal is connected to the distributor.

Spark Plugs

Close inspection of the spark plugs can reveal much about the condition of the engine.

Black sooty plugs may be caused by:

- Engine operated at too low a temperature
- Fuel-air mixture too rich
- Heat range or plug too cold
- If plug is oily — engine consuming oil

Eroded plugs — insulator chipped or blistered from heat may be caused by:

- Heat range of plug too hot.
- Carburetor adjusted too lean, resulting in too much heat in the engine cylinders.

Continued on next page
JB06590,000048A -19-05AUG09-10/32

Good plugs will be relatively free of both deposits and erosion. This means that the plug is operating properly and is serviceable.

It is a good practice to replace all the spark plugs during tune-up if there is any question of their condition. Engine operation can be seriously affected by dirty or eroded plugs.

If the plugs are serviceable, file the electrode surfaces flat and square and adjust the point gap (see illustration) as specified in the engine technical manual.

After the spark plugs have been cleaned and adjusted, check them with a spark plug tester. Replace any which are doubtful.

When installing plugs, use a new gasket and tighten to exactly the specified torque. This is important.

If plugs are not tight enough, the plug can overheat because heat cannot be transferred to the cooling system.

If too tight, the thread area of the plug may stretch. This increases the point gap, resulting in hard starting and poor engine operation under load.

Distributor

Testing

Many modern engines use electronic distributors or even electronic distributorless ignition. Refer to engine technical manual for details on testing.

If possible, the distributor should be checked out on a distributor tester. Testing the distributor often locates troubles that would otherwise not be found.

Cap and Rotor

Remove the distributor cap, rotor, and housing cover. Wipe clean with a rag. Make sure any vent holes in the cap are open. Inspect the posts in the cap for erosion.

Check the rotor for severe erosion at the end of the metal contact. Erosion at this point widens the gap between the end of the rotor and the posts in the distributor cap. This puts an excessive load on the ignition coil and can cause the coil to fail. It also affects engine starting and operation.

Detach the wires from the distributor cap. The ends of the wires may have been burned away, causing a gap which could result in failure of plugs to fire. This makes the engine hard to start.

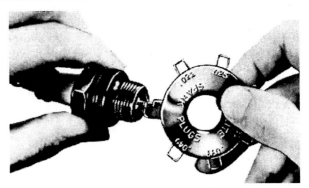

Setting the Spark Plug Gap

When installing the wires into the cap, be sure they are pushed all the way in.

Condenser

Use a reliable meter to check the condenser. Refer to the engine Technical Manual for tests and specifications.

Breaker Points

Examine the condition of the breaker points. If they appear burned, replace them. If there is a spur on one contact surface, a corresponding pit will be found on the other.

If the surfaces are clean and are a light gray in color, merely use a point contact file to true up the surfaces. It is not necessary to remove the pit from one surface, but the spur should be removed.

After the breaker points have been filed or when new points are installed, be sure that the contact surfaces meet squarely. This ensures a full flow of primary current when the points are in contact.

NOTE: Before installing the points, check the operation of the automatic spark advance mechanism (if used). Turn the breaker cam in the normal direction of rotation. It should operate freely without evidence of binding. If not, find the cause and correct it. See the engine technical manual for information.

Avoid using too much cam lubricant. If the grease gets on the contacts, they may prematurely burn out.

Apply a small amount of cam lubricant to the breaker cam.

Continued on next page   JB06590,000048A -19-05AUG09-11/32

Adjust the distributor breaker point gap (see illustration). Follow the instructions in the engine technical manual. Also use a reliable meter to check the cam dwell.

Cam dwell is the period of time the breaker points remain closed until they open to create the spark. If the point gap is too wide, the points may not remain closed long enough for complete buildup of the magnetic field in the ignition coil. This can result in a weak spark and poor engine performance. If the point gap is too small, the points will not remain open long enough for complete collapse of the field. This can cause burning of the points and also a weak spark.

Breaker point tension is very important. If the tension is too great, the rubbing block on the movable point will wear too fast. This will change the cam dwell and affect engine operation. If the tension is too weak, the movable point will bounce, causing erratic engine operation.

On distributors which require lubrication, lubricate according to instructions in the engine technical manual.

Reassemble the distributor making sure all connections are tight and that the spark plug wires are pushed all the way into their sockets.

Adjusting the Distributor Point Gap

A—Feeler Gauge

JB06590,000048A -19-05AUG09-12/32

Distributor Timing

Check the timing of the distributor to the engine using a distributor timing light.

Here is a general guide to distributor timing:

1. Start the engine and warm it up.

IMPORTANT: Disconnect and plug vacuum advance if equipped to avoid false timing results.

2. Stop the engine and connect the timing light as instructed in the engine technical manual. Also locate the timing marks on the engine.

3. Start the engine and operate it as specified.

4. Observe the flash of the timing light when the spark occurs. It should happen when the timing marks on the engine are lined up.

5. If necessary, loosen the distributor mounting and rotate the distributor until the flash occurs at the right instant.

   Rotating the distributor in the direction of cam rotation will delay the spark, while rotating it in reverse will advance the spark.

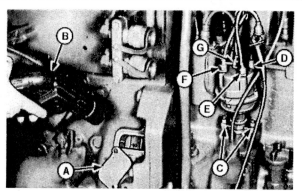

Timing the Distributor

A—Timing Hole Cover	E—No. 2 Cylinder
B—Timing Light	F—No. 3 Cylinder
C—Distributor Clamps	G—No. 4 Cylinder
D—No. 1 Cylinder	

6. Recheck the timing after the distributor mount is tightened again.

Continued on next page   JB06590,000048A -19-05AUG09-13/32

Timing the distributor to the engine should be done with great accuracy. The reason for this can be seen in the illustration.

In a spark-ignition engine the most power from combustion occurs while the piston is traveling the first half of the power stroke. This is because the connecting rod is angled in relation to the crankshaft. The left-hand drawing in the illustration shows why this is true. Notice how the connecting rod drives the crankshaft more directly than during the second half of the power stroke.

If the ignition timing is late, full expansion of the gases will not occur while the piston is within the most effective area of its stroke. This results in a loss of horsepower and greater operating costs.

If the timing is too early, too much expansion takes place before the piston reaches top dead center of its compression stroke and the engine literally works against itself. The result is excessive pinging, loss of horsepower, and higher operating costs.

4. Fuel Systems

Fuel Lines

Check fuel lines for leaks or restrictions. Leaking fuel not only is a fire hazard, but it gathers dirt and wastes fuel.

Checking LP-Gas System for Leaks

Use a soap solution or liquid leak detector to coat all parts of the LP-gas system to check for leaks. Pressurize the system and look for bubbles, which indicate a gas leak.

⚠ CAUTION: Observe all safety rules when working on LP-gas fuel systems (see Chapters 2 and 5).

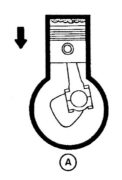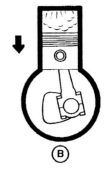

Connecting Rod Position in Relation to Crankshaft

A—First Half of Power Stroke   B—Second Half of Power
                                    Stroke

Checking Natural Gas Systems for Leaks

Open the fuel cylinder manual shutoff valves. Cycle the ignition switch to the ON position for 3 to 4 seconds, then back to the OFF position.

Use a non-ammonia soap and water solution or a commercial leak detector solution and wet all fuel lines, fittings, and other areas to check for leaks. Look for bubbles that indicate leakage from the system.

If a leak is detected, inspect all lines and fittings for damage. Check also for damaged or missing O-rings.

Continued on next page   JB06590,000048A -19-05AUG09-14/32

Fuel Transfer Pump

Remove and clean the fuel pump sediment bowl and strainer (see illustration).

Testing Fuel Transfer Pump Pressure

Install a tee fitting at the fuel pump outlet and connect a low-pressure test gauge. Then operate the engine and check the fuel pressure.

NOTE: On diesel engines, the hand primer lever can be used to build up the pump pressure and to fill the system after service.

Avoid excessive pressures. In spark ignition engines there is a tendency, especially on rough terrain, for the fuel to be forced by the carburetor float valve. This causes the fuel level in the carburetor to be too high, resulting in excess fuel consumption.

If the fuel pump has a line filter, service it as instructed in the engine technical manual.

Carburetor Sump (LP-Gas)

Drain the LP-gas carburetor sump using precautions (see Chapter 5).

Converter (LP-Gas)

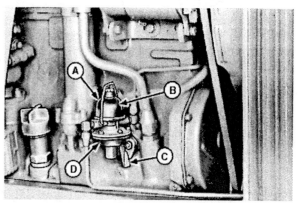

Fuel Transfer Pump

A—Sediment Bowl
B—Fuel Strainer
C—Hand Primer
D—Fuel Pump

Test the LP-gas converter for leaks as follows:

1. Turn on the starter switch and open the vapor withdrawal valve.
2. Remove the radiator cap and check the coolant for bubbles caused by a leaky converter gasket.
3. If bubbles are present, service the converter.

Continued on next page   JB06590,000048A -19-05AUG09-15/32

Carburetor (Gasoline)

Modern engines may also use electronic fuel injection for gasoline engines. Refer to engine technical manual for details on servicing.

Remove the drain plug at the bottom of the carburetor bowl (see illustration) and drain out any water or sediment in the bowl. After draining, replace the drain plug.

Remove and clean the carburetor fuel inlet strainer. Flush the strainer with solvent to remove any dirt that may have worked through it.

If the engine has been in long or severe service, remove and clean the carburetor. Disassemble and clean the metal parts by immersing them in a carburetor cleaning solution. When clean, blow out all passages and dry the parts with compressed air. When reassembling, use a new carburetor repair kit.

In operation, the carburetor float needle and seat wear. This raises the fuel level in the carburetor bowl and increases fuel consumption. It is false economy to make do when new parts will soon pay for themselves in reduced operating costs.

After the carburetor is installed, check the choke disk for proper operation. When the choke is applied, the disk should be centered in the bore and should entirely close the opening. When the choke is open, the disk should be parallel with the opening.

Carburetor fuel adjustment should also be checked later when the engine is tested on the dynamometer.

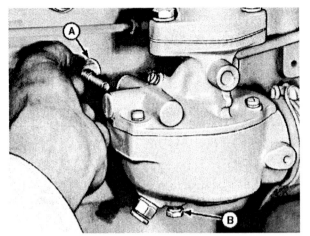

Gasoline Carburetor

A—Inlet Elbow and Strainer   **B—Bowl Drain Plug**

Speed Control Linkage

Check the adjustment of the linkage between the speed control lever and the injection pump or carburetor. Be sure that all linkage operates freely through its entire range. Any binding will cause erratic operation of the engine.

Fuel Filter (LP-Gas)

Check and clean the LP-gas fuel strainer filter element using the precautions in Chapter 5.

JB06590,000048A -19-05AUG09-16/32

Fuel Filter (Natural Gas)

Replace the natural gas filter during tune-up.

Shut off the natural gas supply and perform fuel leak-off procedure using precautions described in Chapter 6.

Crack the fuel supply line fitting to the regulator valve to make certain that internal gas pressure is relieved.

Remove the filter plug (see illustration) and gas filter element from the regulator housing.

Install a new filter element and O-ring. Tighten the filter plug and fuel supply line fitting. Check for gas leaks as previously described in this chapter.

A—Filter Plug

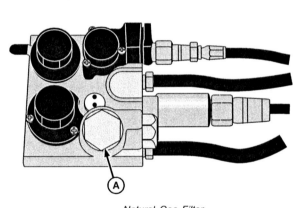

Natural Gas Filter

Continued on next page   JB06590,000048A -19-05AUG09-17/32

Fuel Filters (Diesel)

Diesel fuel must be clean and free from water. Many cases of injection pump or nozzle failure can be traced to dirt or water in the fuel.

Take special care to avoid contaminated diesel fuel — from the supplier, to the storage system, to the fuel tank, and through the injection system.

If the filter has a drain plug, loosen it and drain out any water or sediment. If water is present, be sure to replace the filter. Also find out and eliminate the source of water in the fuel.

On dual-stage filters, if the first-stage filter is extremely dirty or water-soaked, also replace the second-stage filter.

Injection Pump (Diesel)

During major engine service, the injection pump should be removed, cleaned, inspected, and calibrated on an injection pump test stand.

However, it is a good practice to remove the pump and check it during a complete tune-up, if a test stand is available.

Testing the pump ensures that it is operating properly and helps to locate malfunctions which affect engine performance and cause high fuel consumption. It can also reveal whether or not the pump calibration has been tampered with to increase engine horsepower.

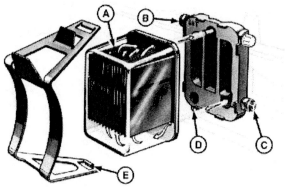

Diesel Fuel Filter

A—Replaceable Filter Element D—Drain Plug
B—Fuel Outlet E—Clip
C—Fuel Inlet

Never calibrate the pump to inject more fuel than specified in the engine technical manual. The engine is designed to produce its rated horsepower at a certain rate of fuel consumption. Any change in this rate puts an undue strain on the working parts of the engine. This can cause early engine failure and always results in higher operating costs.

Excessive smoke from the exhaust is one sign that the injection pump is calibrated too high.

Continued on next page   JB06590,000048A -19-05AUG09-18/32

Injection Nozzles (Diesel)

During major engine service, the injection nozzles should be removed, cleaned, and tested using a nozzle tester (see Chapter 7).

During a complete tune-up, it is also a good practice to remove and service the nozzles (see illustration).

Three factors are of prime importance in nozzle operation:

- Cracking pressure. All nozzles should be about equal. If the pressure is too low, the fuel will not atomize.
- Condition of spray tips. The orifices should be clean and not eroded.
- Spray pattern. The fuel should be finely atomized and spread in an even spray pattern. A bad spray pattern can actually erode metal from the top of the piston. Engines operated with bad nozzles often have low horsepower and excessive exhaust smoke.

Bleeding Air from the Fuel System (Diesel)

After servicing, always bleed air from the diesel fuel system. This will prevent an air lock in the high-pressure injection system.

Usually the fuel transfer pump has a primer lever which helps force air bubbles out of the system at various bleed plugs and loosened connections (see Chapter 7).

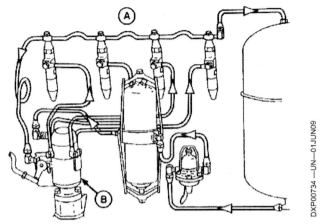

Diesel Fuel Injection Pump and Nozzles

A—Injection Nozzles   **B—Injection Pump**

When all air is bled from the system, the engine will start normally and run without missing.

Continued on next page   JB06590,000048A -19-05AUG09-19/32

Timing the Injection Pump (Diesel)

In a diesel engine, the start of fuel injection occurs before the piston reaches top dead center of its compression stroke. This is necessary because it takes time for the burning fuel to build up pressure. Therefore, the pump is timed so the expanding gases reach peak pressure at about the time the piston reaches top dead center.

If injection is timed too early, expansion occurs before the piston reaches top dead center and the engine is literally working against itself. This results in horsepower loss and decreased engine efficiency.

If injection is timed too late, expansion occurs late, and much power is lost because the connecting rod has a poor angle to the crankshaft.

When the injection is correctly timed, the connecting rod drives the crankshaft more directly and gets full power as shown in illustration.

5. Lubrication System

Pressure Gauge or Indicator Light

Check the operation of the oil pressure gauge or light. Turn on the starter switch to check, then start the engine and check during operation. See the engine technical manual for service information.

Oil Filter

Replace the engine oil filter during tune-up.

If the filter is extremely dirty, consider these likely problems:

- The crankcase oil has not been changed often enough. When oil is used too long in an engine, much of its detergent qualities are neutralized and it can no longer do its job. Either change the oil more often or use a higher quality of oil.
- The oil is of the wrong quality. Check the engine technical manual for the correct quality and weight of oil to use.
- Water or antifreeze is in the oil. Water in the oil is one cause of excessive sludge. Water can leak into the crankcase from around the cylinders through a sand hole in the cylinder block casting, or past the cylinder head gasket. If water is present in the oil, determine the cause and correct it.

Antifreeze can enter the crankcase in any of the above ways. However, antifreeze poses much more of a problem than plain water.

- It forms a deposit on the pistons and rings, causing rings to stick and pistons to score.
- It can cause main and connecting rod bearings to fail.
- It forms a deposit on all surfaces throughout the engine and can cause severe damage to all moving parts.

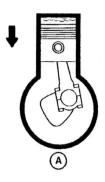
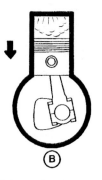

Connecting Rod Position in Relation to Crankshaft

A—First Half of Power Stroke   **B—Second Half of Power Stroke**

If the engine has been run too long with antifreeze in the oil, disassemble the engine. Clean all parts thoroughly and replace any that are damaged.

Crankcase Oil

Remove the oil dipstick and examine the condition of the crankcase oil.

Oil in the engine has a variety of functions:

- Oil's prime purpose is to provide lubrication between the moving parts. This prevents wear and possible damage.
- Oil assists in carrying off heat generated within the engine. The oil is cooled to some extent by the flow of air around the crankcase. Some engines are provided with special coolers to cool the oil. The cooled oil is pumped around the bearing surfaces and thrown around the inside of the engine. This helps to dissipate the engine heat.
- Oil acts as a seal around the piston rings. Oil not only lubricates the pistons and rings, it also acts as a seal between the piston and rings, preventing loss of compression.
- Oil neutralizes acids and prevents corrosion. Modern oils contain additives that neutralize acids formed during combustion and so prevent corrosion within the engine.
- Oil prevents depositing of sludge in the engine. Additives in the oil also enable it to hold in suspension dirt and grit which otherwise would be deposited within the engine as sludge.

From this summary we can see how important it is to use the correct oil.

If the crankcase oil contains water, antifreeze, debris, etc., determine the problem and make the necessary repairs prior to changing the oil and filter.

Refer to the engine operator's manual for draining procedures and for the correct quality and weight of oil to use.

Continued on next page   JB06590,000048A -19-05AUG09-20/32

Oil Pressure

Check the engine oil pressure using a master gauge.

1. Operate the engine until the oil is warmed up to normal. (Also be sure the filter is clean.)
2. Stop the engine and connect a master gauge (see illustration) to the engine block (normally where the pressure sending unit is attached.)
3. Run the engine at the specified speed and check the oil pressure against the listing in the technical manual.
4. If oil pressure is too high or low, adjust it as instructed in the technical manual.

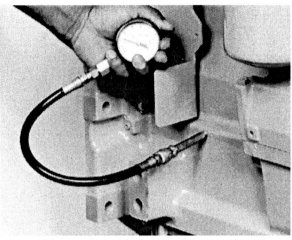

Checking Engine Oil Pressure with a Master Gauge

JB06590,000048A -19-05AUG09-21/32

Problems with Low Oil Pressure Operate the engine at slow idle speed. If the pressure is very low, check for one of these three problems:

- Using too-light oil. Check for correct weight.
- Stuck oil pressure relief valve. Dirt on the valve seat can prevent it from closing. Check and clean the valve and seat.
- Worn-out bearings on the crankshaft or camshaft. Too much oil escapes past worn parts, lowering oil pressure. In this case, recondition the bearings.

Too-low oil pressures can cause the pistons and bearings to wear faster. This is because not enough oil reaches the cylinders for good lubrication (see illustration).

A—Normal Oil Pressure (full oil spray reaches cylinder)

B—Low Oil Pressure (weak oil spray does not reach cylinder)

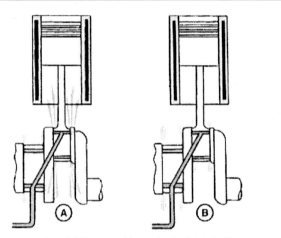

Low Oil Pressure Means Lack of Lubrication

Continued on next page

JB06590,000048A -19-05AUG09-22/32

6. Cooling System

Water Pump

Inspect the water pump. If the pump is leaking or has too much end-play in its shaft, remove the pump and repair it as instructed in the engine technical manual.

Radiator Hoses

Inspect radiator hoses for signs of leakage or rot. Be sure that the inside lining of hoses has not become mushy. This restricts the flow of water. Replace hoses as necessary.

Cleaning and Flushing the Cooling System

1. Run the engine long enough to stir up any rust or sediment in the cooling system.
2. Drain the cooling system. Detach the thermostat housing and remove the thermostat. Reinstall the housing.
3. Fill the cooling system with water and run the engine long enough to warm the water and stir up any rust or sediment.
4. Stop the engine and drain the system at once before the rust or sediment settles.
5. Fill the system with a solution of water and a good commercial radiator cleaner.
6. Install the radiator cap and run the engine until the solution is thoroughly warmed. Place a cardboard over the front of the radiator core for faster warm-up.
7. After several minutes, drain out the solution, fill with clean soft water, start the engine, and let the water circulate for a few minutes.

Water Pump and Thermostat Housing

A—Lower Water Hose
B—Fan Hub
C—Thermostat Housing
D—Upper Water Hose
E—Water Pump

8. Condition the system with a recommended antifreeze (winter) or a coolant conditioner that inhibits rust (summer). Also add a water pump lubricant (when recommended).

JB06590,000048A -19-05AUG09-23/32

Thermostat

The thermostat is the key factor in cooling the engine.

Test the action of the thermostat (see illustration). If it does not open at the temperature given in the engine technical manual or is otherwise defective, replace it.

When the engine is cold, the thermostat is closed. This confines the circulation of the coolant to within the cylinder block, resulting in faster warm-up.

When the engine is warmed up, the thermostat opens to bypass just enough coolant into the radiator core to maintain the proper temperature in the engine cylinder block.

If the thermostat ever sticks shut or fails to open, coolant in the cylinder block will soon overheat, which may damage the engine.

If the thermostat opens at too low a temperature or does not close, the engine will not warm up properly and its efficiency will be lost.

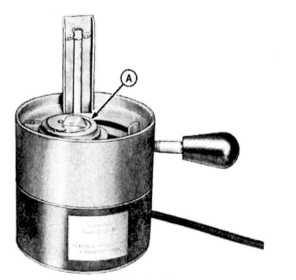

Testing the Thermostat

A—Thermostat

Continued on next page

JB06590,000048A -19-05AUG09-24/32

Radiator

Check the radiator for leaks using a pressure tester (see illustration). Avoid applying too much pressure. A maximum of 15–25 psi (100–170 kPa) should be sufficient.

If any leaks are observed, repair them. See Chapter 10 for details.

Examine all air passages in the radiator core. Blow out any chaff or dirt and straighten any bent fins.

Clean all chaff or dirt from the radiator screens.

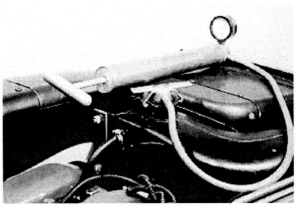

Testing Radiator for Leaks

JB06590,000048A -19-05AUG09-25/32

Radiator Cap

Use a reliable tester to test the radiator cap (see illustration). The pressure valve should open at a certain pressure (see engine technical manual for specifications). If the pressure valve does not open as specified or leaks before reaching opening pressure, replace the gasket or the whole cap as necessary.

Be sure that the cap and radiator are free from leaks. The pressure cap permits pressure to build up in the cooling system and so raises the boiling point of the coolant to approximately 230°F (110°C). This prevents boiling away of the coolant which would occur at a lower temperature.

The pressure cap also prevents loss of coolant through the overflow when traveling over rough terrain.

Fan Belt

Check the fan belt for excessive wear, cracks, or other signs of damage. Replace the belt if any of these conditions exist.

Fan belt tension is checked later during generator inspection.

7. Electrical System

Servicing of electrical systems is covered in the FOS manual, Electronic and Electrical Systems. Refer to that manual for details on the tune-up procedures given here.

Batteries

Cleaning the Batteries

Testing Radiator Pressure Cap

Use a stiff brush and a water and soda solution to thoroughly clean the batteries. Rinse off the batteries with clean water.

If the battery terminals are corroded, disconnect and clean them. Clean the battery posts and the insides of the connectors so they make good electrical contact.

Be sure the batteries are properly installed.

Apply terminal protectant on battery posts and connectors to prevent corrosion.

Continued on next page

JB06590,000048A -19-05AUG09-26/32

Testing The Specific Gravity

Use an accurate battery hydrometer to check the specific gravity of the electrolyte in each battery cell (see illustration).

If the specific gravity of the electrolyte is low, it suggests undercharging.

If the specific gravity shows the battery is fully charged but the electrolyte level is low, the battery may have been charged at too high a rate. This will be discussed later.

Adding Water To Batteries

Fill each battery cell to the proper level. Battery waters are listed below in the order of their quality.

1. Distilled water
2. Water from a dehumidifier
3. Rain water
4. Tap water

The first two waters are best since they contain no minerals.

The third is not so good because rain water can collect minerals from the air.

The fourth is least desirable since it contains chlorine and carbonates that can shorten the life of the battery.

⚠ CAUTION: Avoid adding too much water to a battery during freezing temperatures unless the engine can be run long enough to thoroughly mix

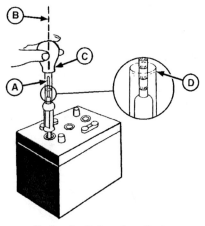

Testing the Battery Specific Gravity

A—Float Must Be Free
B—Hold Tube Vertical

C—Do Not Draw In Too Much Electrolyte
D—Take Reading at Eye Level

the water with the electrolyte — otherwise the water can freeze and cause the battery to burst.

Light Load Test of Batteries

Make a light load test of the batteries to determine their condition.

Recharge or replace the batteries if they fail any part of the light load test. For details, see FOS manual, Electronic and Electrical Systems.

JB06590,000048A -19-05AUG09-27/32

Alternator or Generator

Checking Belt Tension

Check belt for proper tension. The illustration shows a typical belt installation. Refer to the engine technical manual for the correct tension and how to adjust it.

Testing Generating Output

Check the output of the alternator or generator. Refer to the engine technical manual for details.

IMPORTANT: Follow the manual procedure exactly. Making a wrong connection can severely damage an alternator.

Starting Circuit

Safety Starter Switch

Some machines use a safety starter switch to prevent starting the engine when the machine is in gear.

Check the operation of the switch by working the gear shift while starting the engine. Refer to the engine

Belt Tension Adjustment

technical manual and replace or adjust the safety switch as necessary.

Continued on next page JB06590,000048A -19-05AUG09-28/32

Prevent Machine Runaway

⚠ CAUTION: Avoid possible injury or death from machinery runaway. Do not start the engine by shorting across the starter terminals. The machine could start in gear if the circuitry is bypassed.

NEVER start an engine while standing on the ground. Start the engine only from the operator's seat, with the transmission in the neutral or parked position.

Never Bypass the Neutral Start Switch

Starting Motor

With the engine warmed up, check the ampere draw of the starter motor.

Disconnect the battery cable from the starter motor and use an ammeter connected as instructed in the engine technical manual. Record the amperage draw and check it against the specifications.

If the ampere draw is too high, look for worn starter bearings or some drag in the engine.

Typical Starter Motor

JB06590,000048A -19-05AUG09-29/32

8. Clutch Free Travel

Clutch free travel is the distance the pedal travels before it starts to disengage the clutch from the engine flywheel.

Too little clutch free travel may cause the clutch release bearing to operate continuously, resulting in early failure. It can also cause slippage.

Too much free travel may prevent the clutch from completely disengaging and so wear it out.

To check free travel, pull the pedal down until you feel a contact at the flywheel. Measure this distance as shown in the illustration and check it against the engine technical manual. If necessary, adjust as specified.

A—Check Pedal Free Travel

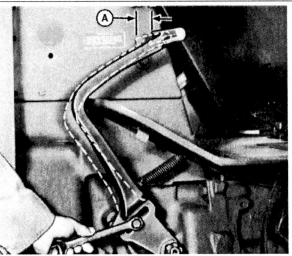

Clutch Free Travel Adjustment

Continued on next page

JB06590,000048A -19-05AUG09-30/32

9. Dynamometer Test

The dynamometer test is the final check of overall engine performance. It will tell you whether the tune-up has been adequate. Compare it with the dynamometer test made before tune-up.

Test for the following things:

- **Engine Horsepower**
- **Exhaust Analysis:**
  - Smoke Analysis (diesel)
  - Carburetor Adjustment (spark-ignition)
- **Fuel Consumption**
- **Crankcase Blowby**

Most dynamometer manufacturers have instruments to be used with the dynamometer for checking the above items.

Use the engine technical manual for procedures and specifications.

NOTE: If the engine fails to produce the desired horsepower, and an air cleaner restriction test was not performed at the beginning, perform test now. It is possible that an air restriction is causing the loss of horsepower.

A completely tuned engine should pass the dynamometer test with no problems.

A Typical Dynamometer Test

However, the engine should not put out more horsepower than it was designed to. Tampering with the engine to get extra horsepower will shorten engine life and raise operating costs. It may also void the engine warranty.

JB06590,000048A -19-05AUG09-31/32

Summary

Tune-up of an engine may seem like a long and expensive ordeal. But actually, most of the items can be checked in a minute or two.

But why check out so many items if the engine has not actually failed?

The answer is that tune-up is preventive maintenance.

Before the engine fails, we keep it tuned up so that causes are corrected early, and possible causes are prevented.

WHY REGULAR TUNE-UP PAYS

Tune-Up Will Not Restore A Badly Worn Engine — Only Major Overhaul Will

BUT:

- Tune-Up Improves the Engine

- And Also Prevents Later Problems

HOW?

- Catching Problems Early Means Fewer Service Calls in the Field

- Shop Service Is Cheaper than Field Calls

- And Shop Service Can Be Scheduled to Avoid Peak Operations

RESULTS:

- Tune-Up Means That the Engine is Ready to Go

- And is Dependable for Long, Productive Hours

JB06590,000048C -19-22JUL09-1/1

Test Yourself

Questions

1. (True or false?) A badly worn engine can be restored by a complete tune-up.

2. Before tuning up an engine, what should be done?
3. When should dynamometer tests be made?

JB06590,000048D -19-22JUL09-1/1

Definitions and Conversions

Definition of Terms and Abbreviations A—K

A

Abrasion

Wearing or rubbing away of a part.

Actuator Solenoid

The solenoid in the actuator housing on the back of the injection pump that moves the fuel control rack as commanded by the Engine Control Unit (ECU).

Adaptive Learn Control

A function of an electronic Engine Control Unit (ECU) that helps to optimize engine operation by acting as a fuel correction memory. When the ECU commands a fuel quantity that is successful in achieving the desired air-fuel ratio, the adaptive learn function remembers that fuel quantity and uses it in the future.

Additive, Coolant

A substance added to engine coolant to give it certain properties. For example, inhibiting additives are added to the coolant to protect the engine against corrosion and cylinder liner erosion or pitting.

Additive, Oil

A substance added to oil to give it certain properties. For example, a material added to engine oil to lessen its tendency to congeal or thicken at low temperatures.

Aftercooler

A heat exchanger that cools the intake air, coming from the turbocharger, before it enters the cylinder. This increases the density of the air, thus creating more power, better economy, and quieter combustion.

Air Cleaner

A device for filtering, cleaning, and removing dust from the air admitted to an engine.

Air-Fuel Mixer

Natural gas "carburetor" that mixes the vaporized natural gas with the intake air to form the air-fuel mixture for combustion.

Ambient Temperature

The temperature of the surrounding medium, such as gas, air, or liquid.

Analog

A signal that has a continuous range of possible voltages. Usually 0 to 5 volts or 0 to 12 volt signals.

Analog-Digital Converter

An integrated circuit within the Engine Control Unit (ECU) that converts analog electrical signals from various sensors into digital signals that the ECU central processing unit can use.

Antifreeze

A material such as alcohol, glycerin, etc., added to water to lower its freezing point.

Antifriction Bearing

A bearing constructed with balls, rollers, or the like between the journal and the bearing surface to provide rolling instead of sliding friction.

Arc Welding

A method of utilizing the heat of an electric current jumping an air gap to provide heat for welding metal.

ASME

American Society of Mechanical Engineers.

Atmospheric Pressure

The weight of air at sea level; about 14.7 pounds per square inch (100 kPa). The pressure decreases at higher altitudes.

Auxiliary Speed Sensor

Engine speed sensor located on the engine timing gear cover. Serves as backup to the primary engine speed sensor for the electronic governor system.

B

Backfire

Ignition of the mixture in the intake manifold by flame from the cylinder such as might occur from a leaking inlet valve or incorrect ignition timing.

Backlash

The clearance or "play" between two parts, such as meshed gears.

Back-Pressure

A resistance to free flow, such as a restriction in the exhaust line.

Baffle or Baffle Plate

An obstruction for checking or deflecting the flow of gases or sound.

Ball Bearing

An antifriction bearing consisting of a hardened inner and outer race with hardened steel balls interposed between the two races.

Bearing

A part in which a journal, pivot, or the like turns or moves.

B.H.P. (Brake Horsepower)

A measurement of the power developed by an engine in actual operation. It subtracts the F.H.P. (friction losses) from the I.H.P. (theoretical horsepower).

Continued on next page

OUO1020,0001972 -19-22JUL09-1/7

100709
PN=470

Blow-By

A leakage or loss of compression past the piston rings between the piston and the cylinder.

Boiling Point

The temperature at which bubbles or vapors rise to the surface of a liquid and escape.

Bore

The diameter of a hole, such as a cylinder; also to enlarge a hole as distinguished from making a hole with a drill.

Braze

To join two pieces of metal using a comparatively high melting-point material. An example is to join two pieces of steel by using brass or bronze as a binder.

Break-In

The process of wearing in to a desirable fit between the surfaces of two new or reconditioned parts.

Burnish

To smooth or polish using a sliding tool under pressure.

Bushing

A removable liner for a bearing.

Bypass

An alternative path for a flow of air or liquid.

C

Calibrate

To determine or adjust the graduations or scale of any measuring instrument.

Calorie

The metric measurement of the amount of heat required to raise 1 gram of water from zero degrees to 1 degree Centigrade.

Calorific Value

A measure of the heating value of fuel.

Calorimeter

An instrument to measure the amount of heat given off by a substance when burned.

Cam-Ground Piston

A piston ground to a slightly oval shape, which becomes round under the heat of operation.

Camshaft

The shaft containing lobes or cams to operate the engine valves.

Carbon Monoxide

Gas formed by incomplete combustion. Colorless, odorless, and very poisonous.

Carbonize

The process of carbon formation within an engine, such as on the spark plugs and within the combustion chamber.

Carburetor

A device for automatically mixing gasoline, LP-gas, or natural gas in the proper proportion with air to produce a combustible vapor.

Carburetor "Icing"

A term used to describe the formation of ice on a carburetor throttle plate during certain atmospheric conditions.

Central Processing Unit (CPU)

Performs the mathematical computations and logical functions for the Engine Control Unit (ECU) that are necessary in controlling air flow, fuel delivery, and ignition timing.

Cetane

Measure of ignition quality of diesel fuel (at what pressure and temperature the fuel will ignite and burn).

Chamfer

A bevel or taper at the edge of a hole.

Chase

To straighten or repair damaged threads.

Choke

A device such as a valve placed in a carburetor air inlet to restrict the volume of air admitted.

Closed-Loop

The Engine Control Unit (ECU) has two operating modes: open-loop and closed-loop. Closed-loop mode is utilized after the engine has been running for 30 to 60 seconds. In closed-loop mode, the ECU has the ability to correct for fuel composition variation and for changing operating conditions.

Coalescing Filter

A filter that removes debris and contaminants from natural gas.

Combustion

The process of burning.

Combustion Chamber

The volume of the cylinder above the piston with the piston on top center.

Compressed Natural Gas (CNG)

Continued on next page

OUO1020,0001972 -19-22JUL09-2/7

Natural gas that is compressed up to 3600 psi (24 821 kPa) and stored in high-pressure tanks.

Compression

The reduction in volume or the squeezing of a gas. As applied to metal, such as a coil spring, compression is the opposite of tension.

Compression Ratio

The volume of the combustion chamber at the end of the compression stroke as compared to the volume of the cylinder and chamber with the piston on bottom center. Example: 9 to 1.

Condensation

The process of a vapor becoming a liquid; the reverse of evaporation.

Condenser

An automotive term that describes a capacitor. It is commonly used in conventional breaker point ignition systems.

Connecting Rod

Rod that connects the piston to the crankshaft.

Contraction

A reduction in mass or dimension; the opposite of expansion.

Convection

A transfer of heat by circulating heated air.

Converter

As used in connection with LP-gas, a device that converts or changes LP-gas from a liquid to a vapor for use by the engine.

Corrode

To eat away gradually as if by gnawing, especially by chemical action, such as rust.

Counterbore

To enlarge a hole to a given depth.

Countersink

To cut or form a depression to allow the head of a screw to go below the surface.

Crankcase

The lower housing in which the crankshaft and many other parts of the engine operate.

Crankcase Dilution

When unburned fuel finds its way past the piston rings into the crankcase oil, where it dilutes or thins the engine lubricating oil.

Crankshaft

The main drive shaft of an engine that takes reciprocating motion and converts it to rotary motion.

Crankshaft Counterbalance

A series of weights attached to or forged integrally with the crankshaft to offset the reciprocating weight of each piston and connecting rod.

Crosshead

Part of the valve train in an engine with two intake and two exhaust valves per cylinder. Permits two valves in the same cylinder to be opened at the same time.

Crude Oil

Liquid oil as it comes from the ground.

Cryogenic Process

A process involving very low temperatures. As an example, natural gas is converted to a liquid by cooling the gas to −259°F (−126°C).

cu in.

Cubic inch.

Cylinder

A round hole having some depth bored to receive a piston; also sometimes referred to as "bore" or "barrel."

Cylinder Block

The largest single part of an engine. The basic or main mass of metal in which the cylinders are bored or placed.

Cylinder Head

A detachable portion of an engine fastened securely to the cylinder block that contains all or a portion of the combustion chamber.

Cylinder Liner

A sleeve or tube interposed between the piston and the cylinder wall or cylinder block to provide a readily renewable wearing surface for the cylinder.

D

Dead Center

The extreme top or bottom position of the crankshaft throw at which the piston is not moving in either direction.

Density

Compactness; relative mass of matter in a given volume.

Detergent

A compound of a soap-like nature used in engine oil to remove engine deposits and hold them in suspension in the oil.

Detonation

Continued on next page

OUO1020,0001972 -19-22JUL09-3/7

A too-rapid burning or explosion of the mixture in the engine cylinders. It becomes audible through a vibration of the combustion chamber walls and is sometimes confused with a "ping" or "knock."

Diagnosis

In engine service, the use of instruments to "troubleshoot" the engine parts to locate the cause of a failure.

Diagnostic Code

A number that represents a problem detected by the Engine Control Unit (ECU). Transmitted for use by onboard displays or a diagnostic reader, so the operator or technician is aware there is a problem and where it can be found in the fuel system.

Diagnostic Scan Tool

Any electronic module that is able to read and clear Diagnostic Trouble Codes (DTCs) stored in the Engine Control Unit's memory, read sensor and actuator data, perform engine tests, and display diagnostic information to the operator or technician.

Diesel Engine

Named after its developer, Dr. Rudolph Diesel. This engine ignites fuel in the cylinder from the heat generated by compression. The fuel is an "oil" rather than gasoline, and no spark plug or carburetor is required.

Digital

A signal that consists of only two volt levels, usually 0 volts and +5 volts.

Dilution

See Crankcase Dilution.

Displacement

See Piston Displacement.

Distortion

A warpage or change in form from the original shape.

Distributor (Ignition)

A device that directs the high voltage of the ignition coil to the engine spark plugs.

Dowel Pin

A pin inserted in matching holes in two parts to maintain those parts in fixed relation one to the other.

Downdraft Carburetor

A type of carburetor in which the air-fuel mixture flows downward to the engine.

Drawbar Horsepower

Measure of the pulling power of a machine at the drawbar hitch point.

Drive-by-Wire

Electronically controlled throttle actuator system that has no mechanical connection between the operator's foot pedal and the throttle valve.

Dynamometer

A test unit for applying a load to an engine to measure the actual power produced by an engine.

E

Eccentric

One circle within another circle but with a different center of rotation. An example of this is a driving cam on a camshaft.

Economizer

A device installed in a carburetor to control the amount of fuel used under certain conditions.

ECU (Engine Control Unit)

An electronic module that controls fuel delivery, diagnostic outputs, backup operation, and communications with other electronic modules.

Electrically Pulsed Fuel Injectors

Natural gas fuel injectors that are controlled by the Engine Control Unit (ECU). The ECU controls the on time (pulse width) of each pair of injectors to deliver the correct amount of fuel to the engine.

Electronic Governor

The computer program within the Engine Control Unit (ECU) that determines the commanded fuel delivery based on throttle command, engine speed, and fuel temperature. Replaces the function of the mechanical governor.

Electronic Ignition System

A system in which the timing of the ignition spark is controlled electronically. Electronic ignition systems have no points or condenser, but instead have a reluctor, sensor, and electronic control unit.

Electronic Unit Injector (EUI)

An electronically controlled injection pump and injector combined. The start of injection and the amount of fuel injected is controlled by a solenoid-actuated spill valve located within the Electronic Unit Injector (EUI) housing.

End Play

The lateral clearance or side-to-side movement of a shaft.

Energy

The capacity for doing work.

Engine

Continued on next page   OUO1020,0001972 -19-22JUL09-4/7

The prime source of power generation used to propel the machine.

Engine Coolant Temperature Sensor

Measures the temperature of the engine coolant.

Engine Displacement

The sum of the displacement of all the engine cylinders. See Piston Displacement.

Evaporation

The process of changing from a liquid to a vapor, such as boiling water to produce steam. Evaporation is the opposite of condensation.

Exhaust Back-Pressure Sensor

Measures the pressure in the exhaust system downstream of the turbocharger.

Exhaust Gas Analyzer

An instrument for determining the efficiency with which an engine is burning fuel.

Exhaust Manifold

The passages from the engine cylinders to the muffler, which conduct the exhaust gases away from the engine.

Expansion

An increase in size. For example, when a metal rod is heated, it increases in length and perhaps also in diameter. Expansion is the opposite of contraction.

F

Ferrous Metal

Made of iron.

F.H.P. (Friction Horsepower)

A measure of the power lost to the engine through friction or rubbing of parts.

Filter (Air, Oil, Water, Fuel)

A unit containing an element, such as a screen of varying degrees of fineness. The screen or filtering element is made of various materials depending upon the type and size of the foreign particles to be eliminated from the air or fluid being filtered.

Flash Point

The temperature at which an oil, when heated, will flash and burn.

Float Level

The height of the fuel in the carburetor bowl, usually regulated by means of a suitable valve and float.

Floating Piston Pin

A piston pin that is not locked in the connecting rod or the piston, but is free to turn or oscillate in both the connecting rod and the piston.

"Flutter" or "Bounce"

In engine valves, refers to a condition where the valve is not held tightly on its seat during the time the cam is not lifting it.

Flywheel

A heavy wheel in which energy is absorbed and stored by means of momentum.

Foot Pedal Position Sensor

Measures the position of the operator's foot (throttle) pedal in the drive-by-wire, electronically controlled throttle actuator system. The Engine Control Unit (ECU) monitors the sensor signals to determine the operator's desired engine load.

Foot-Pound (lb-ft)

This is a measure of the amount of energy or work required to lift 1 pound a distance of 1 foot.

Four-Stroke Cycle Engine

Also known as Otto cycle, where a power stroke occurs every other revolution of the crankshaft. These strokes are (1) intake stroke; (2) compression stroke; (3) power stroke; (4) exhaust stroke.

Fuel Knock

Same as Detonation.

Fuel Lock-Off Device

A valve that shuts off the flow of natural gas to the engine when the engine is not running, even if the ignition switch is in the ON position.

Fuel Pressure Regulator

Reduces the natural gas fuel tank pressure to the working pressure required by the specific engine.

Fuel Temperature Sensor

Measures the fuel temperature. Using this measurement, the Engine Control Unit (ECU) will determine the fuel density, and adjust fuel delivery accordingly.

G

Gas

A substance that can be changed in volume and shape according to the temperature and pressure applied to it. For example, air is a gas that can be compressed into smaller volume and into any shape desired by pressure. It can also be expanded by the application of heat.

Gear Ratio

Continued on next page

OUO1020,0001972 -19-22JUL09-5/7

The number of revolutions made by a driving gear as compared to the number of revolutions made by a driven gear of different size. For example, if one gear makes three revolutions while the other gear makes one revolution, the gear ratio would be 3 to 1.

Glaze

As used to describe the surface of the cylinder, an extremely smooth or glossy surface such as a cylinder wall highly polished over a long period of time by the friction of the piston rings.

Glaze Breaker

A tool for removing the glossy surface finish in an engine cylinder.

Governor

A device to control and regulate speed. May be mechanical, hydraulic, or electrical.

Grind

To finish or polish a surface by means of an abrasive wheel.

H

Heat Exchanger

Sometimes used to describe a Vaporizer.

Heat Treatment

A combination of heating and cooling operations timed and applied to a metal in a solid state in a way that will produce desired properties.

Hone

An abrasive tool for correcting small irregularities or differences in diameter in a cylinder.

Horsepower (hp)

The energy required to lift 550 pounds 1 foot in 1 second.

Hot Spot

Refers to a comparatively thin section or area of the wall between the inlet and exhaust manifold of an engine, the purpose being to allow the hot exhaust gases to heat the comparatively cool incoming mixture. Also used to designate local areas of the cooling system that have attained above average temperatures.

I

I.D.

Inside diameter.

Idle (Fast)

Refers to the engine operating at its highest governed speed with a machine under no load.

Idle (Slow)

Refers to the engine operating at its slowest speed with a machine under no load.

Ignition Control Unit (ICU)

A microprocessor-based ignition control system.

I.H.P. (Indicated Horsepower)

"Pure" horsepower as measured in the combustion chamber before friction and other losses are subtracted.

Inert

A compound that is chemically unreactive with other compounds.

Inertia

A physical law that tends to keep a motionless body at rest or also tends to keep a moving body in motion; effort is thus required to start a mass moving or to retard or stop it once it is in motion.

Inhibitor

A material to restrain some unwanted action, such as a rust inhibitor, which is a chemical added to cooling systems to retard the formation of rust.

Injection Pump (Diesel)

A device by means of which the fuel is metered and delivered under pressure to the injector.

Injector (Diesel)

An assembly that receives a metered charge of fuel from another source at high pressure, then is actuated to inject the charge of fuel into a cylinder or chamber at the proper time.

Intake Manifold

For a spark-ignition engine, the passages that conduct the air-fuel mixture from the carburetor to the engine cylinders. For a diesel engine, the passages that conduct air to the engine cylinders.

Intake Valve

A valve that permits a fluid or gas to enter a chamber and seals against exit.

Integral

The whole made up of parts.

Internal Combustion

The burning of a fuel within an enclosed space.

J

Journal

The area on a crankshaft where bearings and connecting rods are attached.

Continued on next page   OUO1020,0001972 -19-22JUL09-6/7

K

Key

A small block inserted between the shaft and hub to prevent circumferential movement.

Keyway or Keyseat

A groove or slot cut for inserting a key to hold a part on a shaft.

Kilopascal (kPa)

Metric measurement for pressure or volume.

Knock

A general term used to describe various noises occurring in an engine; may be used to describe noises made by loose or worn mechanical parts, preignition, or detonation.

Knurled

Displacing metal on the skirt of a piston to increase the diameter of the piston (on automotive engines). Also can be done to the inside diameter of a valve guide to decrease the clearance between the valve guide and valve stem.

OUO1020,0001972 -19-22JUL09-7/7

Definition of Terms and Abbreviations L—W

L

Lacquer

A solution of solids in solvents that evaporate with great rapidity.

Lapping

The process of fitting one surface to another by rubbing them together with an abrasive material between the two surfaces.

Lean-Burn Combustion

The engine burns an air-fuel mixture that contains more air than is theoretically needed to completely burn the fuel.

L-Head Engine

An engine design in which both valves are located on one side of the engine cylinder.

Liner

Usually a thin section placed between two parts, such as a replaceable cylinder liner in an engine.

Liquefied Natural Gas (LNG)

Natural gas that is converted to a liquid through a cryogenic (freezing) process in which the gas is cooled to −259°F (−126°C).

LP-Gas, Liquefied Petroleum Gas

Made usable as a fuel for internal combustion engines by compressing volatile petroleum gases to liquid form. When so used, it must be kept under pressure or at low temperature in order to remain in liquid form, until used by the engine.

M

Manifold

A pipe or casting with multiple openings used to connect various cylinders to one inlet or outlet.

Manifold Absolute Pressure Sensor

Measures the pressure of the air in the intake manifold.

Manifold Air Temperature Sensor

Measures the temperature of the air in the intake manifold.

Manometer

A device for measuring a vacuum. It is a U-shaped tube partially filled with fluid. One end of the tube is open to the atmosphere and the other is connected to the chamber in which the vacuum is to be measured. A column of Mercury 30 inches high equals 14.7 lb. per square in. (psi), which is atmospheric pressure at sea level. Readings are given in terms of inches of mercury.

Mechanical Efficiency (Engine)

The ratio between the indicated horsepower and the brake horsepower of an engine.

Methane

The primary component of natural gas.

Micrometer

A measuring instrument for either external or internal measurement in thousandths and sometimes tenths of thousandths of inches.

Misfiring

Failure of an explosion to occur in one or more cylinders while the engine is running; may be a continuous or intermittent failure.

Motor

This term should be used in connection with an electric motor and should not be used when referring to the engine of a machine.

Muffler

A chamber attached to the end of the exhaust pipe, which allows the exhaust gases to expand and cool. It is usually fitted with baffles or porous plates and serves to subdue much of the noise created by the exhaust.

Multimeter

An electronic testing device that can be set to read ohms (resistance), voltage (force), or amperes (current) of a circuit.

N

Natural Gas

A naturally occurring gas that is composed primarily of methane with small amounts of ethane, propane, and butane.

Natural Gas Pressure Sensor

Measures the pressure of the natural gas in the fuel metering block.

Natural Gas Tank Pressure Sensor

Measures the pressure of the natural gas in the fuel storage tanks.

Natural Gas Tank Temperature Sensor

Measures the temperature of the natural gas in the fuel storage tanks.

Natural Gas Temperature Sensor

Measures the temperature of the natural gas in the fuel metering block.

Needle Bearing

An antifriction bearing using a great number of thin rollers.

Continued on next page

OUO1020,0001973 -19-23JUL09-1/6

Non-Waste Spark Ignition System

Each engine cylinder has its own ignition coil, and the coil fires on the combustion stroke only.

O

Octane

Measurement that indicates the tendency of a fuel to detonate or knock.

O.D.

Outside diameter.

Oil Pumping

A term used to describe an engine that is using an excessive amount of lubrication oil.

Open-Loop

The Engine Control Unit (ECU) has two operating modes: open-loop and closed-loop. Open-loop mode is utilized during engine start-up and for the first few seconds of run time. Fuel flow rate is determined by the ECU based on calibration data stored in the ECU memory and information from several sensors.

Otto Cycle

Also called four-stroke cycle. Named after the man who adopted the principle of four cycles of operation for each combustion in an engine cylinder. They are 1) intake stroke, 2) compression stroke, 3) power stroke, and 4) exhaust stroke.

Overhead Valve Engine

An engine design in which both valves are located in the cylinder head.

P

Peen

To stretch or clinch over by pounding with the rounded end of a hammer.

Petroleum

A group of liquid and gaseous compounds composed of carbon and hydrogen that are removed from the earth.

Pinion

A small gear having the teeth formed on the hub.

Piston

A cylindrical part closed at one end that is connected to the crankshaft by the connecting rod. The force of the combustion in the cylinder is exerted against the closed end of the piston, causing the connecting rod to move the crankshaft.

Piston Collapse

A condition describing a collapse or a reduction in diameter of the piston skirt due to heat or stress.

Piston Displacement

The volume of air moved or displaced by moving the piston from one end of its stroke to the other.

Piston Head

That part of the piston above the rings.

Piston Lands

Those parts of a piston between the piston rings.

Piston Pin

The journal for the bearing in the small end of an engine connecting rod that also passes through the piston walls; also known as a wrist pin.

Piston Ring

An expanding ring placed in the grooves of the piston to seal off the passages of fluid or gas past the piston.

Piston Ring Expander

A spring placed between the piston ring in the groove to increase the pressure of the ring against the cylinder wall.

Piston Ring Gap

The clearance between the ends of the piston ring.

Piston Ring Groove

The channel or slots in the piston in which the piston rings are placed.

Piston Skirt

That part of the piston below the rings.

Port

The openings in the cylinder block for valves, exhaust and inlet pipes, or water connections. In two-stroke cycle engines, the openings for intake and exhaust.

Position Sensors (Throttle and Foot Pedal)

These sensors are variable resistors (potentiometers). The resistance of the sensor varies depending on the sensor position. A moveable contact slides along a resistor as the sensor position is changed. The Engine Control Unit (ECU) monitors these sensors to determine the throttle opening and the operator's desired engine load.

Potentiometer

A variable resistor used as a voltage divider.

Preignition

Ignition occurring earlier than intended. For example, the explosive mixture being fired in a cylinder as by a flake of incandescent carbon before the electrical spark occurs.

Continued on next page OUO1020,0001973 -19-23JUL09-2/6

Press-Fit

Also known as a force-fit or interference-fit. This term is used when the shaft is slightly larger than the hole and must be forced into place.

Pressure Sensors

These sensors are pressure-sensitive variable resistors. A change in pressure results in a change in sensor resistance. The Engine Control Unit (ECU) monitors the voltage drop across the sensor and translates this voltage to a pressure value.

Primary Speed Sensor

A magnetic pickup that generates voltage pulses to the ECU as the teeth on the speed wheel pass by the tip of the sensor.

Programmable, Read-Only Memory (PROM)

The computer chip that contains the calibration information for the electronic engine control system. This information can only be read, not changed. Information in PROM is retained when battery voltage to the ECU is removed.

PSI

A measurement of pressure in Pounds per Square Inch.

Pulse Width Modulation (PWM)

A digital signal that consists of a pulse generated at a fixed frequency. When an actuator is controlled by a PWM signal, the ON time of the signal is increased or decreased (modulated) to increase or decrease the output of the actuator.

Push Rod

A connecting link in an operating mechanism, such as the rod interposed between the valve lifter and rocker arm on an overhead valve engine.

R

Race

As used with reference to bearings; a finished inner and outer surface in which or on which balls or rollers operate.

Random Access Memory (RAM)

The portion of computer memory within the Electronic Control Unit (ECU) that temporarily stores information as the engine is running. Information in RAM is lost when battery voltage to the ECU is removed.

Rated Horsepower

Value used by the engine manufacturer to rate the power of the engine, allowing for safe loads.

Ratio

The relation or proportion of one number or quantity to another.

Ream

To finish a hole accurately with a rotating fluted tool.

Reciprocating Motion

A back-and-forth movement, such as the action of a piston in a cylinder.

Rocker Arm

In an engine, a lever located on a fulcrum or shaft, one end on the valve stem, the other on the push rod.

Roller Bearing

An inner and outer race upon which hardened steel rollers operate.

Rotary Motion

A circular movement, such as the rotation of a crankshaft.

RPM

Revolutions per minute.

Running-Fit

Where sufficient clearance has been allowed between the shaft and journal to allow free running without overheating.

S

S.A.E.

Society of Automotive Engineers. This group sets the standards for much engine design.

Safety Factor

The degree of surplus strength over and above normal requirements that serves as insurance against failure.

Scale

A flaky deposit occurring on steel or iron. Ordinarily used to describe the accumulation of minerals and metals accumulating in an engine cooling system.

Score

A scratch, ridge, or groove marring a finished surface.

Seat

A surface, usually machined, upon which another part rests or seats. For example, the surface upon which a valve face rests.

Sensor

Devices used by the Engine Control Unit (ECU) to monitor various engine parameters.

Shim

Thin sheets used as spacers between two parts; normally used to adjust the side-to-side clearance or end play between parts.

Shrink Fit

Continued on next page OUO1020,0001973 -19-23JUL09-3/6

Where the shaft or part is slightly larger than the hole in which it is to be inserted. The outer part is heated above its normal operating temperature or the inner part is chilled below its normal operating temperature and assembled in this condition. Upon cooling or warming, an exceptionally tight fit is obtained.

Sliding-Fit

Where sufficient clearance has been allowed between the shaft and journal to allow free running without overheating.

Slip-In Bearing

A liner made to precise measurements that can be used for replacement without additional fitting.

Sludge

A composition of oxidized petroleum products along with an emulsion formed by the mixture of oil and water. This forms a pasty substance that clogs oil lines and passages and interferes with engine lubrication.

Solid Injection

The system used in diesel engines where fuel as a fluid is injected into the cylinder rather than a mixture of fuel and air.

Solvent

A solution that dissolves some other material.

Stoichiometric Combustion

The engine burns an air-fuel mixture that contains just enough air to theoretically burn all the fuel. The stoichiometric air-fuel ratio is approximately 15 parts air to one part fuel for the gasoline engine.

Stress

The force or strain to which a material is subjected.

Stroke

The distance moved by the piston.

Studs

A rod with threads on both ends, such as a cylinder stud that screws into the cylinder block on one end and has a nut placed on the other end to hold the cylinder head in place.

Suction

Suction exists in a vessel when the pressure is lower than the atmospheric pressure; also see Vacuum.

Supercharger

A mechanically driven blower or pump that forces air into the cylinders at higher-than-atmospheric pressure. The increased pressure forces more air into the cylinder, thus enabling more fuel to be burned and more power produced.

Synchronize

To cause two events to occur at the same time. For example, to time a mechanism so that two or more sparks will occur at the same instant.

T

Tachometer

A device for measuring and showing the rotating speed of an engine.

Tap

To cut threads in a hole with a tapered, fluted, threaded tool.

Tappet

The device that follows the profile of the cam on the camshaft, converting the rotary motion of the camshaft into reciprocating motion to actuate the engine's valves.

TDC

Top Dead Center (of the piston).

Temperature Sensors

These sensors are temperature-sensitive variable resistors. As the temperature changes, the sensor resistance changes. The Engine Control Unit (ECU) monitors the voltage drop across the sensor and translates this voltage to a temperature value.

T-Head

An engine design wherein the inlet valves are placed on one side of the cylinder and the exhaust valves are placed on the other.

Thermal Efficiency

A gallon of fuel contains a certain amount of potential energy in the form of heat when burned in the combustion chamber. Some of this heat is lost, and some is converted into power. The thermal efficiency is the ratio of work accomplished to the total quantity of heat in the fuel.

Thermostat

A heat-controlled valve used in the cooling system of an engine to regulate the flow of water between the cylinder block and the radiator.

Throttle Position Sensor

Measures the opening of the throttle in the drive-by-wire electronically controlled throttle actuator system.

Throw

The distance from the center of the crankshaft main journal to the center of the connecting rod journal.

Timing Gears

Any group of gears that are driven from the engine crankshaft to cause the valves, ignition, and other engine-driven components to operate at the desired time during the engine cycle.

Continued on next page   OUO1020,0001973 -19-23JUL09-4/6

Tolerance

A permissible variation between the two extremes of a specification of dimensions. Used in the precision fitting of mechanical parts.

Torque

The effort of twisting or turning.

Torque Wrench

A special wrench with a built-in indicator to measure the applied turning force.

Troubleshooting

A process of diagnosing or locating the source of the trouble or troubles from observation and testing. Also see Diagnosis.

Tune-Up

A process of accurate and careful adjustments to obtain the best engine performance.

Turbine

A series of angled blades located on a wheel against which fluids or gases are impelled to rotate a shaft.

Turbocharger

An air pump that is driven by the heat and volume of exhaust gases that forces more air into the cylinders than could be delivered under naturally aspirated (nonturbocharged) conditions. The turbocharger allows the engine to produce added power without increasing displacement.

Turbocharger Boost

The pressure created in the intake manifold by the action of the turbocharger.

Turbocharger Wastegate

A device that allows a controlled amount of exhaust gas to bypass the turbine wheel to control boost pressure of the turbocharger.

Turbulence

A disturbed, irregular motion of fluids or gases.

Two-Stroke Cycle Engine

An engine design permitting a power stroke once for each revolution of the crankshaft.

U

Universal Exhaust Gas Oxygen (UEGO) Sensor

A sensor that measures the amount of oxygen in the exhaust gas. The Engine Control Unit (ECU) uses the oxygen measurement to determine the correct amount of fuel delivery.

Up-Draft Carburetor

A carburetor type in which the mixture flows upward to the engine.

V

Vacuum

A perfect vacuum has not been created as this would involve the absolute lack of pressure. The term is ordinarily used to describe a partial vacuum; that is, a pressure less than atmospheric pressure.

Vacuum Gauge

An instrument designed to measure the degree of vacuum existing in a chamber.

Valve

A device for opening and sealing the cylinder intake and exhaust ports.

Valve Clearance

The air gap allowed between the end of the valve stem and the valve lifter or rocker arm to compensate for expansion due to heat.

Valve Face

The part of a valve that mates with and rests upon a seating surface.

Valve Grinding

A machining process for mating the valve seat and valve face usually performed with the aid of an abrasive.

Valve Head

The portion of the valve upon which the valve face is machined.

Valve-in-Head Engine

Same as Overhead Valve Engine.

Valve Lifter

A push rod or plunger placed between the cam and the valve on an engine.

Valve Margin

On a poppet valve, the space or rim between the surface of the head and the surface of the valve face.

Valve Seat

The matched surface upon which the valve face rests.

Valve Spring

A spring attached to a valve to return it to the seat after it has been released from the lifting or opening means.

Valve Stem

The portion of a valve that rests within a guide.

Valve Stem Guide

Continued on next page

OUO1020,0001973 -19-23JUL09-5/6

100709
PN=481

A bushing or hole in which the valve stem is placed that allows lateral motion only.

Vapor Lock

A condition wherein the fuel boils in the fuel system, forming bubbles that retard or stop the flow of fuel to the carburetor.

Vaporizer

A device for transforming or helping to transform a liquid into a vapor; often includes the application of heat.

Venturi

Two tapering streamlined tubes joined at their small ends so as to reduce the internal diameter.

Vibration Damper

A device to reduce the torsional or twisting vibration that occurs along the length of the crankshaft used in multi-cylinder engines; also known as a harmonic balancer.

Viscosity

The resistance of flow of an oil.

Volatility

The tendency for a fluid to evaporate rapidly or pass off in the form of vapor. For example, gasoline is more volatile than diesel fuel as it evaporates at a lower temperature.

Vortex

A whirling movement of a mass of liquid or air.

W

Wastegate

A device used to control the amount of boost that the turbocharger is allowed to generate. When the wastegate is closed, the turbocharger generates maximum boost. When the wastegate is opened, a portion of the engine exhaust gases are allowed to bypass the turbocharger turbine, causing a reduction of turbocharger boost pressure.

Wrist Pin

The journal for the bearing in the small end of an engine connecting rod that also passes through the piston walls; also known as a piston pin.

OUO1020,0001973 -19-23JUL09-6/6

Weights and Measures — Metric to U.S.

Metric System					U.S. Equivalent		
Length							
Unit	Abbreviation	Number of Meters					
Kilometer	km	1000			0.62 mile		
Hectometer	hm	100			109.36 yards		
Decameter	dkm	10			32.81 feet		
Meter	m	1			39.37 inches		
Decimeter	dm	0.1			3.94 inches		
Centimeter	cm	0.01			0.39 inch		
Millimeter	mm	0.001			0.04 inch		
Area							
Unit	Abbreviation	Number of Square Meters					
Square Kilometer	sq km or km^2	1,000,000			0.3861 square mile		
Hectare	ha	10,000			2.47 acres		
Are	a	100			119.60 square yards		
Centare	ca	1			10.76 square feet		
Square Centimeter	sq cm or cm^2	0.0001			0.155 square inch		
Volume							
Unit	Abbreviation	Number of Cubic Meters					
Stere	s	1			1.31, cubic yards		
Decistere	ds	0.10			3.53 cubic feet		
Cubic Centimeter	cu cm or cm^3 also cc	0.000001			0.061 cubic inch		
Capacity							
Unit	Abbreviation	Number of Liters	Cubic	Dry	Liquid		
Kiloliter	kl	1,000	1.31 cubic yards				
Hectoliter	hl	100	3.53 cubic feet	2.84 bushels			
Decaliter	dkl	10	0.35 cubic foot	1.14 pecks	2.64 gallons		
Liter	L	1	61.02 cubic inches	0.908 quart	1.057 quarts		
Deciliter	dl	0.10	6.1 cubic inches	0.18 pint	0.21 pint		
Centiliter	cl	0.01	0.6 cubic inch		0.338 fluid ounce		
Milliliter	ml	0.001	0.06 cubic inch		0.27 fluid dram		
Mass and Weight							
Unit	Abbreviation	Number of Grams					
Metric Ton	MT or t	1,000,000			1.1 tons		
Quintal	q	100,000			220.46 pounds		
Kilogram	kg	1,000			2.2046 pounds		
Hectogram	hg	100			3.527 ounces		
Decagram	dkg	10			0.353 ounce		
Gram	g or gm	1			0.035 ounce		
Decigram	dg	0.1			1.543 grains		
Centigram	cg	0.01			0.154 grain		
Milligram	mg	0.001			0.015 grain		
Power							
Unit		Abbreviation					
Kilowatt		kW			1.34 horsepower		
Heat							
Unit		Abbreviation					
Watt		W			3.41 Btu		
Pressure or Vacuum							

Continued on next page   JB06590,00004A5 -19-23JUL09-1/2

Metric System		U.S. Equivalent
Unit	Abbreviation	
Kilopascal	kPa	0.145 pounds per square inch
Torque		
Unit	Abbreviation	
Newton-meter	N·m	142.86 ounce inches 8.85 pound inches 0.74 pound foot
Temperature		
Unit	Abbreviation	
Degrees Celsius	°C	(°C x 1.8) + 32 = °F

JB06590,00004A5 -19-23JUL09-2/2

Weights and Measures — U.S. to Metric

U.S. System			Metric Equivalent
Length			
Unit	Abbreviation	Equivalents in Other Units	
Mile	mi	5280 feet, 320 rods, 1760 yards	1.609 kilometers
Rod	rd	5.50 yards, 16.5 feet	5.029 meters
Yard	yd	3 feet, 36 inches	0.914 meter
Foot	ft or '	12 inches, 0.333 yard	30,480 centimeters
Inch	in. or "	0.083 foot, 0.027 yard	2.540 centimeters
Area			
Unit	Abbreviation	Equivalents in Other Units	
Square Mile	sq mi or m^2	640 acres, 102,400 square rods	2.590 square kilometers
Acre	a	4840 square yards, 43,560 square feet	0.405 hectare, 4047 square meters
Square Rod	sq rd or rd^2	30.25 square yards, 0.006 acre	25.293 square meters
Square Yard	sq yd or yd^2	1296 square inches, 9 square feet	0.836 square meter
Square Foot	sq ft or ft^2	144 square inches, 0.111 square yard	0.093 square meter
Square Inch	sq in. or in^2	0.007 square foot, 0.00077 square yard	6.451 square centimeters
Volume			
Unit	Abbreviation	Equivalents in Other Units	
Cubic Yard	cu yd or yd^3	27 cubic feet, 46,656 cubic inches	0.765 cubic meter
Cubic Foot	cu ft or ft^3	1728 cubic inches, 0.0370 cubic yard	0.028 cubic meter
Cubic Inch	cu in. or in^3	0.00058 cubic foot, 0.000021 cubic yard.	16.387 cubic centimeters
Mass and Weight			
Unit	Abbreviation	Equivalents In Other Units	
Ton Short Ton Long Ton	tn (seldom used)	20 short hundredweight, 2000 pounds 20 long hundredweight, 2240 pounds	0.907 metric ton 1.016 metric tons
Hundredweight Short Hundredweight Long Hundredweight	cwt	100 pounds, 0.05 short ton 112 pounds, 0.05 long ton	45.359 kilograms 50.802 kilograms
Pound	lb or lb av also	16 ounces, 7000 grains	0.453 kilogram
Ounce	oz or oz av	16 drams, 437.5 grains	28.349 grams
Dram	dr or dr av	27.343 grains, 0.0625 ounce	1.771 grams
Grain	gr	0.036 drams., 0.002285 ounce	0.0648 gram
Capacity			
Unit	Abbreviation	U.S. Liquid Measure	
Gallon	gal	4 quarts (231 cubic inches)	3.785 liters
Quart	qt	2 pints (57.75 cubic inches)	0.946 liters
Pint	pt	4 gills (28.875 cubic inches)	0.473 liters
Gill	gi	4 fluid ounces (7.218 cubic inches)	118.291 milliliters
Fluid Ounce	fl oz	8 fluid drams (1.804 cubic inches)	29.573 milliliters
Fluid Dram	fl dr	0.60 minim (0.225 cubic inche)	3.696 milliliters
Minim	min	1/60 fluid dram (0.003759 cubic inche)	0.061610 milliliter
		U.S. Dry Measure	
Bushel	bu	4 pecks (2150.42 cubic inches)	35.238 liters
Peck	pk	8 quarts (537.605 cubic inches)	8.809 liters
Quart	qt	2 pints (67.200 cubic inches)	1.1.01 liters
Pint	pt	1/2 quart (33.600 cubic inches)	0.550 liters
Power			
Unit		Abbreviation	
Horsepower		hp	0.746 kilowatt
Heat			
Unit		Abbreviation	

Continued on next page

JB06590,00004A6 -19-21JUL09-1/2

100709
PN=485

U.S. System		Metric Equivalent
British Thermal Unit	Btu	0.293 watt
Pressure or Vacuum		
Unit	Abbreviation	
Pounds per Square Inch	psi	6.895 kilopascals
Torque		
Unit	Abbreviation	
Ounce Inch	oz/in	0.007 newton-meter
Pound Inch	lb/in	0.113 newton-meter
Pound Foot	lb/ft	1.346 newton-meters
Temperature		
Unit	Abbreviation	
Fahrenheit	°F	(°F - 32) 0.556 = °C

JB06590,00004A6 -19-21JUL09-2/2

Unified Inch Bolt and Screw Torque Values

TS1671 —UN—01MAY03

  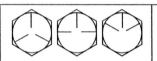

Bolt or Screw Size	SAE Grade 1				SAE Grade 2[a]				SAE Grade 5, 5.1 or 5.2				SAE Grade 8 or 8.2			
	Lubricated[b]		Dry[c]		Lubricated[b]		Dry[c]		Lubricated[b]		Dry[c]		Lubricated[b]		Dry[c]	
	N·m	lb-in	N·m	lb-in	N·m	lb-in	N·m	lb-in	N·m	lb-in	N·m	lb-in	N·m	lb-in	N·m	lb-in
1/4	3.7	33	4.7	42	6	53	7.5	66	9.5	84	12	106	13.5	120	17	150
													N·m	lb-ft	N·m	lb-ft
5/16	7.7	68	9.8	86	12	106	15.5	137	19.5	172	25	221	28	20.5	35	26
									N·m	lb-ft	N·m	lb-ft				
3/8	13.5	120	17.5	155	22	194	27	240	35	26	44	32.5	49	36	63	46
			N·m	lb-ft	N·m	lb-ft	N·m	lb-ft								
7/16	22	194	28	20.5	35	26	44	32.5	56	41	70	52	80	59	100	74
	N·m	lb-ft														
1/2	34	25	42	31	53	39	67	49	85	63	110	80	120	88	155	115
9/16	48	35.5	60	45	76	56	95	70	125	92	155	115	175	130	220	165
5/8	67	49	85	63	105	77	135	100	170	125	215	160	240	175	305	225
3/4	120	88	150	110	190	140	240	175	300	220	380	280	425	315	540	400
7/8	190	140	240	175	190	140	240	175	490	360	615	455	690	510	870	640
1	285	210	360	265	285	210	360	265	730	540	920	680	1030	760	1300	960
1-1/8	400	300	510	375	400	300	510	375	910	670	1150	850	1450	1075	1850	1350
1-1/4	570	420	725	535	570	420	725	535	1280	945	1630	1200	2050	1500	2600	1920
1-3/8	750	550	950	700	750	550	950	700	1700	1250	2140	1580	2700	2000	3400	2500
1-1/2	990	730	1250	930	990	730	1250	930	2250	1650	2850	2100	3600	2650	4550	3350

Torque values listed are for general use only, based on the strength of the bolt or screw. DO NOT use these values if a different torque value or tightening procedure is given for a specific application. For plastic insert or crimped steel type lock nuts, for stainless steel fasteners, or for nuts on U-bolts, see the tightening instructions for the specific application. Shear bolts are designed to fail under predetermined loads. Always replace shear bolts with identical grade.

Replace fasteners with the same or higher grade. If higher grade fasteners are used, tighten these to the strength of the original. Make sure fastener threads are clean and that you properly start thread engagement. When possible, lubricate plain or zinc plated fasteners other than lock nuts, wheel bolts or wheel nuts, unless different instructions are given for the specific application.

[a]*Grade 2 applies for hex cap screws (not hex bolts) up to 6. in (152 mm) long. Grade 1 applies for hex cap screws over 6 in. (152 mm) long, and for all other types of bolts and screws of any length.*
[b]*"Lubricated" means coated with a lubricant such as engine oil, fasteners with phosphate and oil coatings, or 7/8 in. and larger fasteners with JDM F13C zinc flake coating.*
[c]*"Dry" means plain or zinc plated without any lubrication, or 1/4 to 3/4 in. fasteners with JDM F13B zinc flake coating.*

DX,TORQ1 -19-24MAR09-1/1

Metric Bolt and Screw Torque Values

TS1670 —UN—01MAY03

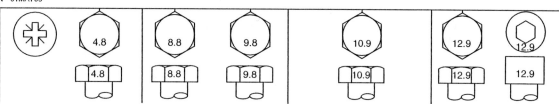

Bolt or Screw Size	Class 4.8				Class 8.8 or 9.8				Class 10.9				Class 12.9			
	Lubricated[a]		Dry[b]		Lubricated[a]		Dry[b]		Lubricated[a]		Dry[b]		Lubricated[a]		Dry[b]	
	N·m	lb-in	N·m	lb-in	N·m	lb-in	N·m	lb-in	N·m	lb-in	N·m	lb-in	N·m	lb-in	N·m	lb-in
M6	4.7	42	6	53	8.9	79	11.3	100	13	115	16.5	146	15.5	137	19.5	172
									N·m	lb-ft	N·m	lb-ft	N·m	lb-ft	N·m	lb-ft
M8	11.5	102	14.5	128	22	194	27.5	243	32	23.5	40	29.5	37	27.5	47	35
			N·m	lb-ft	N·m	lb-ft	N·m	lb-ft								
M10	23	204	29	21	43	32	55	40	63	46	80	59	75	55	95	70
	N·m	lb-ft														
M12	40	29.5	50	37	75	55	95	70	110	80	140	105	130	95	165	120
M14	63	46	80	59	120	88	150	110	175	130	220	165	205	150	260	190
M16	100	74	125	92	190	140	240	175	275	200	350	255	320	235	400	300
M18	135	100	170	125	265	195	330	245	375	275	475	350	440	325	560	410
M20	190	140	245	180	375	275	475	350	530	390	675	500	625	460	790	580
M22	265	195	330	245	510	375	650	480	725	535	920	680	850	625	1080	800
M24	330	245	425	315	650	480	820	600	920	680	1150	850	1080	800	1350	1000
M27	490	360	625	460	950	700	1200	885	1350	1000	1700	1250	1580	1160	2000	1475
M30	660	490	850	625	1290	950	1630	1200	1850	1350	2300	1700	2140	1580	2700	2000
M33	900	665	1150	850	1750	1300	2200	1625	2500	1850	3150	2325	2900	2150	3700	2730
M36	1150	850	1450	1075	2250	1650	2850	2100	3200	2350	4050	3000	3750	2770	4750	3500

Torque values listed are for general use only, based on the strength of the bolt or screw. DO NOT use these values if a different torque value or tightening procedure is given for a specific application. For stainless steel fasteners or for nuts on U-bolts, see the tightening instructions for the specific application. Tighten plastic insert or crimped steel type lock nuts by turning the nut to the dry torque shown in the chart, unless different instructions are given for the specific application.

Shear bolts are designed to fail under predetermined loads. Always replace shear bolts with identical property class. Replace fasteners with the same or higher property class. If higher property class fasteners are used, tighten these to the strength of the original. Make sure fastener threads are clean and that you properly start thread engagement. When possible, lubricate plain or zinc plated fasteners other than lock nuts, wheel bolts or wheel nuts, unless different instructions are given for the specific application.

[a]"Lubricated" means coated with a lubricant such as engine oil, fasteners with phosphate and oil coatings, or M20 and larger fasteners with JDM F13C zinc flake coating.
[b]"Dry" means plain or zinc plated without any lubrication, or M6 to M18 fasteners with JDM F13B zinc flake coating.

DX,TORQ2 -19-24MAR09-1/1

Answers to Test Yourself Questions

Answers to Chapter 1 Questions

1. First blank – "heat." Second blank – "mechanical."
2. Air, fuel, and combustion (or ignition).
3. It heats up. This provides the heat necessary for good ignition of the fuel.
4. Vapor
5. a – 2, b – 1.
6. First blank – "reciprocating" or "up-and-down." Second blank – "rotary."
7. Cylinder bore diameter, length of the stroke, and number of cylinders.
8. Both statements are true. The amount of work performed is the same for both examples.
9. The compression ratio is 8 to 1.
10. Diesel.
11. Because more heat is needed for fuel combustion without a spark in diesel engines. (The more the air is compressed, the hotter it becomes.)
12. Intake, compression, power, exhaust.
13. False. The crankshaft turns two complete revolutions during a four-stroke cycle.
14. a – 1, b – 3, c – 2.
15. Liquid and air.

OUO1020,0001976 -19-21JUL09-1/1

Answers to Chapter 2 Questions

1. True.
2. A job can be done more safely and efficiently if the appropriate tools are accessible and in good working condition.
3. True.
4. False. (Gasoline should NEVER be used as a cleaning solvent.)
5. First blank – "safety goggles." Second blank – "rubber gloves."
6. Disconnect the grounded battery cable first when removing the battery. Connect the grounded cable last when installing the battery.
7. You bypass the neutral start switch by doing so and if the machine is in gear when the engine starts, it could suddenly lurch forward and crush you.
8. False. (Natural gas is lighter than air and will rise to the ceiling in an enclosed space.)
9. First blank – "DANGER." Second blank – "WARNING." Third blank – "CAUTION."
10. Oil, fuel, coolant, brake fluid, solvent, batteries, filters.

OUO1020,0001977 -19-21JUL09-1/1

Answers to Chapter 3 Questions

1. Start at the center of the head and work out on both sides.
2. Upsets metal on the inside wall, making the guide bore smaller. This is a way to recondition some valve guides.
3. By turning the valves, the rotators knock off deposits. This keeps the valves sealing better, running cooler, and cuts down on corrosion.
4. 1/2
5. First blank – "top dead center." Second blank – "compression."
6. "Wet" liners contact the engine coolant, while "dry" liners do not.
7. a. Top inch of ring travel.
8. False. Measure liners before removing them.
9. Because of heat expansion. The thicker top part of the skirt must be slightly narrower since it expands more from the heat of combustion.
10. a – 2, b – 3, c – 1.
11. False. Always replace every piston ring you remove, regardless of its condition.
12. To be sure they are installed in the same cylinder. This gives a better fit, since the parts are already "wear fitted" to their own cylinders.
13. Because of the heavy throws that must be used as counterweights to balance the engine during rotation.
14. The flywheel.
15. True. The lighter flywheel allows faster acceleration and deceleration between various speeds.
16. 1) Cylinder head bolts; 2) valve tappets; 3) engine timing.

OUO1020,0001978 -19-21JUL09-1/1

Answers to Chapter 4 Questions

1. Fuel tank, fuel pump, carburetor
2. To deliver fuel to the carburetor.
3. To mix the proper amounts of fuel and air under all operating conditions.
4. False
5. Natural draft, updraft, downdraft.
6. To provide a richer fuel mixture during starting.
7. To provide extra fuel momentarily during rapid acceleration.
8. The greater the air velocity, the lower the air pressure.
9. Engine camshaft.

OUO1020,0001979 -19-21JUL09-1/1

100709
PN=488

Answers to Chapter 5 Questions

1. False. LP-gas is a vapor when it reaches the carburetor, but gasoline is a liquid.
2. True.
3. True.
4. False! Never fill more than 80% full to allow room for vapor.
5. True.
6. False. Ventilate areas around LP-gas equipment, or use fans and blowers.
7. False. The converter changes liquid to vapor but it lowers the pressure.

OUO1020,000197A -19-21JUL09-1/1

Answers to Chapter 6 Questions

1. True.
2. True.
3. False. Liquefied Natural Gas is produced through a cryogenic process in which the gas is cooled to -259 °F (-126 °C) which converts gas to a liquid.
4. True.
5. Fuel tanks, fuel pressure regulator, fuel metering valve, fuel/air mixer, and throttle assembly.
6. The vaporizer is a heat exchanger located between the fuel tank and the engine. Engine coolant is circulated through the vaporizer to warm and vaporize the liquefied natural gas.
7. a - 2, b - 1.
8. True.
9. The Electronic Control Unit also controls fuel delivery and ignition timing.
10. Open-loop mode is utilized during engine start up and for the first 30 to 60 seconds of run time. Closed-loop mode occurs after the engine has been running for approximately 60 seconds. In closed-loop mode the ECU has the ability to correct for gas composition variation and for changing operating conditions.
11. All natural gas systems operate at high pressures. The natural gas fuel system pressure must be relieved before disconnecting any fuel system component.

OUO1020,000197B -19-21JUL09-1/1

Answers to Chapter 7 Questions

1. False. No spark is used in a diesel engine. Instead, air is compressed until it heats up and ignites the injected shot of fuel.
2. Because one pump can serve all the cylinders.
3. By the diesel fuel circulating through it during operation.
4. First blank – "spring". Second blank – "fuel".
5. Use a brass wire brush. Never use a steel brush as it will scratch the precision tips and erode the spray orifices.
6. In series filters, all the fuel goes through one filter, then through the other. In parallel filters, part of the fuel goes through each filter.
7. The series filters clean fuel the best because the second filter can pick up dirt missed by the first one.
8. Balancing coil and thermostatic.
9. Pump and injector
10. Pumping section, actuator section, injector section.
11. False. Fuel injection begins when the spill valve moves to the closed position.
12. True
13. True
14. 1) The vehicle's digital tachometer; 2) a portable electronic governor tester.
15. False. A thermostatic fuel gauge uses two bimetal blades.
16. True.
17. First blank – "Time." Second blank – "Injector."
18. False. The pressure limiter leaks off excess fuel to the fuel tank if an abnormally high fuel pressure is detected.
19. 19. True.
20. False. The oil pressure sensor is optional and is used for engine protection.

OUO1020,000197C -19-21JUL09-1/1

Answers to Chapter 8 Questions

1. Cylinder feed and crankcase feed.
2. Clean air is a necessity for long engine life. If airborne dust particles are allowed to enter the combustion chamber, premature engine wear and damage will occur.
3. Remove impurities from the air but at the same time allow sufficient volume of air to enter the engine to insure complete combustion of the fuel.
4. The filter element is damaged and must be replaced.
5. Manifold, exhaust pipe, muffler.
6. Removes heat, muffles sound, carries away exhaust gases.
7. Air pumps. The Roots-type supercharger is a positive displacement pump and provides the same amount of air per revolution, regardless of engine speed.
8. Turbochargers are driven by the waste exhaust gases from the engine's exhaust system. Superchargers are mechanically driven by the engine crankshaft.
9. An aftercooler is a heat exchanger that reduces the temperature of the compressed intake air from the turbocharger. This makes the air denser, allowing more to be packed into the combustion chamber.
10. High-engine RPM.

OUO1020,000197D -19-21JUL09-1/1

Answers to Chapter 9 Questions

1. Any three of these: 1) Reduces friction; 2) absorbs heat; 3) seals the piston rings; 4) cleans and flushes moving parts; 5) helps deaden the noise.

2. Bypasses oil around the filter if the filter clogs. This assures a supply of lubricating oil to the engine at all times.
3. Additives.
4. False. Oil loses many of its good lubricating qualities as it gets dirty and its additives wear out.

OUO1020,000197E -19-21JUL09-1/1

Answers to Chapter 10 Questions

1. Radiator, radiator cap, fan and fan belt, water pump, engine water jacket, thermostat, connecting hoses, liquid or coolant.
2. True.

3. The BELLOWS-type consists of a short length of circular corrugated copper tube closed at both ends and filled with a liquid having a low boiling point. The BIMETALLIC-type of thermostat consists of a spiral of bimetallic strip. This is a strip of steel welded to a strip of bronze.
4. Water pump.
5. False.

OUO1020,000197F -19-02JUL09-1/1

Answers to Chapter 11 Questions

1. Governors can do any of three jobs: maintain a selected speed, limit the slow and fast speeds, or shut down the engine when it overspeeds.
2. Speed droop is the change in governor speed required to cause the throttle rod to move from full-open to full-closed or full-closed to full-open.

3. Vacuum governor and centrifugal governor.
4. Constant speed governors are isochronous governors. An isochronous governor is able to maintain a constant speed without correcting it.
5. True
6. It performs governor (speed control) and fuel system diagnostic functions.

OUO1020,0001980 -19-02JUL09-1/1

Answers to Chapter 12 Questions

1. Testing tools and servicing tools.

2. Testing tools, which locate the trouble which the servicing tools are then used to help correct.
3. Special tools designed for servicing each model of engine.

OUO1020,0001981 -19-02JUL09-1/1

Answers to Chapter 13 Questions

1. Know the system, 2) Ask the operator, 3) Inspect the engine, 4) Operate the engine, 5) List the possible causes, 6) Reach a conclusion, 7) Test your conclusion.

2. During none of these steps. Do all these things before you begin repairing the engine.
3. The dynamometer test.
4. Fuel-air mixture, compression, and ignition.

OUO1020,0001982 -19-02JUL09-1/1

Answers to Chapter 14 Questions

1. False. Tune-up can restore an engine which needs minor checks and adjustments, but it cannot make up for overdue repairs. Only a major overhaul can do that.
2. A visual inspection should be made to find out the condition of the engine and to see whether a tune-up or an overhaul is necessary.

3. Both before and after engine tune-up. Testing before tune-up gives an idea of the engine's condition and whether tune-up will do the job. Testing after tune-up gives a final check of engine performance to see if the tune-up has been successful.

OUO1020,0001983 -19-02JUL09-1/1

100709
PN=490

Index

Continued on next page

Continued on next page

Continued on next page

Continued on next page